Estuaries of the World

Series Editor
Jean-Paul Ducrotoy

Fishing village in the southwest part of El Caimanero coastal lagoon

Estuaries are amongst the most endangered areas in the world. Pollution, eutrophication, urbanization, land reclamation; over fishing and exploitation continuously threaten their future. The major challenge that humans face today is managing their use, so that future generations can also enjoy the fantastic visual, cultural and edible products that they provide. Such an approach presupposes that all users of the environment share views and are able to communicate wisely on the basis of robust science. The need for robust science is pressing. Over the last decade there have been numerous advances in both understanding and approach to estuaries and more and more multidisciplinary studies are now available. The available scientific information has come from a multiplicity of case studies and projects local and national levels. Regional and global programs have been developed; some are being implemented and some are in evolution. However, despite the rapidly increasing knowledge about estuarine ecosystems, crucial questions on the causes of variability and the effects of global change are still poorly understood. Although the perception of politicians and managers of coasts is slowly shifting from a mainly short-term economic approach towards a long-term economic—ecological perspective, there is a need to make existing scientific information much more manageable by non-specialists, without compromising the quality of the information. The book series includes volumes of selected invited papers and is intended for researchers, practitioners, undergraduate and graduate students in all disciplines who are dealing with complex problems and looking for cutting-edge research as well as methodological tools to set up truly transversal science and technology projects, such as the restoration of damaged habitats.

More information about this series at http://www.springer.com/series/11705

Felipe Amezcua · Brian Bellgraph
Editors

Fisheries Management of Mexican and Central American Estuaries

Springer

Editors
Felipe Amezcua
Universidad Nacional Autónoma de México
Instituto de Ciencias del Mar y Limnología
Mazatlán
Sinaloa
Mexico

Brian Bellgraph
Battelle Memorial Institute
Pacific Northwest National Laboratory
Richland
Washington
USA

ISSN 2214-1553 ISSN 2214-1561 (electronic)
ISBN 978-94-017-8916-5 ISBN 978-94-017-8917-2 (eBook)
DOI 10.1007/978-94-017-8917-2
Springer Dordrecht Heidelberg New York London

Library of Congress Control Number: 2014940081

Printed on acid-free paper

Springer is part of Springer Science+Business Media (www.springer.com)

Felipe dedicates this book to his parents for their unconditional support through the years

Brian dedicates this book to his wife, who encourages him to pursue his dreams, and his daughters, who inspire him to be a better person

Prologue

Historically, our knowledge of how aquatic systems function has advanced faster in both the freshwater and marine realms than where these two interface. It may be that estuaries are more complicated and/or more variable than freshwater and marine systems; they certainly vary widely across the globe in their ecological, hydrological and physical/chemical characteristics. At least in the past, however, it may also be that studying estuarine systems was a bit intimidating to some aquatic scientists. Because of the ever-increasing interactions of humans with the earth's coasts, its water, and its fishery resources, our need to understand how estuaries work, much less how humanity is impacting them (rarely in a good way) is now paramount.

In the past and still within today's scientific community, studies on natural systems often remain fragmented, isolated from each other rather than being integrated into a coherent (and more useful) whole. The process of designing (and funding), implementing, analyzing, and publishing a study can often take quite a bit of time. During that process, much of the emerging information is not shared among the various scientific groups working on a common topic. That unproductive situation is exacerbated by our increasingly globalized society; scientists working in a given region may come from different places around the globe, making communication among them difficult. Fisheries Management of Mexican and Central American Estuaries, edited by Felipe Amezcua and Brian Bellgraph (another entry in Springer Verlag's series, Estuaries of the World), breaks that mold, pulling together a number of current studies, successfully synthesizing a great deal of work into a single volume.

To help organize the synthesis of ideas, the volume is organized into three sections, Physical/Chemical Considerations, Ecological Considerations, and Socioeconomic Considerations. The first section has four chapters, three that cover trophic relationships within the estuarine community, how the estuarine ecosystem is being impacted by human residential and industrial wastes, including mercury, and one chapter that lays out management strategies based on nutrient carrying capacity. The second section also has four chapters that together cover how changing hydrologic regimes (and human impacts on them) affect coastal fisheries, especially in the mangrove forest complex, and argue for an ecosystem-based approach to fisheries management, including the use of fishing zones to manage artisanal fisheries. The third section has two papers that nicely outline the use of areas of responsible fishing for marine systems and interesting strategies for protecting wetland areas. Each of the chapters is nicely conceived and well-delivered, and together, they make a really nice whole.

The value of this volume goes beyond a synthesis just for scientists working in Mexican and Central American estuaries. It provides a great source of current information on estuaries for scientists around the globe and serves as a strategic model for how and why to take the next step of pulling multiple studies into a single synthesis. I applaud the efforts of this group of dedicated scientists and recommend Fisheries Management of Mexican and Central American Estuaries to all aquatic scientists.

Chair Fisheries Conservation Foundation —Dr. David P. Philipp

Preface

It was during my participation at the 6th World Fisheries Congress in Edinburgh when the idea to edit this book started to take form. At the time I met Alexandrine Cheronet, the Publishing Editor of Environmental Sciences from Springer, I was enquiring about another book from the same editorial (Nagelkerken. 2009. Ecological Connectivity among Tropical Coastal Ecosystems). After going through this book, and discovering that it was the kind of book I was looking for—considering my current projects on estuarine systems and their linkages with the adjacent marine environment in the SE Gulf of California—I started to chat with Alexandrine about this and other similar books from Springer. Once I was back at home, she told me about a new series called "Estuaries of the World" that was being developed by Springer, and it was immediately something that interested me as I was looking for more information on estuarine systems from Mexico and Central America. However, Alexandrine told me that a book covering Mexico and Central America had not been planned, but encouraged me to contribute with a volume on estuarine systems from this region of the world. After giving this idea some thought, and doing an initial research among my colleagues that work on these systems (in order to have an idea on how many contributed chapters the volume could have), I decided to go ahead with this volume and presented an initial idea to Jean-Paul Ducrotoy, the series editor. I came to realize that there were, in general, few books and published works dealing with fisheries management in estuarine systems from Mexico and Central America that also included aspects such as the ecology, hydrology, contamination, etc.–topics related to fisheries management that are not always considered when management policies are formed. Considering also the rapid degradation of estuarine systems in Mexico and Central America, I realized that a book with these characteristics could provide a vision, and a path forward, to help conserve and properly manage these ecosystems. So, several emails later, the idea took form and over the following months, thirty three authors mainly from Mexico, but also from Costa Rica, the USA, Canada and France worked extra hours to make this book possible. As usual, there were some individuals that promised a chapter and at the last minute had to withdraw their contribution; however, besides those incidents, and with the help of all the authors, my first experience in editing a book was free of frustrations. Part of this success was also thanks to several editors, reviewers and peer reviewers that provided a critical, but mainly cursory review of the book chapters: Gabriela Anaya-Reyna, Gastón Bazzino, Claire Coiraton, Elaine Espino, Francisco Flores-Verdugo, Francisco Gutiérrez-Mendieta, Germán Ramirez, Jorge R. Ruelas-Inzunza, Cleve Steward, Carlos Suarez and Piotr Szefer. I am also very thankful to David P. Philipp who managed to write the foreword for the book in spite of his busy and tight agenda. Finally, after realising that a book in English language written mainly by non-English speakers would need an Editor whose mother language was English, I invited Brian Bellgraph to help me co-edit this book. Brian did a fantastic job, as he not only edited the written English but also made very critical comments and observations that undoubtedly helped to improve the quality of this book. Now looking at the final version of the book, I hope that this material is useful for fisheries ecologists, students, managers and others interested in the fisheries and estuarine systems of Mexico and Central America.

Felipe Amezcua

About the Editors

Felipe Amezcua is a research professor at the Institute of Marine Science and Limnology of the National University of Mexico and is based at a station of the Institute in Mazatlan. He specializes in the ecology of estuarine and coastal fish of the Gulf of California as well as small-scale fisheries. He completed his undergraduate studies in Biology at the Faculty of Science from the National University of Mexico and received his Ph.D. in Marine Biology from the University of Liverpool (UK).

Felipe has extensive experience in the fin fish fauna of the coastal areas of the SE Gulf of California, and has worked and participated in projects and workshops on conservation, and fisheries management, and regulation with governmental entities such as the National Fisheries Institute (INAPESCA) and the National Commission for Fisheries and Aquaculture (CONAPESCA), among others.

His project experience includes research on fisheries and the carrying capacity of reservoirs and estuarine areas to develop management plans; the effects of the shrimp trawl fishery on the fish fauna, trophic ecology and ecological connectivity of estuarine areas and the adjacent marine environments; and more recently, the use of otoliths and vertebrae to determine migration patterns of fish and sharks.

He is actively involved with fisheries associations and has served as President of the Mexican Fisheries Society and the Mexico Chapter and the International Fisheries Section of the American Fisheries Society. He is also a member of other societies such as the Japanese Society of Fisheries Sciences, the Fisheries Society of the British Isles, and the Société Française d'Ichtyologie.

Brian Bellgraph is a fisheries scientist for Battelle Memorial Institute at the Pacific Northwest National Laboratory (PNNL), one of nine United States Department of Energy federal research laboratories. He primarily studies fish ecology and behavior and the impacts of energy production on fisheries sustainability, but has additional interests in artisanal and subsistence fisheries. Brian obtained a Master's of Science degree in Fish and Wildlife Management from Montana State University and a Bachelor's of Science in Fisheries and Wildlife Management from Michigan State University.

A majority of his experience has been in the Pacific Northwest U.S. studying the impacts of hydroelectric dams on juvenile fish survival, adult fish behavior, and the impacts of human development on lake fishes. Brian also has experience evaluating ecological interactions among riverine fishes, assessing the effects of water temperature changes on fish behaviour and survival, and building individual-based computer simulation models.

Brian's love of Latin America culture, his practice of the Spanish language for 19 years (although he has a lot of learning left to do), and his respect for the aquatic world all led to his interest in editing a book about Mexico and Central America's estuaries. When he is not working, Brian enjoys spending time with his family, international travel, and cooking.

Contents

Contributors

Angela Aguiar-Moreno Posgrado en Ciencias del Mar y Limnología, Universidad Nacional Autónoma de México, México, DF, México

Felipe Amezcua Instituto de Ciencias del Mar y Limnología, Universidad Nacional Autónoma de México, Mazatlan, Sinaloa, México

Manuel Blanco-Correa Investigación y Posgrado, Universidad Autónoma de Nayarit, Tepic, NAY, México

Omar Calvario-Martínez Av. Sabalo-Cerritos s/n, Centro de Investigación en Alimentación y Desarrollo, Mazatlán, Sinaloa, México

José Luis Carbajal-Pérez Laboratorio de Hidrobiología, Departamento de Zoología, Instituto de Biología, Universidad Nacional Autónoma de México, México, México

Alejandro Castillo-López Pronatura Noroeste A.C., Programa de Conservación Marina y Pesca Sustentable, Ensenada, B.C., Mexico

Guadalupe de la Lanza-Espino Laboratorio de Hidrobiología, Departamento de Zoología, Instituto de Biología, Universidad Nacional Autónoma de México, México, México

Caroline Downton-Hoffmann Centro Intercultural de Estudios de Desiertos y Océanos (CEDO), Puerto Peñasco, SON, Mexico

Ofelia Escobar-Sánchez Laboratorio de Metales Pesados, Instituto Tecnológico de Mazatlán, Mazatlán, Sin., México

Luc Fargier UMR 7266 LIENSs (Research Center for Coasts, Environment and Societies), Université de La Rochelle (ULR), La Rochelle, France

Francisco Flores-Verdugo Instituto de Ciencias del Mar y Limnología, Universidad Nacional Autónoma de México, Mazatlan, Sinaloa, México

Hans J. Hartmann UMR 7266 LIENSs (Research Center for Coasts, Environment and Societies), Université de La Rochelle (ULR), La Rochelle, France

Salvador Hernández-Pulido Laboratorio de Hidrobiología, Departamento de Zoología, Instituto de Biología, Universidad Nacional Autónoma de México, México, México

Martín E. Jara-Marini Departamento de Ciencia de los Alimentos, Centro de Investigación en Alimentación y Desarrollo, A.C., Hermosillo, SON, México

Programa de Posgrado en Ciencias del Mar y Limnología, Instituto de Ciencias del Mar y Limnología, Unidad Académica Mazatlán, Universidad Nacional Autónoma de, México, Mazatlán, Sinaloa, México

John M. Kovacs Department of Geography, Nippising University, North Bay, ON, Canada

Carlos H. Lechuga-Devéze Av. Sabalo-Cerritos s/n, Centro de Investigación en Alimentación y Desarrollo, Mazatlán, Sinaloa, México

Iván Martinez-Tovar Centro Intercultural de Estudios de Desiertos y Océanos (CEDO), Puerto Peñasco, SON, Mexico

Helena Molina-Ureña CIMAR (Research Center in Marine Sciences and Limnology), Universidad de Costa Rica (UCR), San José, Costa Rica

Hem Nalini Morzaria-Luna Centro Intercultural de Estudios de Desiertos y Océanos (CEDO), Puerto Peñasco, SON, Mexico

Adrian Munguia-Vega Conservation Genetics Laboratory, School of Natural Resources and the Environment, University of Arizona, Tucson, AZ, USA

Federico Páez-Osuna Instituto de Ciencias del Mar y Limnología, Universidad Nacional Autónoma de México, Mazatlán, Sinaloa, México

Hiram Peña-Bonilla Centro Intercultural de Estudios de Desiertos y Océanos (CEDO), Puerto Peñasco, SON, Mexico

Mauricio Ramírez-Rodríguez Centro Interdisciplinario de Ciencias Marinas, Instituto Politécnico Nacional, La Paz, BCS, México

Jorge R. Ruelas-Inzunza Laboratorio de Metales Pesados, Instituto Tecnológico de Mazatlán, Mazatlán, Sin., México

David Alberto Salas-de-León Instituto de Ciencias del Mar y Limnología, Universidad Nacional Autónoma de México, México, D.F., México

Miguel A. Sánchez-Rodríguez Av. Sabalo-Cerritos s/n, Unidad Mazatlán, Centro de Investigación en Alimentación y Desarrollo, Mazatlán, Sinaloa, México

Norma Sánchez-Santillán División de Ciencias Biológicas y de la Salud, Departamento del Hombre y su Ambiente, Universidad Autónoma Metropolitana-Xochimilco, Coyoacan, Mexico

David Serrano Facultad de Ciencias del Mar, Universidad Autónoma de Sinaloa, Mazatlán, SIN, México

Martín F. Soto-Jiménez Instituto de Ciencias del Mar y Limnología, Universidad Nacional Autónoma de México, Mazatlán, Sinaloa, México

Hitandehui Tovar-Vázquez Centro Intercultural de Estudios de Desiertos y Océanos (CEDO), Puerto Peñasco, SON, Mexico

Peggy J. Turk-Boyer Centro Intercultural de Estudios de Desiertos y Océanos (CEDO), Puerto Peñasco, SON, Mexico

Paloma A. Valdivia-Jiménez Centro Intercultural de Estudios de Desiertos y Océanos (CEDO), Puerto Peñasco, SON, Mexico

Román Rodolfo Vera-Mendoza Instituto de Ciencias del Mar y Limnología, Universidad Nacional Autónoma de México, México, D.F., México

1 Introduction on Managing Fisheries in Estuarine Systems of Mexico and Central America

Felipe Amezcua

Abstract

The contributed papers in this book provide research undertaken in estuarine systems of Mexico and Central America that aim to provide a scientific basis for proper estuarine management. The book is divided in three parts that cover topics associated with fisheries management and regulations: physicochemical studies, ecological studies and socioeconomic studies. This introduction outlines the contents of this book in relation to the management of coastal ecosystems, which have a high socioeconomic importance for a large proportion of the population in this area of the world. Rather than be a definitive suite of papers for the regulation of these habitats, the goal of this book is to outline the urgent need to continue research of threatened coastal ecosystems of Mexico and Central America.

Keywords

Fisheries management · Estuaries · Mexico · Central America · Anthropogenic impacts

Many books and papers highlight the economic and ecologic importance of tropical estuarine systems (e.g., Harborne et al. 2006; Manson et al. 2005; Meynecke et al. 2008; Aburto-Oropeza et al. 2008; UNEP 2006; Nagelkerken 2009). Others have documented the degradation these systems have incurred as a consequence of human impacts such as pollution, eutrophication, sedimentation, deforestation and habitat destruction (Short and Wyllie-Echeverria 1996; Alongi 2002; Hughes et al. 2003). The potential negative consequences of these habitat alterations on fisheries yields have also been discussed (Aburto-Oropeza et al. 2008).

It is generally recognized that tropical wetlands and estuarine systems have received much less scientific attention than estuarine systems in temperate zones, rainforests and other upland forests in tropical areas (Ellison 2004). Additionally, most of the available publications on tropical wetlands and estuarine systems come from experimental studies undertaken by researchers from developed countries such as the USA, Australia, Germany, Belgium, Denmark, etc. (Nagelkerken 2009).

In Mexico and Central America, wetland and estuarine ecosystems cover large areas of the region's coastline. For example, approximately 1,370,000 ha of mangrove forest grow along the coast of Mexico and Central America (FAO 2008). Mexico has some of the largest mangrove and wetland areas in the world; there are 137 estuarine systems (92 on the Pacific coast and 45 on the Gulf of Mexico and Caribbean) that cover an area of 1,567,000 ha (Contreras-Espinoza and Zabalegui 1988; Contreras-Espinoza and Castañeda 1993). These 137 estuarine systems contain 655,667 ha of mangrove forests (CONABIO 2008). In fact, more than one third of the regional mangrove area occurs in Mexico (FAO 2008). The largest wetland/mangrove complex of the American Pacific is also located in Mexico (i.e., Marismas Nacionales) and includes a large number of seasonal flood plains, coastal lagoons and tidal channels that are located among 175,300 ha of mangrove wetlands (DOF 2010).

F. Amezcua (✉)
Instituto de Ciencias del Mar y Limnología, Universidad Nacional Autónoma de México, Av. Joel Montes Camarena s/n, C.P. 82040, Mazatlan, Sinaloa, México
e-mail: famezcua@ola.icmyl.unam.mx

F. Amezcua, B. Bellgraph (eds.), *Fisheries Management of Mexican and Central American Estuaries*, Estuaries of the World, DOI 10.1007/978-94-017-8917-2_1, © Springer Science+Business Media Dordrecht 2014

In Central America, wetlands have received a paucity of attention from the scientific and management communities (Ellison 2004) and consequently, the exact total areal extent of wetlands in Central America still remains unknown (Davidson and Gauthier 1993; CCAD 2002). Ecological studies in this region have centered primarily on rain forests (Leigh et al. 1982; McDade et al. 1994). However, it is estimated that forested estuarine wetlands account for at least 6,500 km^2 (Groombridge 1992) and possibly as much as 12,000 km^2, or 2.2 % of the total land area of Central America (Spalding et al. 1997). The area covered by mangrove forests is about 488,000 ha; the country of Panama contains the largest mangrove complex in Central America (35 % of the total; FAO 2008).

Estuarine and coastal lagoon systems in Mexico and Central America are important fishing grounds for a large number of small-scale fishers in the region. As it happens in other regions of the world, small-scale fishing communities in this part of the world have a complex interrelationship with estuarine environments (Blaber 2011; Jennings et al. 2001; Nagelkerken et al. 2008). Small-scale fisheries in Mexico and Central America are also considered to be one of the most valuable natural resources in terms of ecosystem services per hectare (Costanza et al. 1997). Considering the large extent of these systems in the region and the large number of small-scale fishers operating in these systems, fisheries landings are high. In fact, with the exception of Panama, artisanal landings are more important in all Central American countries than industrial landings; in Mexico, artisanal landings represent 20 % of the total landings (SAGARPA 2011). Thus, the socioeconomic importance of these systems and the fisheries resources is paramount.

Management of fisheries and the associated estuarine habitat is very difficult to implement because it is a complex activity that involves not only the biology of the exploited resources, but also environmental and anthropogenic factors. Local authorities in the region are increasingly recognizing the importance and benefits of healthy mangrove forests, wetlands and estuarine systems due to their aesthetic, ecological and economic value (e.g., tourism and fisheries). However, adequate legislation for the protection and conservation of mangroves and estuarine areas is not a common practice in the region, and only very few countries have specific laws for the conservation of these ecosystems (i.e. Costa Rica). In some countries these ecosystems are included on the Ramsar List of Wetlands of International Importance and therefore designated as Protected Areas (e.g., Terraba-Sierpe National Park (Costa Rica), Jeanette Kawas National Park (Honduras), Marismas Nacionales (Mexico)). Even with the presence of laws and regulations, poaching and limited knowledge of the fishing laws are two common problems among fishers and other inhabitants of the region (Villaseñor and Amezcua 2013).

In general, there is a lack of proper management programs for these ecosystems in Mexico and Central America. Several areas of mangroves and coastal lagoons have been modified or destroyed for aquaculture, agriculture, urbanization and tourism development in Mexico. In Honduras, El Salvador and Panama, the shrimp and salt production industries have been the main cause of the loss of more than 160,000 ha of mangrove forests. Agricultural encroachment and use of land for livestock grazing are additional causes of the loss of these ecosystems (FAO 2008). The lack of management has also indirectly resulted in high poaching activity in the region; it is estimated that 60 % of fisheries landings in Mexico come from poaching, mainly from estuarine areas (IMCO 2013). It is likely that a similar scenario is happening in Central America, although information on this subject is not available.

There are many issues that affect estuarine systems in Mexico and Central America including population growth and poverty; limited technical and management capacity; limited possibility of sustainable development of the fishing industry; poor infrastructure; low income and/or ignorance of the fishermen and associated problems; limited possibilities to obtain assistance and training in new technologies; poor governance and corruption; lack of institutional collaboration; weak governmental institutions; a focus on solving single issues; lack of stakeholder participation; mismatch between the issue and the geographic scale of management; lack of an ecosystem perspective; ineffective governance and management; lack of awareness of the consequences of human activities; and a lack of information for decision making (Duda and Sherman 2002; Villaseñor and Amezcua 2013).

In particular, this book addresses the last issue—that a lack of information on the dynamics and functioning of these systems is available for decision making in this region. Developing adequate management measures for an activity as complex as artisanal fisheries in estuarine areas requires information on the biology and ecology of the exploited resources, as well as on the environmental, social, economic and other anthropogenic factors. Studies that consider all aspects of fishing activity in estuarine areas are needed. These sort of interdisciplinary studies have seldom been undertaken in this region, or they have only been published as memos or reports that are not widely available to managers, academic and scientific staff. Thus, this book compiles a series of studies with information that will be useful for proper estuarine management.

The present book consists of three parts, each covering a different topic associated with fisheries management of estuarine systems in the region: physicochemical studies, ecological studies and socioeconomic studies. All of the studies were undertaken in estuarine systems of the region with the participation of local institutions. The first section deals with

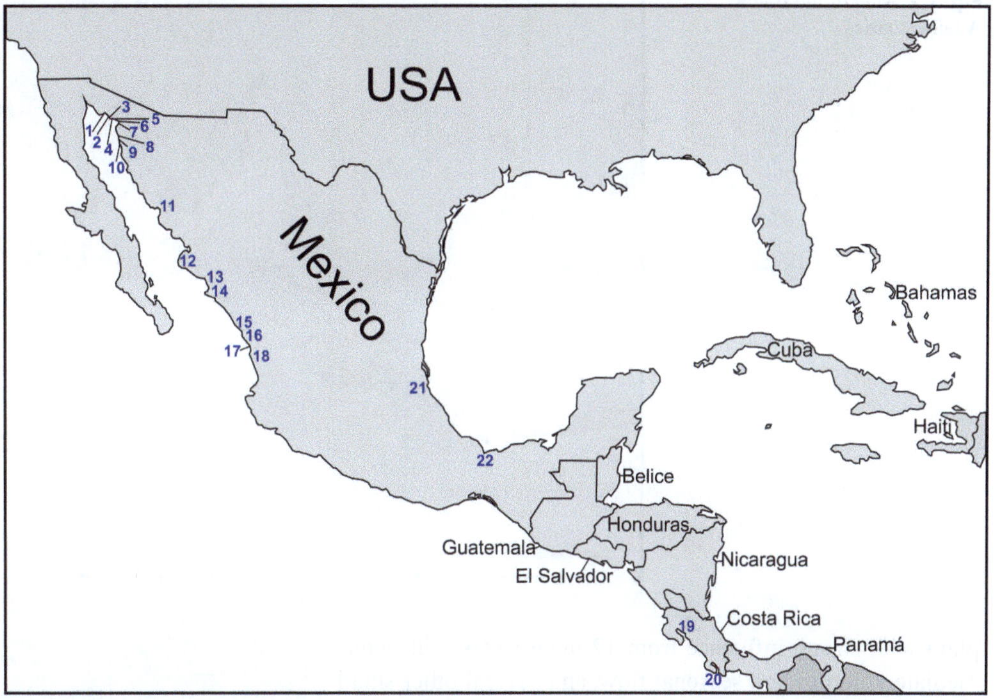

Fig. 1.1 Locations of the estuarine systems studied in this book. Numbers correspond to the systems discussed in Sect. 1.1

physicochemical factors and their relation to the living resources in estuarine zones. Chapter 2 examines the effect of human discharges on estuarine systems and the consequent effect to biota and fisheries. Chapter 3 reviews the presence of mercury (Hg), which is present in the edible parts of commercially exploited fishes and invertebrates captured by artisanal fishermen, and consequently poses a human health risk. Chapter 4 discusses the importance of determining the carrying capacity and water quality of coastal lagoons for management purposes. Chapter 5 examines the estuarine food web structure through the use of carbon and nitrogen stable isotopes, establishes the importance of the macrobenthos for exploited fish species, and provides essential information for an ecosystem-based management approach to these systems.

In the second section, four chapters review different aspects of the ecology of estuarine systems. Chapter 6 examines how hydrological changes in coastal lagoons affect mangrove forests and associated biota, and deals also with the hypothesis that more mangroves does not necessarily mean that fisheries will be more abundant, but rather there is an intermediate amount of mangrove coverage that produces maximum fishery yields. Chapter 7 reviews the effects of estuarine hydrodynamics on zooplankton and how hydrodynamics could affect zooplankton abundance and therefore, the abundance of commercially exploited species. Chapter 8 proposes a way to manage small-scale fisheries in these ecosystems using available information, rather than expecting to acquire all needed information to start proposing management plans. However, it is recognized that more information would improve management plans. Chapter 9 proposes an ecosystem-based approach to fisheries management of an extensive wetland region in the upper Gulf of California based on extensive information gathered in the area, and deals with the fact that such management is possible given that the necessary information is available.

The last section includes two chapters that discuss the human aspect of fisheries management in these areas. Chapter 10 reviews the small-scale fisheries management situation in Central America with emphasis on Costa Rica and proposes the creation of "Marine Areas for Responsible Fishing" as a management tool. Finally, Chap. 11 reviews the legal tools needed to achieve long-term protection of estuarine systems using existing policy instruments.

Although similar studies to those presented in this book have been undertaken in similar systems throughout the world, the results cannot be readily applied to the specific situations found in Mexico and Central America. The dynamics surrounding the specific ecological diversity and socioeconomic factors of the region are particular, as they are in other regions, and consequently local studies are essential to develop adequate management programs. A general description of the characteristics of the estuarine areas of Mexico and Central America covered in this book can be found in Sect. 1.1.

1.1 Estuarine Systems Studied in this Book (Fig. 1.1)

Mexican Pacific All of the estuarine systems from the Mexican Pacific mentioned in this book belong to the Eastern Coastal Plain of the Gulf of California. It is a fluvial-deltaic

Fig. 1.2 Map of the Bahía Adair estuaries

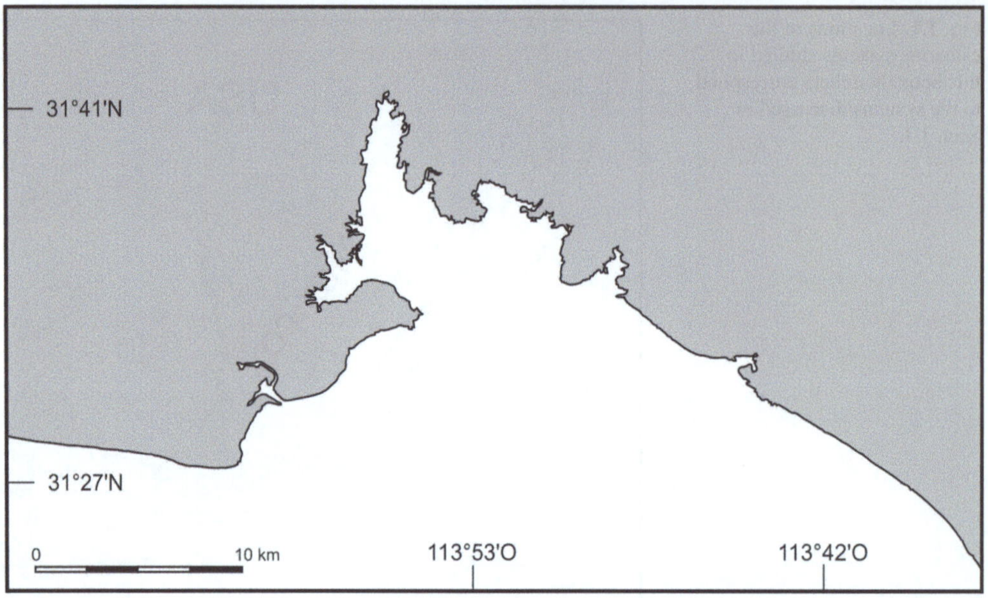

plain with strong influence from 12 main rivers with intermediate volumes and seasonal flow, and several other small tributaries with small, ephemeral drainage basins (Lankford 1977). The predominant sediment type is sedimentary rock and volcanic igneous rock from the Cenozoic. It has a wide depositional continental shelf. The weather in the north is dry with precipitation that ranges from 0 to 600 mm per year. To the south of the Urias estuarine system, the weather becomes hot and sub humid with an annual precipitation of 600 to 2000 mm. Geomorphologically, the coasts in this area are primary coasts formed mostly by non-marine agents according to Shepard (1984). Principally, the estuarine systems present in this area are subaerial deposition coasts created by rivers and wind (Contreras-Espinosa 1993). Also present are coastal mountain ranges with intermediate- to high-relief flank, wide to narrow coastal plains, and narrow watersheds (Lankford 1977). Precipitation occurs mostly in summer and increases with elevation and to the south. The continental shelf is usually narrow and irregular from 5–25-km wide, but widens to 70 km in north. Wave energy is highest along delta fronts and increases to the south near the mouth of the Gulf of California and tidal energy varies from intermediate in the south with high ebb velocity to extremely high in the north with high flood velocity (Lankford 1977).

1. Bahía Adair Estuaries (31°20′ N—115°36′ W to 31°34′ N—114°02′ W; Chaps. 8 and 10, Figs. 1.2 and 1.3) This area includes the Las Lisas and San Judas estuaries. Lankford (1977) classified these estuarine systems as Type III-A (I-C) coastal lagoons (i.e., they have a barred inner shelf with a Gilbert de Beaumont lagoon). The total area of these estuaries is 29226 ha, and the main habitats are halophyte marsh, mud flats, coastal dunes, salt flats and freshwater springs. The tidal regime is macrotidal and is mixed semidiurnally. The continental topography is regular and is characterized by broad, gently sloping plains that stretch from the sea to the mainland with some high points. There are dunes formed by wind erosion and sedimentary rocks are of various origins. The estuaries were formed by the interaction of extreme tides and intermittent deposits and have lacustrine and eolian deposits that form dunes. There are also artesian wells formed by deposits of sand and gravel from ancient rivers that have since been covered by clay dunes and salt flats that make them relatively impermeable. In these estuaries, 12 species of animals are under special protection from the Mexican government under the Norm NOM-059-SEMARNAT 2010, and/or have been listed by the Convention on International Trade in Endangered Species of Wild Fauna and Flora (www.cites.org) e.g., *Cyprindon*

Fig. 1.3 View of the Bahia Adair and its habitats

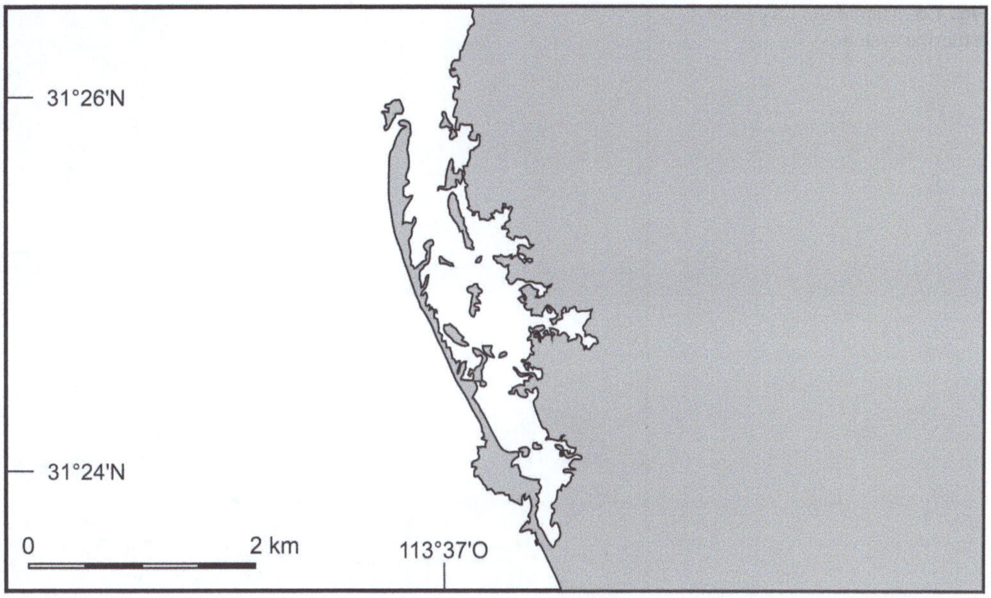

Fig. 1.4 Map of the Cerro Prieto estuarine system

macularis and *Lepidochelys olivacea*. There are 5 species of endemic plants and animals (e.g., *Leuresthes sardina* and *Distichlis palmeri*) and several commercial species that are recruited in these estuaries (e.g., *Litopenaeus stylirostris* and *Callinectes bellicosus*). The region connects bird populations in North America as part of the Pacific Flyway. At a landscape level, it is an intermediate zone between the Sonoran Desert and the marine ecosystem of the Gulf of California.

2. Cerro Prieto (31°27′ N—113°37′ W to 31°24′ N—113°37′ W; Chaps. 8 and 10, Figs. 1.4 and 1.5) This estuary has a total area of 1987 ha and the main habitats are halophyte marsh and mud flats. The tidal regime is macrotidal, mixed semidiurnal. There is no information available regarding its geomorphology and sedimentary processes, or information available on its ecological and biological characteristics.

3. La Cholla (31°22′ N—113°36′ W to 31°20′ N—113°37′ W; Chaps. 8 and 10, Figs. 1.6 and 1.7) La Cholla estuary is a small system of 235 ha, inhabited by halophyte marsh and mud flats; coquina reefs (consolidated shells) are also present. The tidal regime is macrotidal, mixed semidiurnal. There is no information available regarding its geomorphology and sedimentary processes, or its ecological and biological characteristics.

4. Morua (31°17′ N—113°28′ W to 31°16′ N—113°23′ W; Chaps. 8 and 10, Figs. 1.8 and 1.9) This estuarine system has a total area of 1097 ha, which is primarily composed of halophyte marsh and mud flat habitats; the tidal regime is

Fig. 1.5 View of the Cerro Prieto estuary and its habitats

macrotidal, mixed semidiurnal. The estuary is an elongated lagoon about 19.3-km long and 4.8-km wide at the mouth. A stabilized dune system separates it from the open coast. Morúa and La Pinta were previously part of a larger estuarine system fed by the Sonoyta River. There are over 140 resident and migratory bird species including the endangered willow flycatcher (*Empidonax traillii extimus*). There are avifauna and shorebirds present in the tidal flat habitats and lagoons such as the endangered least tern (*Sternula antillarum*). Commercial shrimp and blue crab (*Callinectes bellicosus*) use these wetlands during specific stages of their life cycle. About 10 to 15 species of halophytes inhabit these estuaries. There is also an endemic fish fauna, which includes such species as *Gillichthys seta*.

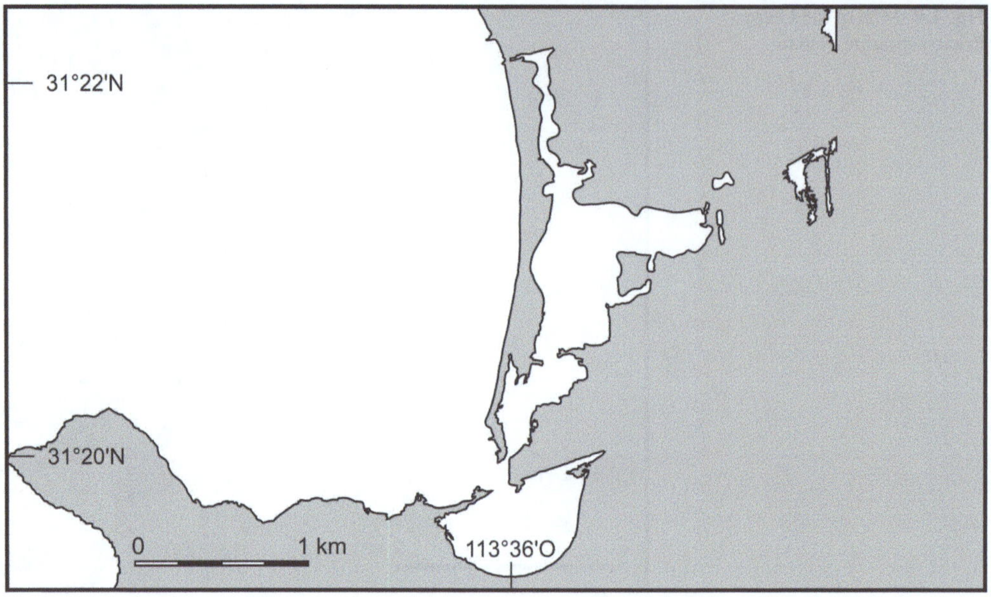

Fig. 1.6 Map of the La Cholla estuarine system

5. La Pinta (31°16′ N—113°16′ W to 31°13′ N—113°12′ W; Chaps. 8 and 10, Figs. 1.10 and 1.11) This estuarine system has a total area of 3338 ha and the main habitats are halophyte marsh and mudflats. In terms of geomorphology and sedimentary processes, this system is macrotidal, mixed semidiurnal. There is no information available on its geomorphology and sedimentary processes, or its ecological and biological characteristics.

6. Bahía San Jorge (30°58′ N—113°02′ W to 31°14′ N—113°10′ W; Chaps. 8 and 10, Figs. 1.12, 1.13 and 1.14) Bahia San Jorge includes the estuaries of Almejas and La Salina. The total area of these estuaries is 9874 ha and the main habitats are halophyte marsh, mudflats, coastal dunes, and salt flats. Lankford (1977) classified these estuarine systems as coastal lagoon type III-A (i.e., they have a barred inner shelf with a Gilbert de Beaumont lagoon). The tidal regime is macrotidal, mixed semidiurnal. The area contains igneous, metamorphic and granitic rocks. The continental topography includes unconsolidated plains and coastal landform tips that were transported by intermittent streams composed of alluvial silt, sand and gravel. There is a large sand bar about 10-km long composed of unconsolidated dunes. The sandbar separates the La Salina estuary from the open ocean. In this area there are 23 species of animals under special protection from the Federal Norm NOM-059-SEMARNAT 2010 and/or have been listed in CITES (www.cites.org) e.g., *Sterna antillarum*.
There are 7 species of endemic plants and animals in the estuaries (e.g., *Gillichthys seta* and *Suaeda puertopenascoa*).

7. San Francisquito (30°58′ N—113°05′ W to 30°56′ N—113°05′ W; Chaps. 8 and 10, Figs. 1.15, 1.16 and 1.17) This estuarine system has a total area of 543 ha and

Fig. 1.7 View of the La Cholla estuary and its habitats

the main habitats are halophyte marshes; the tidal regime is macrotidal, mixed semidiurnal. There is no further information regarding the geology and ecology of this area.

8. Delta del Rio Asunción (30°33′ N—113°W to 30°30′ N—112°57′ W; Chaps. 8 and 10, Figs. 1.18 and 1.19) This estuary has a total area of 9233 ha and the main habitats are brackish marsh vegetated by mesquites (*Prosopis* spp.) and halophyte marsh. The tidal regime is mixed semidiurnal. Information on the geomorphology, sedimentary processes, ecology and biology is not available for this system.

9. Los Tanques (30°27′ N—112°53′ W to 30°26′ N—112°52′ W; Chaps. 8 and 10, Figs. 1.20 and 1.21) The total area of this estuary is 543 ha, the main habitat is halophyte marsh and the tidal regime is mixed semidiurnal.

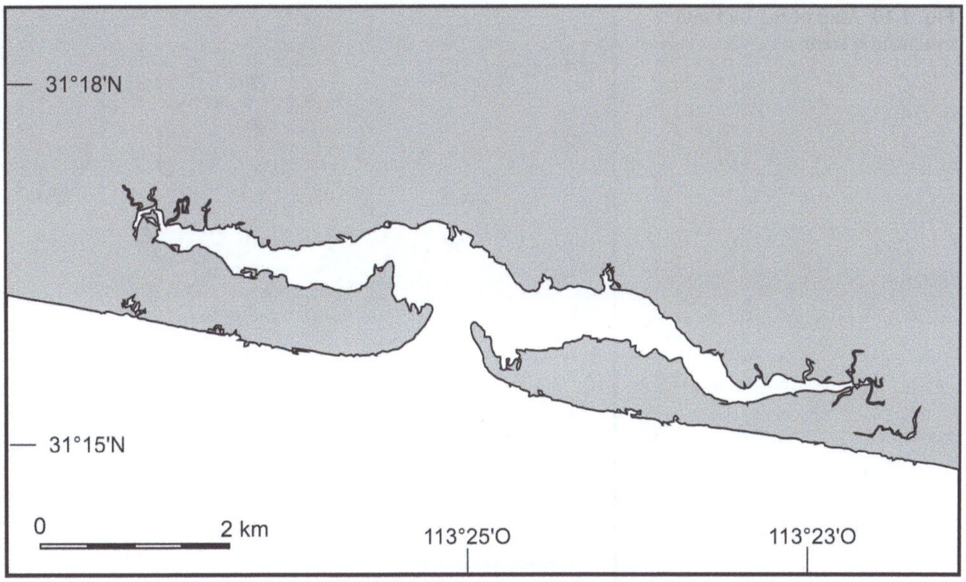

Fig. 1.8 Map of the Morua estuarine system

Information on the geomorphology, sedimentary processes, ecology and biology is not available for this system.

10. Puerto Lobos (30°16′ N—112°51′ W to 30°16′ N—112°50′ W; Chaps. 8 and 10, Figs. 1.22 and 1.23) This is a small system of only 4 ha. The main habitats are halophyte marsh and mangrove forest (black mangrove, *Avicennia germinans*) and the tidal regime is mixed semidiurnal. Information on the geomorphology, sedimentary processes, ecology and biology is not available for this system.

11. Guaymas (27°54′ N—110°48′ W to 27°59′ N—110°55′ W; Chap. 2, Figs. 1.24, 1.25 and 1.26) This estuary has a total area of 2,800 ha, the main habitat is mangrove forest (red mangrove, *Rhizophora mangle*; black mangrove, *Avicennia germinans*; and white mangrove, *Laguncularia racemosa*) and the tidal regime is mixed semidiurnal. In terms of the geomorphology and sedimentary processes, this system is bordered with tertiary acidic extrusive igneous rocks. Lankford (1977) classified this estuarine system as a type III-B (I-E) coastal lagoon (i.e., it has a barred inner shelf with a cuspate lagoon). There is no further information on other aspects of this system.

12. Topolobampo-Navachiste estuarine complex (25°45′ N—109°25′ W to 25°17′ N—108°30′ W; Chap. 7; Figs. 1.27, 1.28, 1.29, 1.30 and 1.31) This estuarine complex includes the estuarine systems of Colorado, Santa María, Topolobampo, Ohuira, San Ignacio and Navachiste, which is part of the El Fuerte River delta. Lankford (1977) classified these estuarine systems as type II-A (I-C) coastal lagoons (i.e., they have an intra-deltaic depression with a flooded valley and a barrier separating the system from the sea). The delta is formed of alluvium of the Pleistocene. The sediments come from the weathering of volcanic rocks from the Cretaceous period (Ayala-Castañares et al. 2003). The watered area of the entire complex is 20,840 ha. The annual precipitation is 241 mm (Berlanga-Robles and Ruiz-Luna 2003) and the mean depth ranges between 2 and 4 m, although there are channels up to 20-m deep. This estuarine complex does not receive runoff from rivers, but only discharge from irrigation channels, which are the main source of pollutants in these systems. Other sources of contaminants are from domestic effluents of the city of Los Mochis and surrounding settlements. The water temperature oscillates from 20.2 to 33.5 °C and salinity ranges from 30.3 to 37.9 ppm.

Fig. 1.9 View of the Morua estuary and its habitats

The vegetation of the area is typical of deserts. Mangrove forests begin in the plains and are composed by red mangrove (*Rhizophora mangle*) and black mangrove (*Avicennia germinans*) species.

The main species of mollusks are *Crassinella pacifica*, *Lucina approximata* and *Chione compta*. The most representative

Fig. 1.10 Map of the La Pinta estuarine system

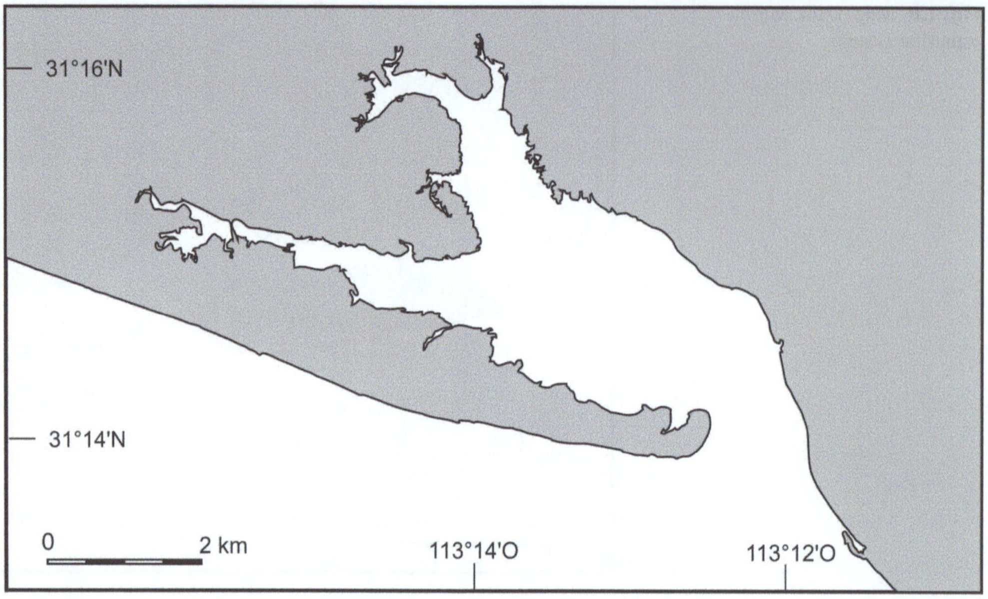

crustaceans are from the Penaeidae and Portunidae families. The most representative fish species are from the families Mugilidae, Sciaenidae and Scombridae.

13. Santa María La Reforma (25°17′ N—108°30′ to 24°41′ N—108°01′ W; Chap. 7; Figs. 1.32, 1.33 and 1.34) This is the largest estuarine system of the state of Sinaloa, and has a high flora and fauna diversity; the watered area is about 51,172 ha and is divided in 2 sections due to the presence of 2 large islands. It is classified as a type III-A (III-C) system, which is an inner-shelf coastal lagoon with permanent communication to the sea and semiparallel orientation to the coast (Lankford 1977). The coastal lagoon is separated from the sea by a 32-km long sand barrier; this barrier has 10-m high dunes, and at each end of this barrier there are 2 channels that connect this system to the sea. Depth generally varies from 12 to 20 m with a mean depth of 7 m; however, maximum depth is 24 m (Zamora-Arroyo et al. 2000). The dominant sediment types are sand and silty sand with some silt and clay; however, in the central part of the estuary there is gravel where the tidal waves converge (de la Lanza and Cáceres 1994). Mean annual precipitation ranges from 400 to 500 mm and is highest during summer and lowest during winter. The main fresh water input comes from irrigation channels used for agricultural purposes. The lower basin of this system has an adjacent human population of almost 170,000 inhabitants. Agricultural activities are commonplace and the most important crops include corn, wheat, peas and beans. At the margins of the system there are approximately 7,700 ha of shrimp ponds.

The vegetations is dominated by mangrove forests consisting of black mangrove (*Avicennia germinans*), button mangrove (*Conocarpus erectus*) and white mangrove (*Laguncularia racemosa*) as well as halophyte vegetation, which are important for a large community of migratory birds (Flores-Verdugo et al. 1993). One-hundred ninety-one fish species have been recorded and the most abundant are the moonfish, *Selene peruviana,* and the puffer, *Sphoeroides annulatus* (Amezcua and Amezcua-Linares 2014)

Fig. 1.11 Aerial view of the La Pinta estuarine system

14. Altata—Ensenada del Pabellón (24°41′ N—108°01′ W′ to 23°55′ N—106°58′ W; Chaps. 2 and 7, Figs. 1.35, and 1.36) This estuary has a total area of 36,000 ha, the main habitats are mangrove forests and halophyte marshes, and the tidal regime is mixed semidiurnal. Lankford (1977) classified these estuarine systems as type III-C (I-D) coastal lagoons (i.e., they have a barred inner shelf with strand plain depressions). In terms of the geomorphology and sedimentary processes, this system has a grain size that ranges from fine grain sediments to sands although sandy sediments predominate in the navigation channel. In terms of ecological and biological characteristics, the highest primary

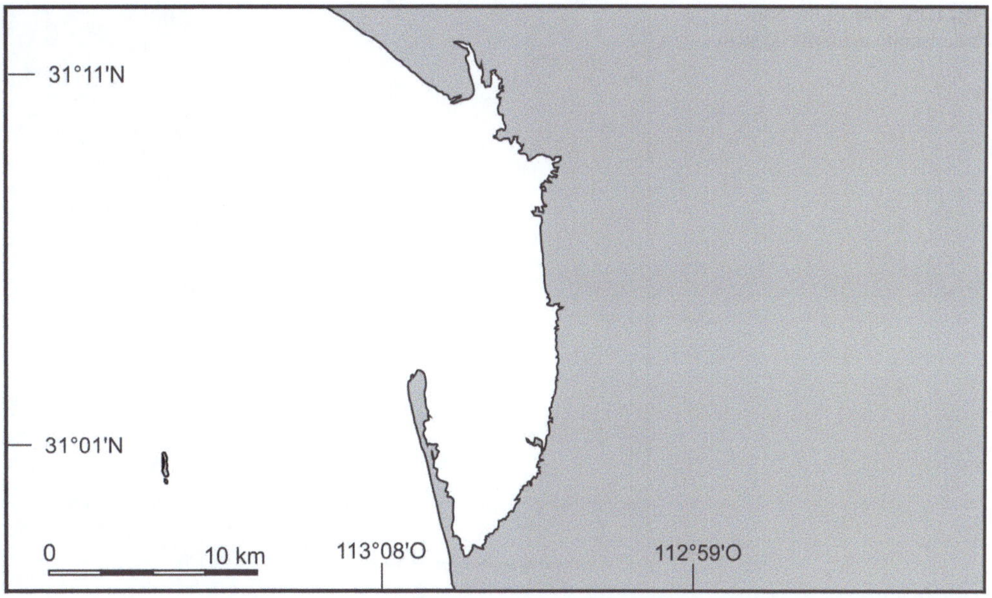

Fig. 1.12 Map of the Bahia San Jorge estuaries

Fig. 1.13 Aerial view of the Bahia San Jorge estuaries

Fig. 1.14 Halophyte marsh habitat of the Bahia San Jorge

production of this system occurs during spring and summer (5.34 gC m^{-3} d^{-1}). Diatoms and dinoflagellates predominate the phytoplanktonic community. Among benthic macroalgae, *Rhizoclonium* sp, *Hydrocoleum* sp. and *Chaetomorpha* sp. predominate.

15. Urias (23°55′ N—106°58′ W to 23°07′ N—106°19′ W; Chaps. 2, 4 and 7; Figs. 1.37, 1.38, 1.39, 1.40 and 1.41) The Urias estuarine system is a coastal body of water classified as a coastal lagoon with an internal platform barrier. The lagoon is tidally dominated and characterized by a mixed tide with an average range of about 1.0 m. Because of limited fresh water discharges through some streams during the rainy seasons, the salinity is usually high during the year in the range of 25.8–38.4 ‰. Water depth varies between 1 and 3 m except in the navigation channel where it is up to 13 m. The exchange of lagoon and ocean waters occurs in this channel, which is periodically dredged to maintain the required depth. Prevailing winds are associated with weather systems from the northwest. A strong tidal current (up to 0.60 m/s) within the navigation channel characterizes the flow of water into the lagoon during the flood tide (t=6 h and t=18 h), while the flow reverses its direction during the ebb (t=12 h and t=24 h). In the inner part of the lagoon, low velocity tidal currents occur (<0.30 m/s) as the lagoon becomes very shallow (1–4 m water depth) (Montaño-Ley et al. 2008).

This estuarine system is classified as a type II coastal lagoon, with a differential terrigenous sedimentation associated with fluvial/deltaic systems produced by irregular sedimentation and/or surface subsidence due to compaction/loading effects. It has a sandy barrier, and the tidal energy is low except in the navigation channel (Lankford 1977).

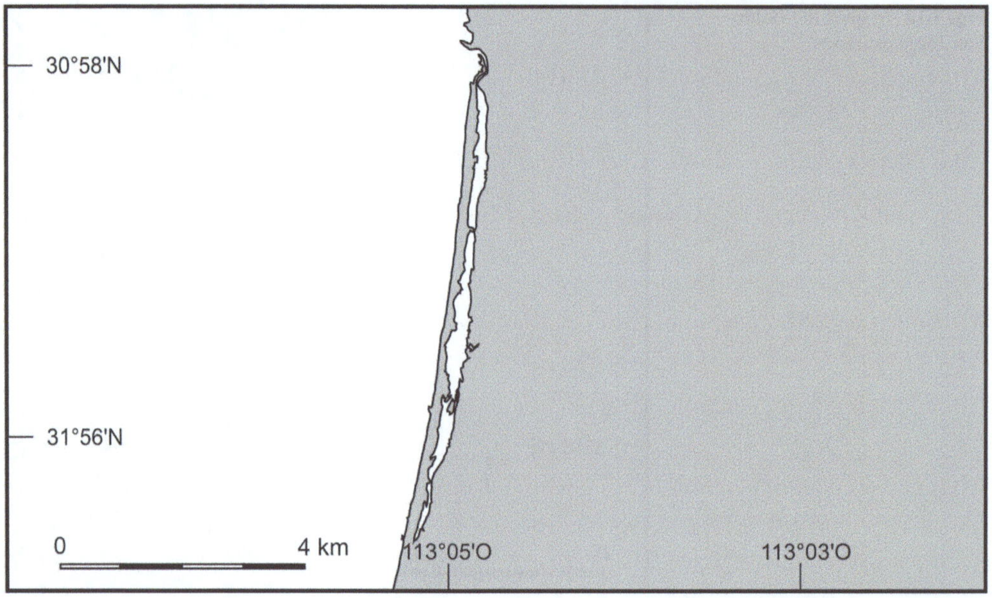

Fig. 1.15 Map of the San Francisquito estuarine system

Fig. 1.16 View of the San Francisquito estuary and its habitats

Fig. 1.17 Aerial view of the San Francisquito estuary

The Urias estuary is occasionally affected by tropical storms originating in the northeast Pacific; these storms generate waves that are mostly received from a west-southwest direction. Dominant littoral transport in the study area is toward the northwest along the coastline, with movement of sands toward the beach line and open sea (Montaño-Ley et al. 2008).

Sediments in the studied sites were dominated by mud, ranging from 86.2 to 92.5%, and organic matter content varying from 3.75 to 7.86%. High mud content and levels of organic matter in sediments are related to the low energy from waves and channel currents in the study area; mangrove root systems trap floating detritus and reduce tidal flow, which allows the settling of suspended clay and silt particles. Others zones of the lagoon are characterized by a predominance of sands with variable organic carbon (ranging from 0.6 to 5.2%) (Soto-Jiménez and Páez-Osuna 2001).

This system is characterized by the presence of mangroves (*Rhizophora mangle, Laguncularia racemosa* and *Avicennia germinans*) along the margins. They possess high biological diversity and a rich and complex food chain. They are also a large organic matter reserve and create a diverse habitat that creates optimal niches for numerous endemic and migratory aquatic species that utilize these areas as refuge and/or breeding grounds (Flores-Verdugo et al. 1987). The most representative species of this system are the crustaceans *Litopenaeus vannamei* and *Callinectes arcuatus* (Hendrickx et al. 2002), the fishes *Gerres cinereus*, *Haemulopsis leuciscus* and *Lutjanus argentiventris* (Fischer et al. 1995) and the birds *Tringa semipalmata*, *Numenius*

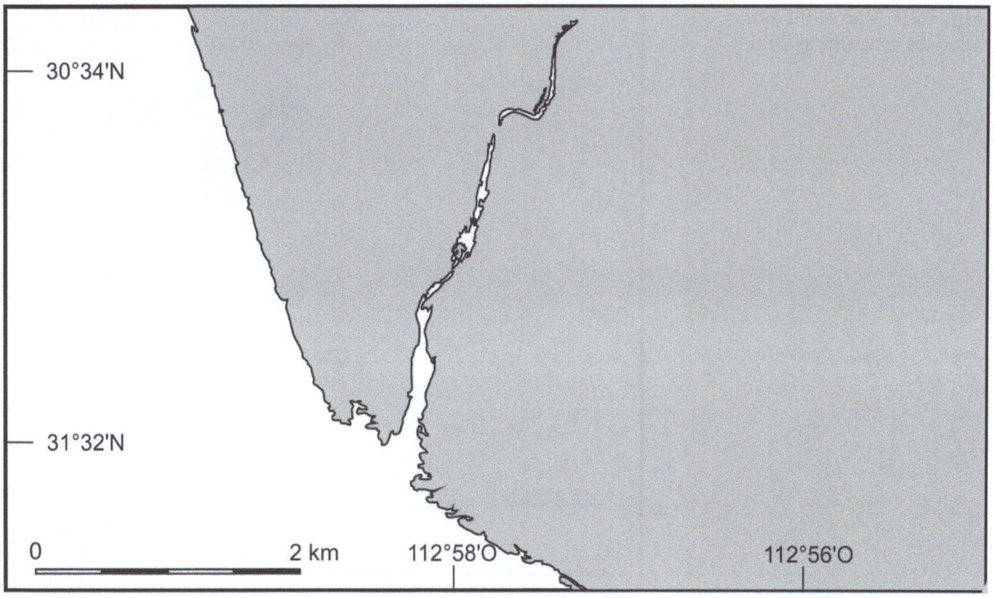

Fig. 1.18 Map of the Delta del Rio Asunción estuarine system

phaeopus, *Numenius americanus* and *Limosa fedoa* (Sauma-Castillo 2010).

16. Huizache-Caimanero and Baluarte estuary (23°07′ N—106°19′ W to 22°49′ N—106°00′ W; Chaps. 2 and 7, Figs. 1.42, 1.43, 1.44, 1.45) This estuarine system has an area of 17,100 ha approximately. It is a type III-A coastal lagoon (i.e., it has a barred inner shelf with a Gilbert de Beaumont lagoon). There are many small currents that flow from the channels towards the main lagoon body, and flow velocities are higher during the rainy season. There is a sand barrier that limits the lagoon extension and its morphology varies throughout its length. In the inner part, there are irregularly shaped sand dunes. The external part of the barrier has parallel sand dunes. The part of the barrier that is adjacent to the rivers has an aggregation of alluvial deposits originating from the river's delta. Most of the sediments from this system are silty clays with some zones of sand and silt. At the barrier the sediments are mainly sands with some silt and clay (Ayala-Castañares et al. 1970).

During the dry season, the water level decreases dramatically and some parts of the lagoon become totally dry, which creates a flood plain.

The main types of vegetation include mangrove forests and deciduous rain forest. At the margins of the main water body there are halophyte associations. A type of sea grass (*Ruppia* sp) covers the bottom of the lagoon.

17. Teacapán (22°49′N—106°00′W to 22°32′N—105°45′ W; Chap. 7; Figs. 1.46, 1.47, 1.48 and 1.49) This system is the boundary between the States of Sinaloa and Nayarit. It is a type III-A (III-C) coastal lagoon with a permanent connection to the sea (Lankford 1977), and includes a complex of estuaries and minor coastal lagoons with a size of approximately 5,500 ha. This is a shallow system that connects to the sea through a 1-km wide channel and a depth that ranges from 3 to 9 m. The system is influenced by semidiurnal tides with high (spring) tides reaching 2 m. Stratigraphic analysis shows the existence of both palustrian and lacustrian sediments of the Quaternary consisting of sand, silt, mud and peat deposits. The soils of the alluvial plain, seasonal flood plains and mangroves are solonchak gleyco, solonchak ortic, and solonchak takiric (Flores-Verdugo et al. 1997). Mean annual relative humidity, at 14:00 hr., varies from 65 to 70 %. The mean annual temperature exceeds 18 °C with dominant northwesterly winds in winter and a system of breezes from the southeast during summer (Garcia and Trejo 1990). This system is influenced by the discharge of 6 rivers and several streams, but only 2 have permanent water flow throughout the year. Average annual rainfall is 1,459 mm and the evaporation rate is 1,991 mm/year (Flores-Verdugo et al. 1997).

There is a human settlement of approximately 97,000 inhabitants at sub basin drainage associated with this system,

Fig. 1.19 Aerial view of the Delta del Rio Asunción estuarine system

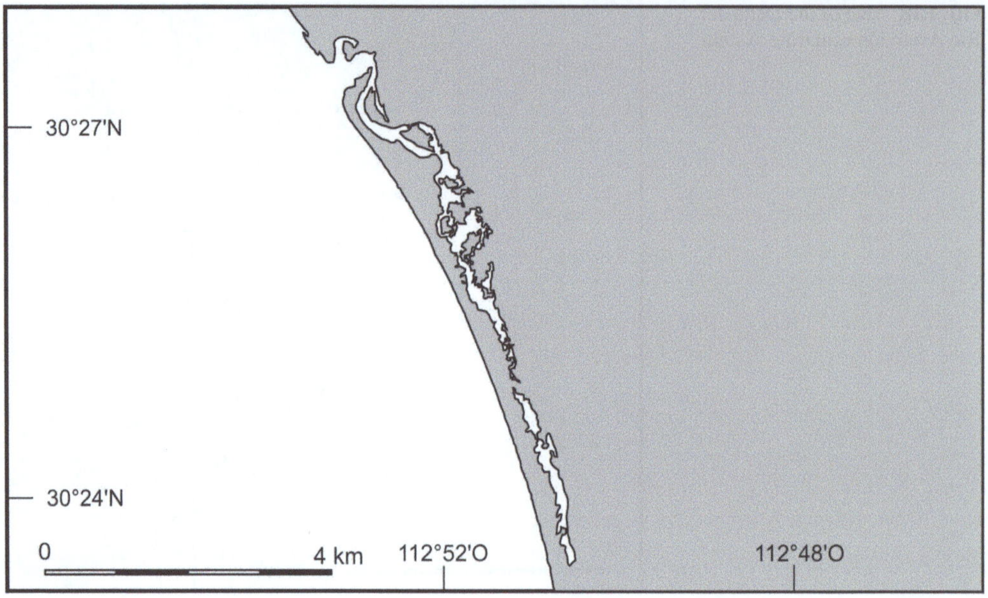

Fig. 1.20 Map of the Los Tanques estuarine system

Fig. 1.21 Aerial view of the Los Tanques estuary

which hosts the largest mangrove forest along the Pacific coast in Mexico with an extension of 83,000 ha, accounting for 15 to 20% of the total area of mangrove forests in Mexico (De la Lanza et al. 1993).). It is inhabited by white mangrove, red mangrove, black mangrove and button mangrove. This system hosts a high diversity of fishes, crustaceans, and molluscs (Flores-Verdugo et al. 1990) and provides an important site for migratory wild1ife and endangered species such as the jaguar (*Felix onca*) and the crocodile (*Crocodylus acutus*). It has been chosen as a priority region for mangrove conservation in Mexico, and is listed with the Ramsar Convention on Wetlands.

18. Marismas Nacionales (22°50′ N—106°01′ W to 21°31′ N—105°17′ W; Chaps. 3 and 5; Figs. 1.50, 1.51, 1.52, 1.53, 1.54 and 1.55) The mangrove-estuarine complex of Marismas Nacionales comprises more than 120,000 ha of mangrove wetlands, coastal lagoons, tidal channels, mud flats, seasonal flood plains and freshwater swamps amongst river floodplains. This complex also contains sand barriers consisting of approximately 280 sub-parallel ridges that were formed by successive accretion of narrow beach ridges during the Holocene Transgression, from 3,600 to 4,750 years BP. This Transgression was locally balanced by a continuous deposition of sand along the shore, causing a progradation in the coast that was colonized by mangroves while the eustatic sea level was still rising. Currently, this is the most extensive mangrove region (80,000 ha) of the American Pacific. It contains riverine mangroves as tall as 15 m, fringe mangroves of 7–8 m height, overwash mangroves up to 4 m high, and a dense forest of dwarf mangroves. It is

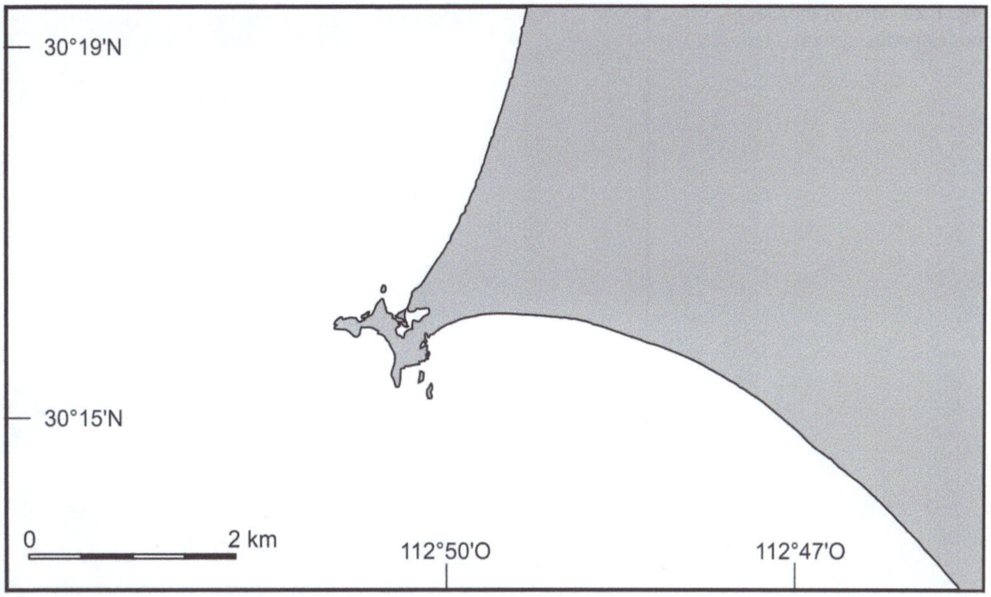

Fig. 1.22 Map of the Puerto Lobos estuarine system

Fig. 1.23 Aerial view of the Puerto Lobos estuary

considered a RAMSAR site (80% of the pacific trend for migratory birds use the region during winter) and a significant part was declared a Biosphere Reserve. The region has high biodiversity, which includes endangered species such as jaguar and crocodiles.

The presence of more than 3,000 shell middens ("conchales") of *Tivela* spp and oysters suggest that fishing was an important activity since prehispanic times (5,000 years BP). After the classical violent expeditions of Nuño de Guzmán between 1530–1532, the Spanish soldiers mentioned an indigenous populated region that produced crops of maize in the alluvial plains. They also described high abundances of fish, crustaceans, mollusks and a good hunting ground for deer, hare and ducks.

In 1974 an artificial inlet was opened to improve the fisheries in the region; however, this caused a notorious hydrological change from predominately brackish-freshwater wetlands to marine-hyperhaline waters. This change negatively affected the area and killed approximately 8,000 ha of mangroves. The artificial inlet expanded from about 40-m wide and 3-m deep to more than 1500-m wide and 12-m depth. To this day, the channel continues to grow in width and depth.

Central American Pacific (Costa Rica) Geographically and biologically, the 1254-km long Pacific coast of Costa Rica features diverse habitats such as headlands, cliffs, peninsulas, bays, islands, coral reefs, beaches, sandspits and estuaries with sandy-muddy coasts and mangrove wetlands due to its tectonic origin and the many rivers draining the central mountains (Allen and Robertson 1998; Bergoeing 1998; Robertson and Allen 2008). Species diversity in the area is the highest in the Eastern Pacific region, with at least 750 fish species associated with a high level of endemism (Mora and Robertson 2005a, b). The Nicoya and Osa peninsulas encompass two major gulfs, the Gulf of Nicoya dominating the central-northern Pacific sector and Golfo Dulce in the South Pacific Zone near the Panamanian border. Precipitation increases from north to south, from 1600 mm and distinct seasonality in Puntarenas (Golfo de Nicoya) to 4500 mm in Golfito (Golfo Dulce).where the dry season is comparatively short (Quesada-Alpízar and Cortés 2006 for Golfo Dulce). Other geographic and climactic traits of Costa Rica include high-relief flank of coastal mountain ranges; narrow watersheds; numerous rivers with small drainage basins; a coastal sub-humid climate that becomes very humid to the southeast with precipitation in summer; small-volume rivers with marked seasonal flow that may become dry in winter; a narrow continental shelf varying from 5–10 km that is wider towards the southeast; very high wave energy on open, exposed coasts; and high tidal energy with high ebb velocity (Lankford 1977).

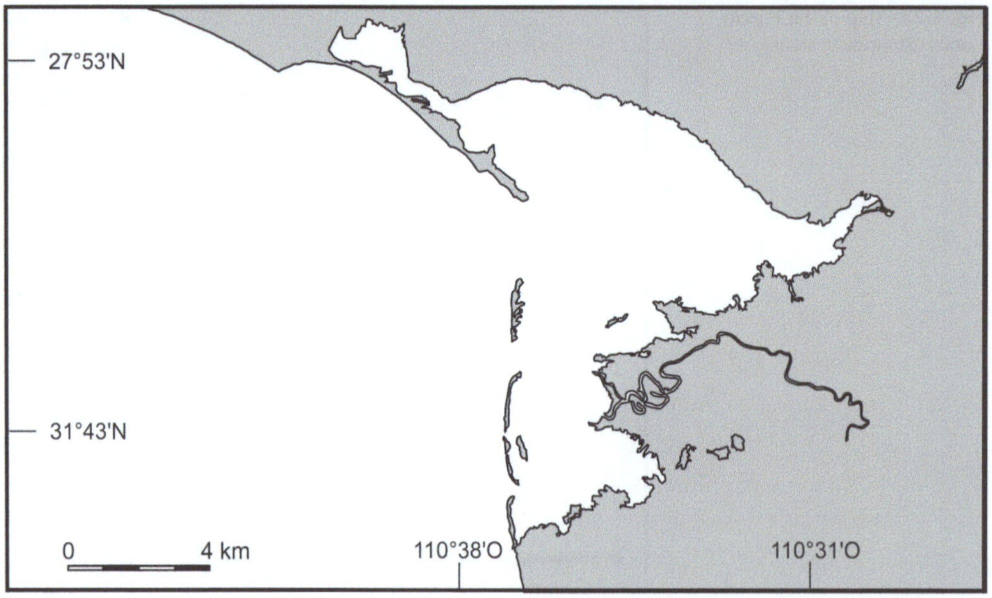

Fig. 1.24 Map of the Guaymas estuarine system

Fig. 1.25 View of the entrance of the Guaymas estuarine system

Fig. 1.26 View of the Gulf of California from a mountain within the Guaymas estuary

19. Golfo de Nicoya (9°30′ N—84°32′ W to 10°12′ N—85°14′ W; Chap. 9; Figs. 1.56, 1.57 and 1.58) The Gulf of Nicoya, on the northern Pacific coast of Costa Rica, represents one of the largest estuaries of Central America. It is about 80-km long from its narrowest northern part (the mouth of the Tempisque River) to its 55-km-wide border with the open ocean (Epifanio et al. 1983). The area of the system is approximately 1,380,000 ha and is divided into two zones. The shallow upper gulf (with a depth of 25 m or less), north of Puntarenas, includes five large islands and is fringed by extensive mangrove stands (about 15,200 ha, Jiménez 1999), sandy-muddy flats and fine, organically enriched sea-bottom sediments (Maurer and Vargas 1984). Two major freshwater inputs come from the Tempisque River on the northwest end, and the Estero Chomes on the eastern edge. The latter is an estuary that gathers water from several moderately sized rivers. Biological cycling, estuarine circulation and land run-off are important nutrient sources (Gocke et al. 2001b). Additional food web energy sources are supplied by epiphytobenthos and mangrove detritus (Wolff et al. 1998, Gocke et al. 2001a, b). The inner gulf is highly productive (net 610 $Cm^{-2}d^{-1}$), with an excess of primary phytoplankton production that is exported to the outer gulf (Gocke et al. 2001b). The outer gulf, dominated by the outflow of the Grande de Tárcoles River, deepens to over 200 m. Together with the Tempisque River and the Estero Chomes, the catchment basin is about 9844 km^2. Its coastline is characterized by rocky shores and sandy beaches. A majority of nitrogen entering the outer system comes from offshore deep water, which is upwelled into the gulf (Voorhis et al. 1983). The

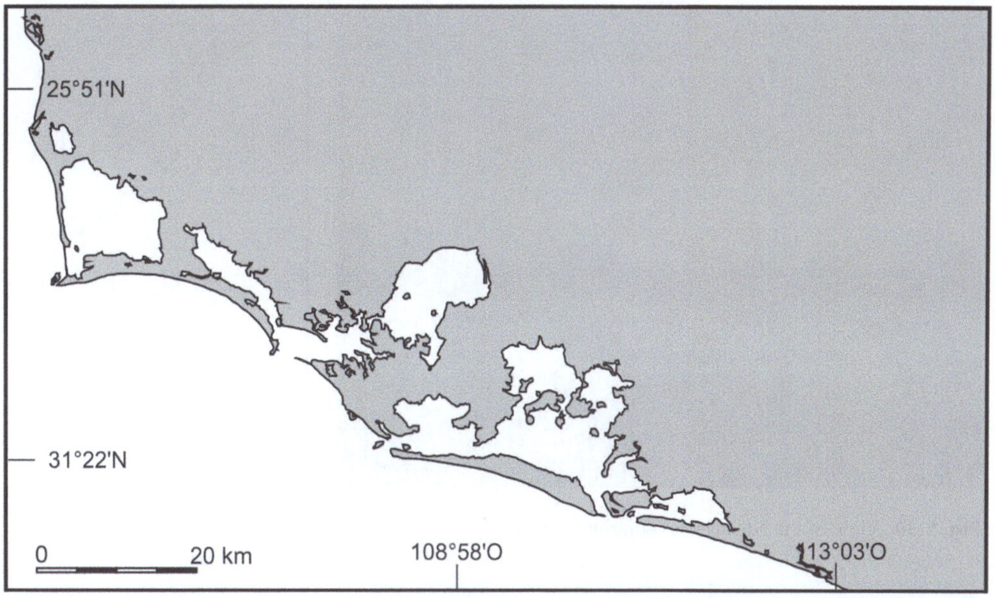

Fig. 1.27 Map of the Topolobampo-Navachiste estuarine complex

Fig. 1.28 View of Bahia Ohuira (Ohuira Bay)

Fig. 1.29 View of the Ohuira estuarine system

plume of the Tárcoles River often expands towards the gulf's mouth from the southeastern areas outside the gulf. Due to its high productivity and diversity of habitats, the Golfo de Nicoya is the most important fishing area in the country in terms of fish production and the number of artisanal fishermen that have historically operated there. Tides are on a semidiurnal regime, with spring tides averaging 280 cm in Golfo de Nicoya (Lizano 2006).

20. Golfo Dulce (8°22′ N—83°87′ W to 8°43′ N—83°29′ W; Chap. 9; Figs. 1.59, 1.60 and 1.61) Golfo Dulce on the southern Pacific coast of Costa Rica is a periodically anoxic fjord-like embayment, one of only four anoxic basins known to exist in the Tropics and the only one along the Pacific coast of the Americas. With a length of 50 km and a width of 10–15 km, the area of the water mirror is approximately 700,000 ha and is divided into 2 sections. A steeply sloped deep inner basin in the northern waters attain depths of 215 m and are sheltered against the open Pacific by an outer part with gentler slopes and a shallow sill (60 m), which allows limited water exchange with the ocean. The rugged coastline of the inner basin is covered by parts of the last primary rainforest on the Pacific coast of Central America (Allen 1977); a few sandy beaches and estuaries alternate throughout the landscape. The coast of the outer basin of Golfo Dulce features alluvial deposits interspersed with low mountains, where rocky headlands and cliffs alter with extensive sandy beaches and estuarine mudflats. Due to this topography, mangroves are not important (<1,000 ha; Jiménez 1999). The climate is humid tropical with a marked rainy season from May to November. Monthly average precipitation varies from a maximum of above 700 mm in

Fig. 1.30 View of the Navachiste estuarine system

Fig. 1.31 Small scale fishers' skiffs in the Navachiste estuarine system

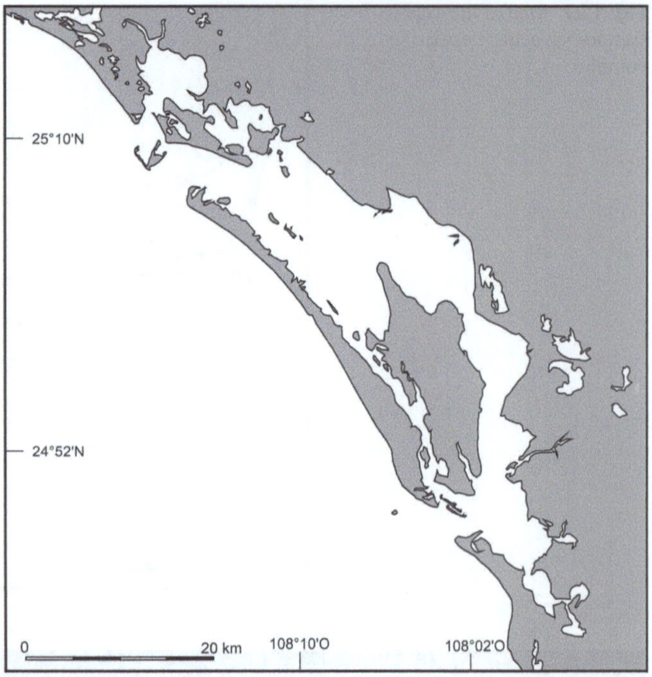

Fig. 1.32 Map of the Santa María La Reforma estuarine system

Fig. 1.33 View of one of the islands of the Santa María La Reforma estuarine system

Fig. 1.34 A view of the La Reforma fishing village

October to a minimum of about 100 mm in February resembling the dominant Pacific Central. Precipitation exceeds evaporation during at least eight months of the year (Herrera 1985). The total rainfall per year is more than 4.5 m. The main source of freshwater is from four rivers: the Coto, Tigre, Esquinas and Rincón, which drain a catchment basin of about 3200 km^2 (Quesada-Alpízar and Cortés 2006). A large proportion of non-biogenic material (>90%) in sediments reflect the dominance of terrigenous inputs to Golfo Dulce. Biogenic components such as organic carbon and carbonate are also supplied from terrigenous sources (Hebbeln and Cortés 2001). The morphology of the Golfo Dulce Basin, as in temperate fjords, and the effects of El Niño-Southern Oscillation (ENSO), lead to a strong vertical stratification of the water column (Quesada-Alpízar and Morales-Ramírez 2004), which restricts circulation in the inner basin and promotes the development of deep anoxic conditions (Richards et al. 1971; Dalsgaard et al. 2003). Nutrient inputs to surface waters are primarily from river runoff. The Golfo Dulce's net primary productivity is comparatively low as compared to

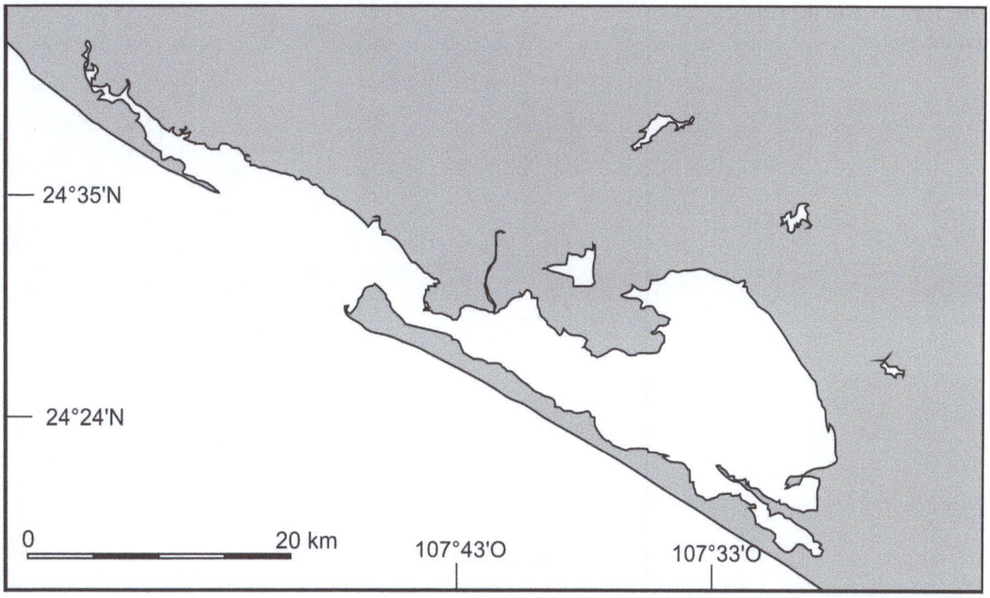

Fig. 1.35 Map of the Altata—Ensenada del Pabellón estuarine system

the highly productive upper gulf of Nicoya (Jiménez 1999). Due to its low productivity and the geographic isolation of this border region, which is far from major urban centers, the fisheries productivity and number of artisanal fishers in Golfo Dulce is much smaller than in the Nicoya gulf where more than half of Costa Rica's artisanal fishers live. Below 50-m depth, fishery production is negligible (HMU, Pers. Obs.), partially due to hypoxic and anoxic conditions.

Golfo Dulce is an impacted estuary with relatively high metal concentrations as compared to other sites (García-Céspedes et al. 2004). Wave energy in the gulf is low, but increases off the open, exposed coasts of the Gulf's mouth. Tides are on a semidiurnal regime, with spring tides averaging 289 cm in inner parts of Golfo Dulce (Lizano 2006). The barotropic tidal component is weakest in the inner deeper part, ~0.1 m s^{-1}, while in the shallow sill area it is strong (0.5 m s^{-1}). Wind conditions are characterized by a monsoon like pattern.

Mexican Gulf of Mexico All of the estuarine systems from the Gulf of Mexico mentioned in this book belong to the Coastal Plain of the Gulf of Mexico and Yucatan Peninsula. The estuarine systems are located on a large fluvial-deltaic plain with influence from a large number of major rivers with large drainage basins; some of the largest rivers influence this area including the Tuxpan, Pánuco and Coatzacoalcos. The continental shelf in this region is typically narrow but varies from 10 to 150 km, and the predominant sediment type is sedimentary rock and volcanic igneous rock from the Cenozoic. Low-relief coastal plains that are 30- to 150-km wide are typical and low sea cliffs may also occur. The weather in the region is hot and humid with yearly precipitation that ranges from 600 to more than 4,000 mm;

Fig. 1.36 Aerial view of the Altata—Ensenada del Pabellón estuarine system

however, it is higher during summer and increases markedly with elevation. Geomorphologically, both estuaries are primary coasts formed mostly by non-marine agents according to Shepard (1984). Principally, the estuarine systems present in this area are subaerial deposition coasts from rivers and wind (Contreras-Espinosa 1993).

Wave energy is low to intermediate except for summer hurricanes and winter "northers", which are waves associated with polar air mass movement. Tidal energy is low except for storm surges (Lankford 1977).

21. Tampamachoco Lagoon (20°18′ N—97°′ W to 20°18′ N—98°W; Chap. 1; Figs. 1.62 and 1.63) Tampamachoco lagoon is located in the northern region of the State of Veracruz, with an area of 1,500 ha and about 1.5-m

Fig. 1.37 Map of the Urias estuarine system

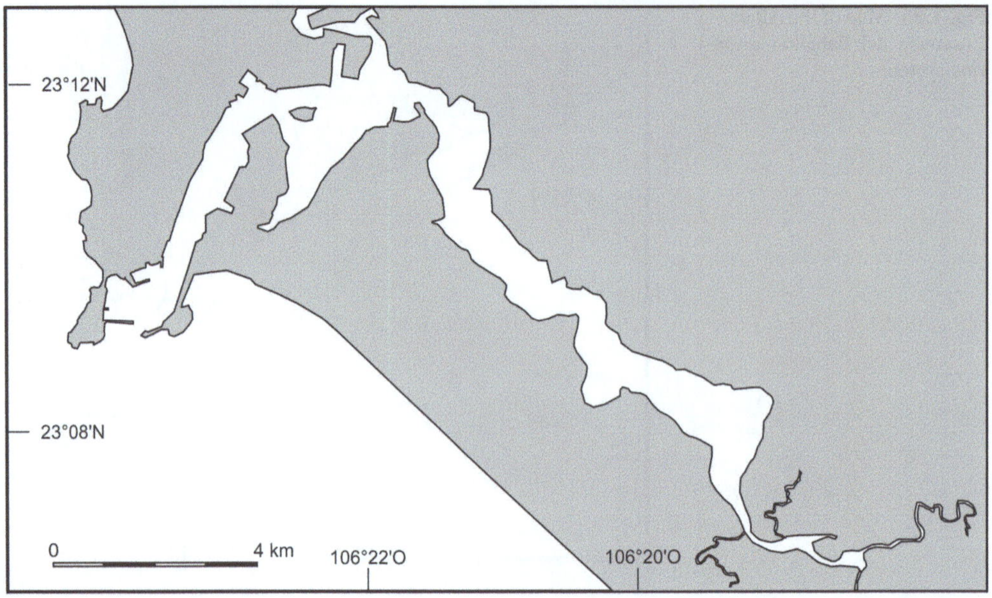

Fig. 1.38 Mangrove forest in the Urias estuarine system

Fig. 1.40 Pelicans landing in front of Mazatlan's shrimp fishing fleet shelter in the Urias estuarine system

Fig. 1.39 Thermoelectric complex along the shore of the Urias estuarine system

Fig. 1.41 Tourist cruise harbour in the Urias estuarine system

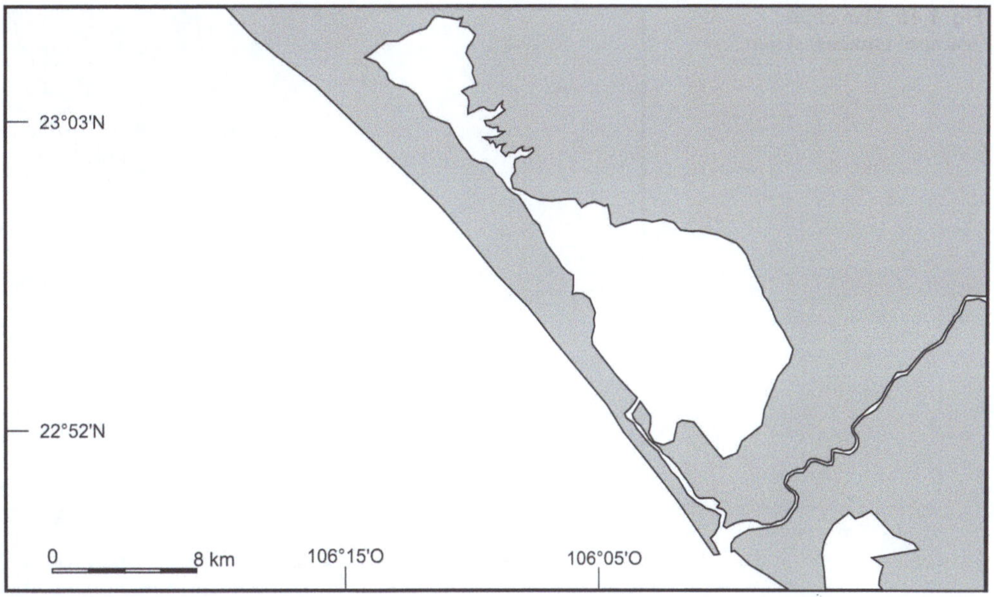

Fig. 1.42 Map of the Huizache-Caimanero estuarine system

Fig. 1.43 Artisanal fishing boat in the Caimanero estuary

Fig. 1.45 View of a "tapo" (fish barrier) in the Huizache estuary

Fig. 1.44 View of the Caimanero estuary and its habitats

depth. It is a type II coastal lagoon (Lankford 1977), which corresponds to a coastal lagoon associated with fluvial deltaic systems formed by irregular sedimentation or surface subsidence that causes compaction of the bed load. Sandy barriers are formed rapidly and surround very shallow marginal or intra-deltaic depressions; the deltas contain a small supply of sediment that may be shallow and is frequently short-lived, causing formation of elongated lagoons between beach hills. The barriers of the deltaic depressions can be composed of mud, sand, or mangroves and are commonly bed by direct runoff from rivers and tributaries. This causes slow changes in shape and bathymetry, though some changes may be rapid. Tidal energy is typically very low, except in the canals. Salinity is also low, but may vary with river input. The Tuxpan River, which empties into the lagoon, has a drainage area of 4441 km^2, a length of 150 km, an annual

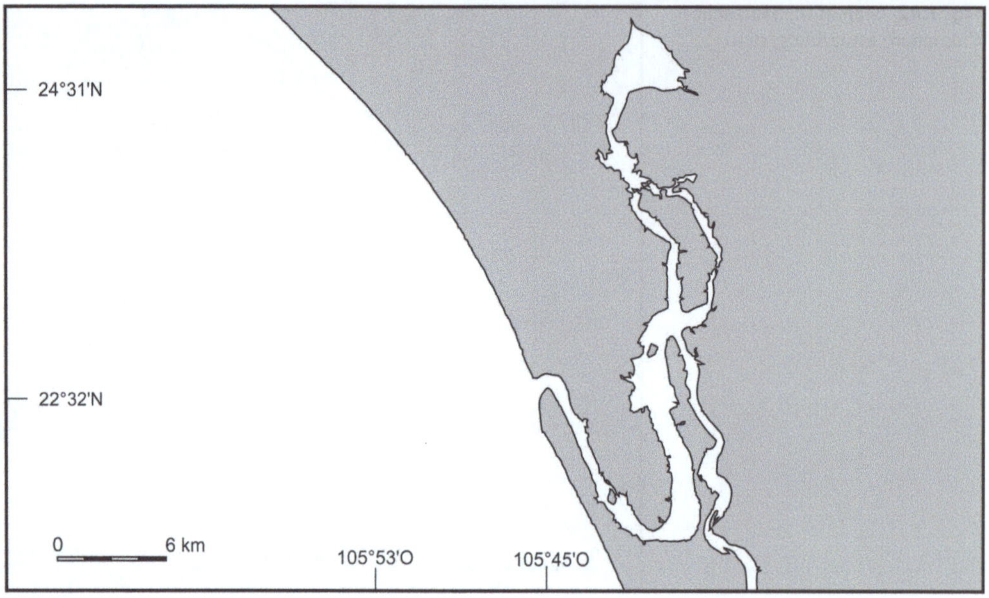

Fig. 1.46 Map of the Teacapán estuarine system

Fig. 1.47 La Brecha fishing village in the Teacapán estuarine system

Fig. 1.48 Mangrove forest of the Teacapán estuary

mean runoff volume of 2076 hm^2, a velocity over 0.5 m s^{-1}, and a maximum discharge of 452.75 m^3 s^{-1}.

The average high tide is 0.22 m and the average low tide is −0.28 m. The tide is non-symmetrical with low amplitude and is predominantly mixed diurnal. The wind generates mixing throughout the water column with differential velocities between seasons. For example, 8.1 m s^{-1} is possible during the Northern wind season, 6.5 m s^{-1} is typical during the dry season and 7.5 m s^{-1} is likely during the rainy season (Sánchez-Santillán and de la Lanza-Espino 1996).

The geomorphology of the basin of Tampamachoco lagoon is very restricted and close to the coast with altitudes less than 20 m above sea level and slopes of 1° or less. The shallowness of the lagoon and the high density of vegetation characterize Tampamachoco as a deposit basin without erosion. The lagoon also receives pollutant inputs from the Tuxpan River; however, the ebb and flow of the tides affects the permanence of the pollutants.

Tampamachoco lagoon has silty sediments with no or minimal variation in texture throughout the year. Coincidentally, the low tidal energy allows the basin to collect silt and pollutants. The silty mud is terrigenous and the calcareous mud is generated by biological activity which, at times, is greater than the terrigenous input. The presence of quartz indicates geologic stability of the basin (Márquez-García 1996).

Tampamachoco is a benthic environment characterized by low biological diversity, but with high specific densities (Quintana-Molina 1981). Light coloured fine sediment is predominant in the sediment-water interphase, and dark-silty-sandy sediment is predominant in the underlying layers, although the physiography of the lagoon affects the distribution of these percentages (Quintana-Molina 1981). Cruz-Abrego and Solís-Weiss (1990) identified 39 species of mollusks (38 bivalves and 1 gastropod) and 20

Fig. 1.49 View of fishing skiffs and the town of Teacapán at the entrance of the Teacapán estuarine system

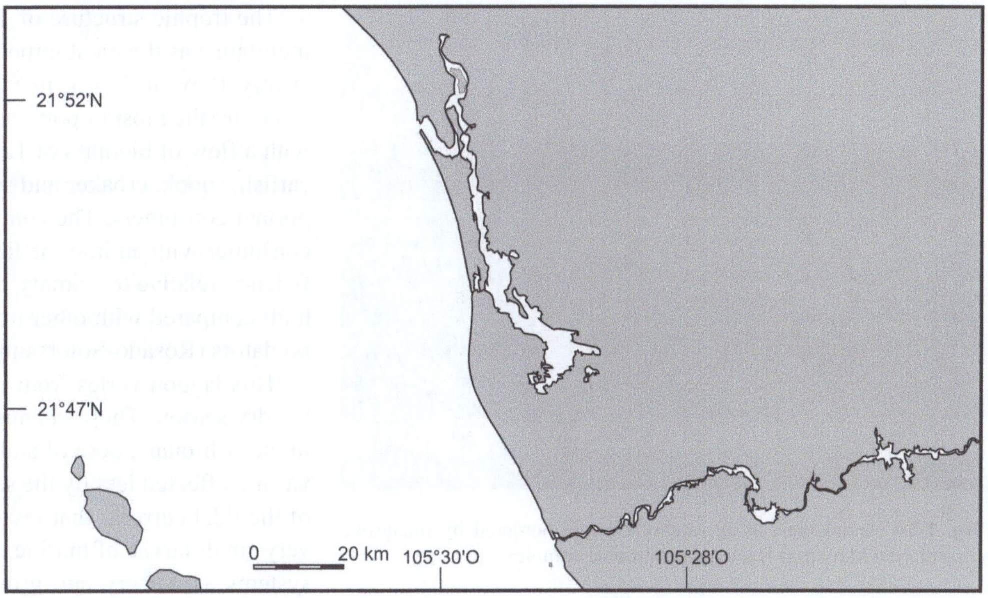

Fig. 1.50 Map of the estuarine complex of Marismas Nacionales (National flood plains)

Fig. 1.51 Aerial view of the artificially opened Cuautla channel connecting the Agua Brava lagoon with the Gulf of California

Fig. 1.52 Freshwater swamp of the Marismas Nacionales estuarine complex

Fig. 1.53 A racoon in a freshwater swamp in the Marismas Nacionales estuarine complex

Fig. 1.55 Detailed view of the coastal area of the Marismas Nacionales estuarine complex

Fig. 1.54 Aerial view of an estuary channel bordered by mangrove forest in the Marismas Nacionales estuarine complex

species of fauna associated with the mangrove roots of *Rhizophora mangle*. The percentages of the most representative species are 47.77 % for *Crassostrea virginica*, 20.42 % for *Brachidontes recurvus*, 19.49 % for the barnacle *Balanus eburneus*, 8.44 % for tube annelids and 1.85 % for other groups like decapods, gastropods, isopods and amphipods.

Bulit and Signoret (1988–1987) recorded that the nanophytoplanctonic fraction contributed over 90 % to the concentration of chlorophyll a in Tampamachoco, with a maximum of 16.06 mg m^{-3} in September. This indicates that primary production had ample temporal variation, from heterotrophy to autotrophy, and was dominated by the diatoms *Nitzschia longissima* and *Asterionella japonica*. The macroalgae composition of Tampamochoco includes the families Chlorophyceae, Phaeophyceae and Rhodophyceae (Dreckmann and Pérez 1994).

The trophic structure of Tampamachoco lagoon includes meiofauna as the most important primary consumer, with an energy flow of 57.8 g m^{-2} $year^{-1}$ dry weight. The crustaceans are the most important group of secondary consumers, with a flow of biomass of 12.8 g m^{-2} $year^{-1}$ dry weight. The catfish, snook, croaker and swimming crab are the most important consumers. The commercial fishery is an additional consumer with an intermediate trophic level of 3.08 and efficiency relative to primary productivity of 0.0087, which is high compared with other tropical fisheries based on higher predators (Rosado-Solorsano and Guzmán del Proo 1998).

This lagoon varies from oligohaline in July to marine in the dry season. These changes are accompanied by changes in the fish fauna, both of adults and larvae, although the larvae are affected less by the salinity and more by the velocity of the tidal currents that favors the immigration of eggs and very small larvae of marine organisms that use the estuarine systems as nursery and growing areas (Ríos-Salazar et al. 1988).

22. Coatzacoalcos estuary (18°09′ N—94°24′ W to 18°05′ N—94°25′ W; Chaps. 2 and 6; Figs. 1.64 and 1.65) The Coatzacoalcos River is one of the largest rivers in Mexico and has a variable discharge volume. The estuary has a total area of 1,500 ha, the main habitats are mangrove forests, grasses and swamps, and the tidal regime is mixed semidiurnal. In terms of the geomorphology and sedimentary processes, the sediments are mostly composed of sands in the mouth of the system. In terms of ecological characteristics, 31 fish species, ten crustaceans species and three mollusk species are reported in this system. The average annual rainfall in the river basin is 2,780 mm. The water contribution of the Coatzacoalcos River fluvial moves an annual volume of 18.381 billion m^3, which can range from 410 m^3 s^{-1} in the dry season to 6737 m^3 s^{-1} in rainy season (Riverón Enzástiga 2008). Its basin covers

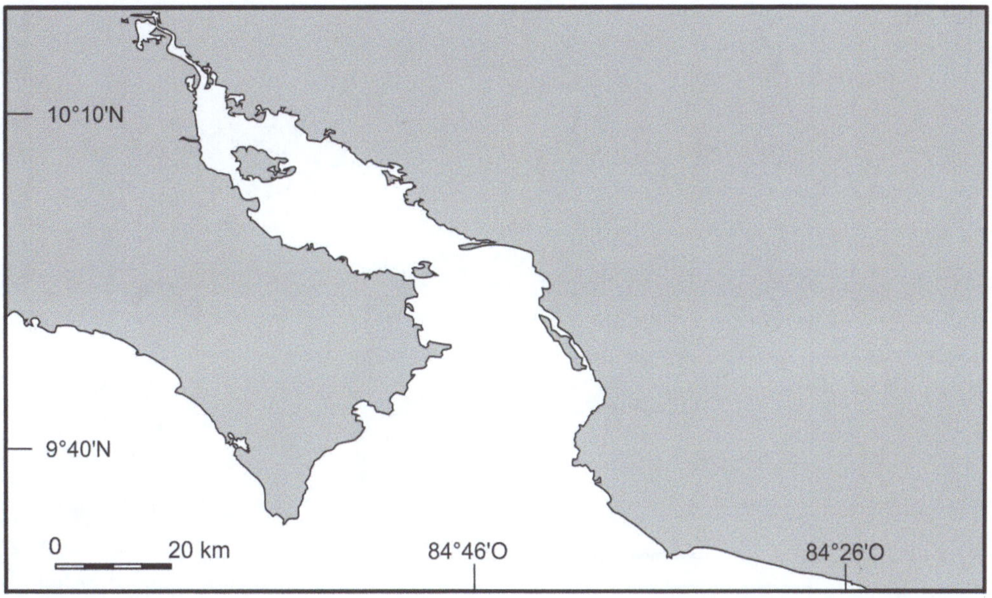

Fig. 1.56 Map of the Golfo de Nicoya

Fig. 1.57 Aerial view of the Golfo de Nicoya

Fig. 1.58 Fishing village within the Golfo de Nicoya

almost the entire northern side of the Isthmus of Tehuantepec. Born in Sierra Atravesada in the state of Oaxaca and after running westward, it then bends north and northeast to empty into the southern Gulf of Mexico.

In the maritime region of Coatzacoalcos, the continental shelf extends to the north and reaches a width of about 65 km with a steep slope between 100 and 200 m deep. This is part of the Bay of Campeche whose circulation is predominantly cyclonic (Vázquez de la Cerda 1979; Monreal Gómez and Salas de León 1990) and is caused by the formation of a spin on the east side of the bay during the months of August and September, which continues until December and moves westward (Vázquez de la Cerda 1979). The sedimentology of the maritime area is mainly coarse sand from the upper parts of the basin area, where there are intrusive igneous rocks, volcanic sediments and clastic (Gutiérrez-Estrada and Galaviz-Solis 1983).

The temperature at the sea surface is between 24 and 29 °C, with no marked seasonal differences (Villalobos and Zamora 1975; Padilla-Pilotze et al. 1986). During the fall and winter, surface layer mixture is caused by the effect of "norther" winds (local name of polar air masses), which cause a decrease in temperature from the surface to 150-m depth (Monreal Gómez and Salas de León 1990; Gio Argáez 2000). In the summer, the depth of the mixed layer established in the fall and winter is thinned by high surface temperatures, creating a strong vertical gradient in addition to the presence of a strong thermocline (Licea and Luna 1999)

The variation in salinity is more intense than that of temperature. Low salinity waters reach the Campeche Bank in spring and summer from the northeast to the Yucatan Current (Bogdanov 1969).

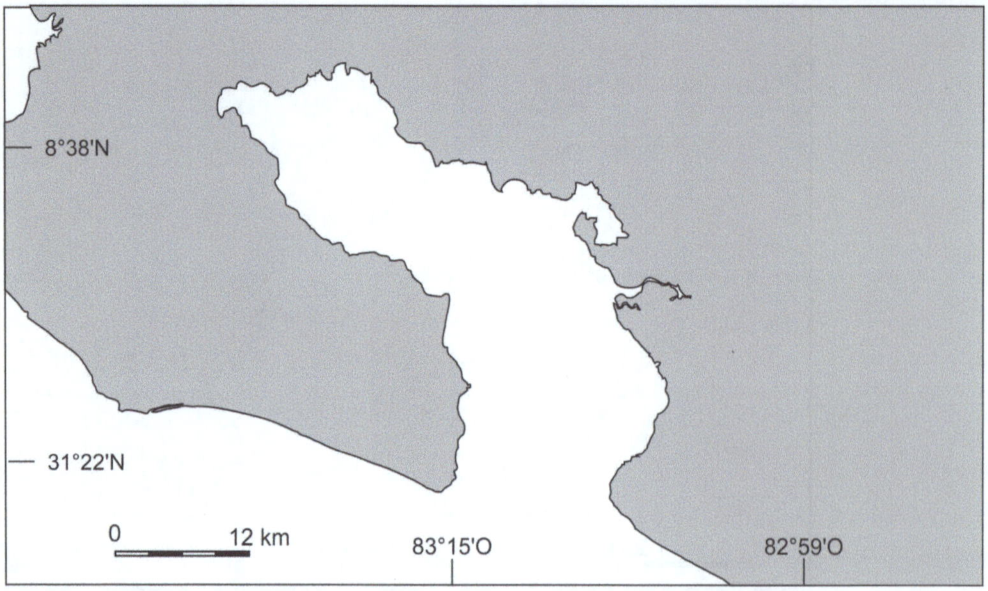

Fig. 1.59 Map of the Golfo Dulce

Fig. 1.60 Aerial view of the Golfo Dulce

Fig. 1.61 Fishing boat in the Golfo Dulce

The weather in the region is sub-humid with summer rains and an average annual rainfall of 1100–2000 mm (García 1973). The winds have a predominant direction of east–southeast throughout the year with speeds 4 m s^{-1}, except in the months of November to March when winds from the north–northwest predominate due to polar air masses known locally as "northers" whose speeds can range from 25 to 36 m s^{-1} (Gutiérrez-Estrada and Galaviz-Solis 1983).

Tides are mixed semidiurnal. From 1999 to 2004, the maximum high tide was recorded in the months of September-October and the minimum tide occurred during the months of June-July.

The topics covered within this book are not an exhaustive compendium of all the issues needed for proper fisheries management in the region. For example, topics such as climate change, invasive species and their influence on fisheries, bycatch and classic works on stock assessment are not included in this book. Also, many countries and regions are not mentioned. However, this book contains information on the natural and socioeconomic factors that can serve as guidance to achieve conservation and management of estuarine systems in Mexico and Central America. According to Gladstone (2009), achieving long-term success in conservation and management requires considering the exploited species, the ecosystem as a whole and the people who use these resources; and to manage them through the establishment of very specific goals. These goals should include maintaining ecosystem resilience by allowing connectivity and water quality of the estuarine ecosystems, recovering endangered and threatened species, conserving biodiversity that is representative of the area to be managed, understanding the socioeconomic context within these ecosystems, including stakeholders as participants, educating the public, and managing resources at the appropriate spatial scale.

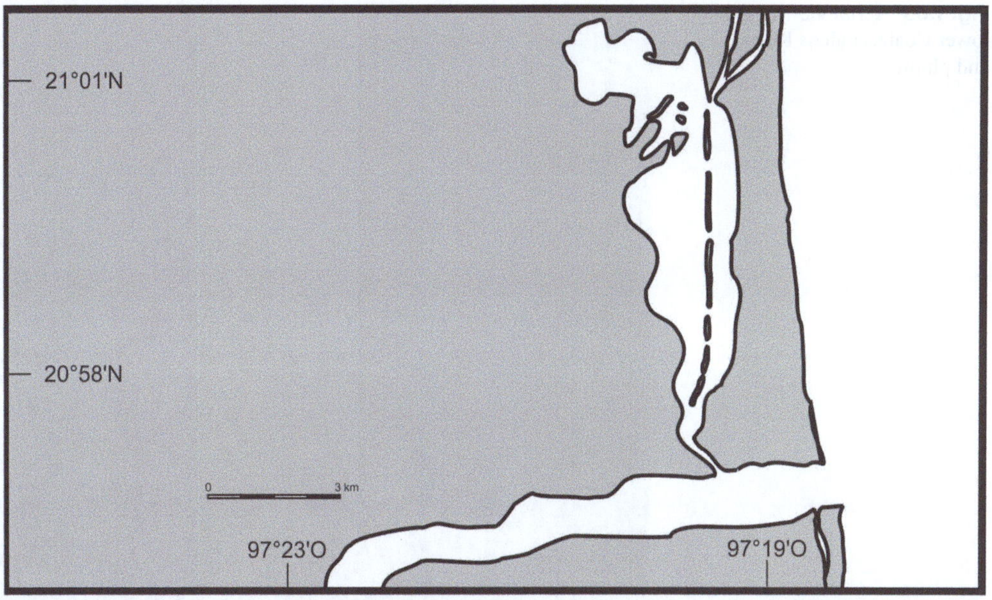

Fig. 1.62 Map of the Tampamachoco Lagoon

Fig. 1.63 View of a fishing village in the Tampamachoco Lagoon

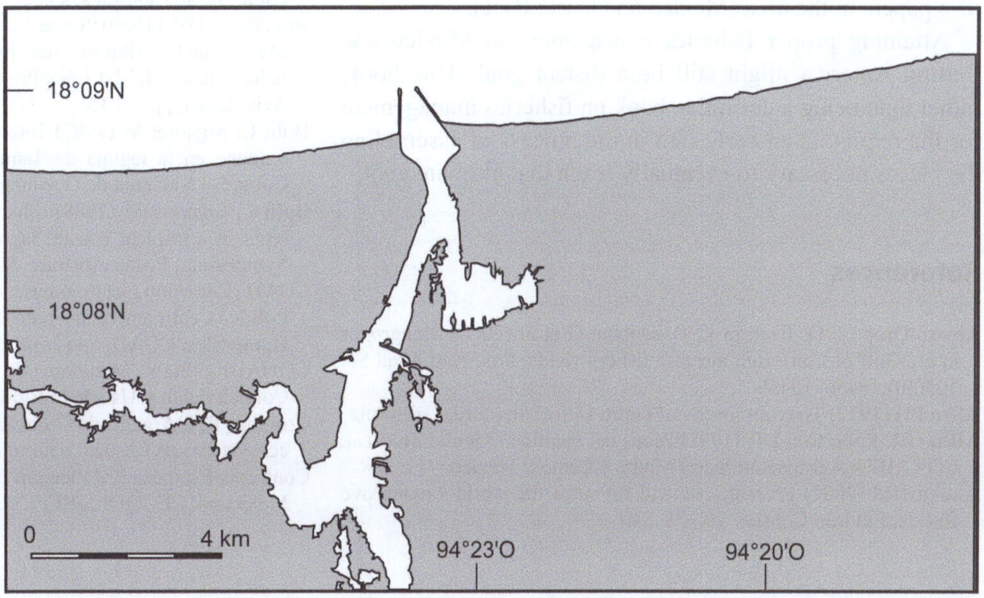

Fig. 1.64 Map of the Coatzacoalcos estuarine system

Fig. 1.65 Aerial view of the lower Coatzacoalcos River and plume

This book covers most of these goals using specific case studies from the region—recovery of endangered species and conservation of local biodiversity were not covered—although considering the large number and extent of estuarine systems in Mexico and Central America, similar studies are needed for other sites not covered here. A problem in this regard is that the number of researchers in fishery-related sciences in these countries is scarce. In fact, there is a lack of scientists and scientific publications in general (SCImago 2007). From the eight countries present in this region, five are among the countries with the fewest number of published scientific documents in the world (i.e., Belize, Guatemala, El Salvador, Honduras and Nicaragua). Thus, there is no surprise that the information and research available from these countries in regard to fisheries science is scare or nonexistent; all of the research covered in this book comes from the two countries with the highest number of published scientific papers in the area (Mexico and Costa Rica).

Attaining proper fisheries management in Mexico and Central America might still be a distant goal. This book, rather than being a definitive book on fisheries management for the region, is an early step in the process of assembling the pieces necessary to eventually reach that ultimate goal.

References

Aburto-Oropeza O, Ezcurra E, Danemann G et al (2008) Mangroves in the Gulf of California increase fishery yields. Proc Natl Acad Sci 105(30):10456–10459

Allen PH (1977) The rain forests of Golfo Dulce. Stanford, California

Allen GR, Robertson DR (1998) Peces del Pacifico Oriental Tropical. CONABIO. Agrupación Sierra Madre y Cemex, México

Alongi DM (2002) Present state and future of the world's mangrove forests. Environ Conserv 29:331–349

Amezcua F, Amezcua-Linares F (2014) Seasonal changes of fish assemblages in a subtropical lagoon in the SE Gulf of California. Sci World J 2014:1–15. doi:10.1155/2014/968902

Ayala CA, Guttierrez AM, Malpica UM (1970) Informe final de los estudios de geologia marina en las regiones de Yavaros (Sonora), Huizache-Caimanero (Sinaloa) y Agiabampo (Sinaloa) durante la primera etapa. Instituto de Biologia, Universidad Nacional Autonoma de México, pp 3–90

Ayala CA, Phleger FB, Gutiérrez Estrada M (2003) Bases para el Estudio, la Interpretación y el Manejo de las Lagunas Costeras. In: Cifuentes Lemus JL, Gaxiola López J (eds) Atlas de los Ecosistemas de Sinaloa. El Colegio de Sinaloa, Culiacán, 481 p (pp 127–174)

Bergoeing JP (1998) Geomorfología de Costa Rica. Instituto Geográfico Nacional, San José

Berlanga Robles C, Ruiz Luna A (2003) Los Sistemas Acuáticos Costeros de Sinaloa. In: Cifuentes Lemus JL, Gaxiola López J (eds) Atlas de los Ecosistemas de Sinaloa. El Colegio de Sinaloa, Culiacán, 481 p (pp 197–214)

Blaber SJM (2011) Removals (wild harvesting) of the biological resources from systems. In: Eric W, Donald M (eds) Treatise on Estuarine and soastal science. Academic, Waltham, pp 253–275

Bogdanov DV (1969) Some oceanographic features of the Gulf of Mexico and Caribbean Sea. In: Bogdanov AS (ed) Soviet-Cuban-fishery research, 1st edn. Program for Scientific Translations, Tel Aviv, Israel, pp 13–15

Bulit C, Signoret M (1987) Biomasa fitoplanctónica y producción de oxígeno en la laguna de Tampamachoco, Veracruz. Resumen VII Congreso Nacional de Oceanografía

Bulit C, Signoret M (1988). Phytoplankton distribution and tidal currents in a tropical coastal lagoon. 23rd European Marine Biology Symposium. Poster Abstract. Swansea, 5th–9th September 4

CCAD (Comisión Centroamericana de Ambiente y Desarrollo) (2002) Política Centroamericana para la Conservación y Uso Racional de los Humedales. CCAD, San Jose

CONABIO (2008) Manglares de México. Comisión Nacional para el Conocimiento y Uso de la Biodiversidad, México

Costanza R, d'Arge R, de Groot R et al (1997) The value of the world's ecosystem services and natural capital. Nature 387:253–260

Contreras-Espinosa F, Zabalegui L (1988) Aprovechamiento del Litoral Mexicano. CECODES-SEPESCA, Mexico

Contreras-Espinosa F, Castañeda L (1993) Ecosistemas costeros mexicanos. Universidad Autónoma Metropolitana-Iztapalapa, Comisión Nacional para el Conocimiento y Uso de la Biodiversidad, Mexico

Cruz-Abrego FM, Solís Weiss V (1990) Influencia del abanico costero del río Tuxpan (oeste del Golfo de México) sobre las poblaciones de moluscos. Resumen II Congreso de Ciencias del Mar

Dalsgaard T, Canfield DE, Petersen J, Thamdrup B, Acuna-Gonzalez J (2003) N2 production by the anammox reaction in the anoxic water column of Golfo Dulce, Costa Rica. Nature 422:606–608

Davidson I, Gauthier M (1993) Wetland conservation in Central America. Report 93-3 of the North American Wetlands Conservation Council (Canada), Ottawa, Ontario

De LLanza-EspinoG (1993) Ambientes y pesquerías en el litoral pacífico mexicano. Instituto Nacional de Estadística, Geografía e Informática, México

De laLanza-EspinoG, Cáceres M (1994) Lagunas costeras y el litoral mexicano. Universidad Autónoma de Baja California Sur, La Paz

DOF (2010) Decreto por el que se declara como área natural protegida, con el carácter de reserva de la biosfera, la región conocida como Marismas Nacionales Nayarit, localizada en los municipios de Acaponeta, Rosamorada, Santiago Ixcuintla, Tecuala y Tuxpan en el Estado de Nayarit. Presidencia de la República, Estados Unidos Mexicanos

Dreckmann KM, Pérez HMA (1994) Macroalgas bentónicas de la laguna de Tampamachoco, Veracruz, México. Rev Biol Trop 42(3):715–717

Duda AM, Sherman KS (2002) A new imperative for improving management of large marine ecosystems. Ocean Coast Manag 45:797–783

Ellison AM (2004) Wetlands of Central America. Wetl Ecol Manag 12:3–55

Epifanio CE, Maurer D, Dittel A (1983) Seasonal changes in nutrients and dissolved oxygen in the Gulf of Nicoya, a tropical estuary on the Pacific coast of Central America. Hydrobiologia 101:231–238

FAO (2008) The world's mangroves 1980–2005. FAO Forestry Paper 153. Food and Agriculture Organization of the United Nations, Rome

Fischer W, Krupp W, Schneider W et al (1995) Vertebrados. Guía FAO para la identificación de especies para los fines de la pesca. Organización de las Naciones Unidas para la Agricultura y la Alimentación. Volumen II. FAO, Roma, pp 647–1200

Flores-Verdugo F, Day JW, Briseño-Dueñas R (1987) Structure, litterfall, decomposition, and detritus dynamics of mangroves in a Mexican Coastal lagoon with an ephemeral inlet. Mar Ecol Prog Ser 35:83–90

Flores-Verdugo EJ, González-Farias F, Amezcua F et al (1990) Mangrove ecology, aquatic pri-mary productivity and fish community dynamics in the Teacapán-Agua Brava lagoon-estuarine system (Mexican Pacific). Estuaries 13:219–230

Flores-Verdugo F, Rodríguez-García J, González-Farias F (1993) Variación espacio temporal de la productividad primaria acuática en la Bahía de Mazatlán, Sinaloa. Ecología de los Manglares, productividad acuática y perfil de las comunidades en ecosistemas lagunares estuarinos de la costa noroccidental de México. Parte I, Ensenada del Pabellón, Bahía de Altata y Bahía de Mazatlán. Informe Técnico. DGPA (IN202389)

Flores-Verdugo FJ, Gonzalez-Farias F, Blanco-Correa M et al (1997) The Teacapan-Agua Bra-va-Marismas Nacionales mangrove ecosystem on the Pacific coast of Mexico. In: Kjerfve B, Drude de Lacerda L, Salif-Diop EH (eds) Mangrove ecosystem studies in Latin America and África. United Nations Educational, Scientific and Cultural Organization, Paris, pp 35–46

García E (1973) Modificaciones al sisema de clasificación climática de Köppen. Universidad Nacional Autónoma de México, México

García E, Trejo RL (1990) Causas de la precipitación en Nayarit. Proc. XII Congreso Nacional de Geogr. March 27, 1990. Sociedad Mexicana de Geografía y Estadística. Tepic, Nayarit. Mexico, pp 234–243

García-Céspedes J, Acuña-González J, Vargas-Zamora J (2004) Metales traza en sedimentos costeros de Costa Rica. Rev Biol Trop 52(2):51–60

Gio-Argáez FR (2000) Distribución y sistemática de los ostrácodos de la Bahía de Campeche. Dissertation, Universidad Nacional Autónoma de México

Gladstone W (2009) Conservation and management of tropical coastal ecosystems. In: Nagelkerken I (ed) Ecological connectivity among tropical coastal ecosystems. Springer, Dordrecht, pp 565–605

Gocke K, Cortés J, Murillo MM (2001a) Planktonic primary production in a tidally influenced mangrove forest on the Pacific coast of Costa Rica. Rev Biol Trop 49(2):279–288

Gocke K, Cortés J, Murillo MM (2001b) The annual cycle of primary productivity in a tropical estuary: the inner regions of the Golfo de Nicoya, Costa Rica. Rev Biol Trop 49(2):289–306

Groombridge B (1992) Global biodiversity status of the Earth's living resources. Chapman and Hall, London

Gutierrez-Estrada M, Galaviz-Solis A (1983) Morfología y sedimentos recientes de las lagunas El Carmen, Pajonal y La Machona, Tabasco, México. An Inst Cienc Mar Limnol 10(1):249–270

Harborne AR, Mumby PJ, Micheli F et al (2006) The functional value of Caribbean coral reef seagrass and mangrove habitats to ecosystem processes. Adv Mar Biol 50:57–189

Hebbeln D, Cortés J (2001) Sedimentation in a tropical fjord: Golfo Dulce, Costa Rica. Geo-Marine Lett 20:142–148

Hendrickx ME, Brusca RC, Ramírez-Reséndiz G (2002) Biodiversity of macrocrustaceans in the Gulf of California. In: Hendrickx ME (ed) Contributions to the study of East Pacific crustaceans. UNAM, México, pp 349–369

Herrera W (1985) Clima de Costa Rica. Universidad Estatal a Distancia, San José

Hughes TP, Baird AH, Bellwood DR et al (2003) Climate change, human impacts, and the resilience of coral reefs. Science 301:929–933

IMCO (2013) La pesca ilegal: una barrera a la competitividad. http://imco.org.mx/wp-content/uploads/2013/5/pescailegal_ppt.pdf. Accessed 23 Aug 2013

Jennings S, Kaiser MJ, Reynolds JD (2001) Marine fisheries ecology. Blackwell Science, London

Jimenez JA (1999) Ambiente, distribucíon y características estructurales en los manglares del Pacífico de Centro América: contrastes climáticos. In: Yañez-Arancibia A, Lara-Dominguez AL (eds) Ecosistemas de manglar en América Tropical. Instituto de Ecología A.C. México, UICN/ORMA, Costa Rica, NOAA/NMFS Silver Spring MD USA, 380 p

Lankford RR (1977) Coastal lagoons of Mexico: their origin and classification. In Wiley M (ed) Estuarine processes. Academic, New York, pp 182–215

Leigh EG, Rand AS, Windsor DM (eds) (1982) The ecology of a tropical forest: seasonal rhythms and long-term changes. Smithsonian Institution Press, Washington, DC

Licea S, Luna R (1999) Spatio-temporal variation of phytoplankton on the continental margin in the SW Gulf of Mexico. Rev Soc Mex Hist Nat 49:89–99

Lizano O (2006) Algunas características de las mareas de la costa Pacífica y Caribe de Centroamérica. Ciencia y Tecnología 24(1):51–64

Maurer D, Vargas JA (1984) Diversity of soft-bottom benthos in a tropical estuary: Gulf of Nicoya, Costa Rica. Mar Biol 81(1):97–106

Manson FJ, Lorenagan NR, Skilleter GA et al (2005) Evaluation of the evidence for linkages between mangroves and fisheries: a synthesis of the literature and identification of research directions. Oceanogr Mar Biol Annu Rev 43:483–513

Márquez-García AZ (1996) Evaluación Geoquímica Ambiental y Diagnosis de la Zona Costera de Veracruz: Lagunas de Tamiahua, Pueblo Viejo y Tampamachoco. Informe Final del Pro-yecto de Investigación. Universidad Autónoma Metropolitana Unidad Iztapalapa-Consejo Nacional de Ciencia y Tecnología, México, pp 29–82

McDade LA, Bawa KS, Hespenheide HA, Hartshorn GS (1994) La Selva: ecology and natural history of a neotropical rain forest. University of Chicago, Chicago

Meynecke JO, Lee SY, Duke NC (2008) Linking spatial metrics and fish catch reveals the im-portance of coastal wetland connectivity to inshore fisheries in Queensland, Australia. Biol Conserv 141:981–996

Monreal-Gómez MA, Salas-de-León DA (1990) Simulación de la circulación de la Bahía de Campeche. Geofis Int 29(2):101–111

Montaño-Ley Y, Peraza-Vizcarra R, Páez-Osuna F (2008) Tidal hydrodynamics and their implications for the dispersion of effluents in Mazatlán Harbor: an urbanized shallow coastal lagoon. Water Air Soil Pollut 194:343–357

Mora C, Robertson DR (2005a) Causes of latitudinal gradients in species richness: a test with the endemic shorefishes of the Tropical Eastern Pacific. Ecology 86:1771–1782

Mora C, Robertson DR (2005b) Factors shaping the range-size frequency distribution of the endemic fish fauna of the Tropical Eastern Pacific. J Biogeogr 32:277–286

Nagelkerken I (2009) Ecological connectivity among tropical coastal ecosystems. Springer, Dordrecht

Nagelkerken I, Blaber SJM, Bouillon S et al (2008) The habitat function of mangroves for terrestrial and marine fauna: a review. Aquat Bot 89:155–185

Padilla-Pilotze AR, Alatorre MA, Ruíz F et al (1986) Observaciones recientes de la estructura termohalina en el sur del Golfo de México. In: Urrutia-Fucugauchi J (ed) Memoria de la reunión anual 1986 de la Unión Geofísica Mexicana, Morelia

Quesada-Alpízar MA, Cortés J (2006) Los ecosistemas marinos del Pacífico sur de Costa Rica: estado del conocimiento y perspectivas del manejo. Rev Biol Trop 54(1):101–145

Quesada-Alpízar MA, Morales-Ramírez A (2004) Comportamiento de las masas de agua en el Golfo Dulce, Costa Rica durante El Niño (1997–1998). Rev Biol Trop 52(2):95–103

Quintana-Molina J (1981) Estudio preliminar de las poblaciones bentónicas de la laguna Tam-pamachoco, Veracruz, México. Memorias del VII Simposio Latinoamericano de Oceano-grafía Biológica

Richards FA, Anderson JJ, Cline JD (1971) Chemical and physical observations in Golfo Dulce, an anoxic basin on the Pacific coast of Costa Rica. Limnol Oceanogr 16:43–50

Ríos-Salazar H, Marmolejo A, de la Campa S (1988) Análisis ecológico del ictioplancton de la laguna de Tampamachoco, Veracruz y de la zona nerítica adyacente. Resumen III Sociedad Mexicana de Planctología

Riverón-Enzástiga ML (2008) Análisis de la variabilidad de las descargas de agua de los principales ríos de la cuenca del Golfo de México. Dissertation, Universidad Nacional Autónoma de México

Robertson DR, Allen GR (2008) Shorefishes of the Tropical Eastern Pacific online information system. Smithsonian Tropical Research Institute, Balboa, Panamá

Rosado-Solorsano R, Guzmán del Proo SA (1998) Preliminary trophic structure model for Tampamachoco lagoon, Veracruz, Mexico. Ecol Model 109:141–154

SAGARPA (2011) Anuario estadístico de acuacultura y pesca. Secretaría de Agricultura, Gana-dería, Desarrollo Rural, Pesca y Alimentación. CONAPESCA, Mexico

Sánchez-Santillán N, de la Lanza-Espino G (1996) The influence of the regional climate in a lagoon in northeastern Mexico. Atmósfera 9:201–222

Sauma-Castillo LF (2010) Uso de hábitat y comportamiento alimenticio de tres especies de aves playeras en el Estero de Urías, Sinaloa, México, durante una temporada no-reproductiva. Master Science Thesis. Posgrado en Ciencias del Mar y Limnología. Unidad Académica Mazatlán. Universidad Nacional Autónoma de México

SCImago (2007) SJR—SCImago journal & country rank. http://www.scimagojr.com. Accessed 23 Aug 2013

Shepard FP (1984) Submarine geology. Harper Collins, New York

Short FT, Wyllie-Echeverria S (1996) Natural and human induced disturbance of seagrasses. Environ Conserv 23:17–27

Soto-Jiménez MF, Páez-Osuna F (2001) Distribution and normalization of heavy metal concentrations in mangrove and lagoonal sediments from Mazatlán Harbor (SE Gulf of California). Estuar Coast Shelf S 53:259–274

Spalding MD, Blasco F, Field CD (1997) World Mangrove Atlas. International Society for Mangrove Ecosystems, Okinawa

UNEP (2006) Marine and coastal ecosystems and human well-being: a synthesis report based on the findings of the Millennium Ecosystem Assessment. United Nations Environment Programme, Nairobi

Vázquez-de la Cerda AM (1979) Circulación del agua en el sureste del Golfo de México. Secretaría de Marina, Dirección General de Oceanografía, Mexico

Villalobos A, Zamora ME (1975) Importancia biológica de la Bahía de Campeche. In: Memo-rias del I Simposio Latinoamericano de Oceanografía Biológica, Secretaria de Marina. 25–29 November 1974, Mexico

Villaseñor R, Amezcua F (2013) Mexican fisheries: their management and administration to regulate impacts to fish habitat. In: LeBlanc P (ed) Managing impacts to fish and fish habitat. American Fisheries Society Symposium 78, Taylor & Francis, pp 1–19

Voorhis A, Epifanio CE, Maurer D, Dittel A, Vargas JA (1983) The estuarine character of the Gulf of Nicoya, an embayment on the Pacific coast of Central America. Hydrobiologia 99:225–237

Wolff M, Koch V, Chavarria JB, Vargas JA (1998) A trophic flow model of the Golfo de Ni-coya, Costa Rica. Rev Biol Trop 46(6):63–79

Zamora Arroyo F, Egido Villareal J, Albar Reynoso MP, Villalba Loera A, Rodríguez G (2000) Programa Piloto de Monitoreo Ecológico de Bahía Santa María La Reforma. Centro de Conservación para el Aprovechamiento de los Recursos Naturales (CECARENA). Conservación Internacional México (CIMEX), Universidad Autónoma de Sinaloa

Part I
Physicochemical Considerations

Water Quality Effects on Fish Larvae in a Tropical Coastal Lagoon of the Gulf of Mexico

2

Guadalupe de la Lanza-Espino, Salvador Hernández-Pulido, Norma Sánchez-Santillán and José Luis Carbajal-Pérez

Abstract

Human settlements and industrial activities located along rivers and coastal lagoon margins have led to the discharge of untreated waste effluents into proximate waters, a situation that has affected the biota, fisheries and man himself. Many examples of this phenomenon exist throughout the world, including along the coast of Mexico. This study analyzed the physicochemical water quality parameters of a coastal lagoon in the northwest Gulf of Mexico during four sampling seasons in 2009 and 2010 that included dry, rainy and north-wind seasons; results were compared to conditions of the lagoon in 1983. Anthropogenic discharges along rivers and lagoons in the study area were correlated with slight increases in ammonium, total nitrogen and phosphorus starting 30 years ago, with concentrations remaining stable during this time period. Intermittent decreases in these nutrients occurred during heavy rains. Residence time of these nutrients varied from 19–40 days and depended on the depth of the lagoon. Results suggest that water quality does not differ greatly between historic and present times, suggesting that these fluvial-lagoon systems do not currently require environmental management. However, controlling urban discharge, as the human population increases will be necessary to minimize the impact of anthropogenic discharges. Fish larvae were only affected by the variation of temperature and salinity of the lagoon water.

Keywords

Tropical costal lagoon · Water-quality behavior · Estuarine physicochemistry · Fish larvae · Gulf of Mexico

2.1 Introduction

Wetlands and the environmental services they provide have attracted humans and their activities, which has resulted in environmental runoff and unintended nutrient inputs. It is for these reasons that many ancient ethnic groups settled along the margins of both inland water bodies like lakes and rivers, and coastal systems such as lagoons, estuaries, salt marshes and bays. Monitoring water quality in coastal wetlands, including riverine lagoon systems, is fundamental to discriminating between natural changes in physicochemical characteristics and those changes generated by human activities. This information can then be used to establish management recommendations and guide decision making, with conservation as the ultimate goal.

G. de la Lanza-Espino (✉) · S. Hernández-Pulido · J. L. Carbajal-Pérez
Laboratorio de Hidrobiología, Departamento de Zoología, Instituto de Biología, Universidad Nacional Autónoma de México, Tercer Circuito Exterior s/n, Ciudad Universitaria, 04510 México, México
e-mail: gdlle@unam.mx

S. Hernández-Pulido
e-mail: fshp@ib.unam.mx

J. L. Carbajal-Pérez
e-mail: jlcarbajal54@hotmail.com

N. Sánchez-Santillán
División de Ciencias Biológicas y de la Salud, Departamento del Hombre y su Ambiente, Universidad Autónoma Metropolitana-Xochimilco, Calzada del Hueso 1100, 04960 Coyoacan, Mexico
e-mail: santilla@correo.xoc.uam.mx

F. Amezcua, B. Bellgraph (eds.), *Fisheries Management of Mexican and Central American Estuaries*, Estuaries of the World, DOI 10.1007/978-94-017-8917-2_2,

Water quality monitoring programs in countries around the world have led Mexico to analyze basic parameters to determine the productivity and trophic state associated with phytoplankton in both inland and coastal waters. Aquatic environments provide food and refuge to a great variety of organisms as a result of both functional and biochemical traits. These organisms represent a greater and more diverse array of species than in terrestrial systems, as wetlands provide a greater number of habitats as well as a buffer against extreme climate variations (Sánchez-Santillán and de la Lanza-Espino 2012).

There are slightly more than 600 coastal water bodies in Mexico (Contreras Espinosa 2010). The chemical composition of the freshwater that flows into these water bodies is determined by the geochemical composition of the lands through which the rivers flow, and typically results in the high productivity of coastal eco-systems. However, agricultural activities occurring within these basins may increase nutrient concentrations to levels that could have profound impacts on the historic food web.

Laguna Tampamachoco, in the state of Veracruz, is a prime example of a coastal lagoon that has been impacted by anthropogenic modification and monitored through time. The lagoon has been studied since 1980 when the first physicochemical attributes were measured (Contreras-Espinosa 1983, 1985). Several years later an environmental impact study was conducted (CFE 1994). Calva and Torres-Alvarado (2000) evaluated the organic-matter content of three lagoons in Veracruz and recorded the greatest content of nitrogen in the sediments of Laguna Tampamachoco. Further, de la Lanza-Espino et al. (1996) completed a temporal and spatial analysis on the content of silicates in the lagoon. In 2009, as part of an interdisciplinary project, de la Lanza-Espino et al. (2012) carried out three surveys to determine the water quality of Laguna Tampamachoco by analyzing basic physicochemical parameters during one dry season (March 2009) and two rainy seasons (August 2009 and September 2010). The aim of the current study was to compare the changes in water quality of Laguna Tampamachoco since the original study in 1980; a period of more than 30 years. Notably, human population abundance has grown considerably along the margins of Laguna Tampamachoco and consequently, anthropogenic nutrient inputs to this coastal system have also increased.

2.2 Study Area

Laguna Tampamachoco is located north of the city of Veracruz at 20°18′-21°02′ N and 97°19′-97°22′ W. It has an area of 1,500 ha and its main source of freshwater is the Tuxpan River. The lagoon has an average depth of 0.6–0.8 m; the estuary inlet is about 3-m deep in March and 5-m deep in

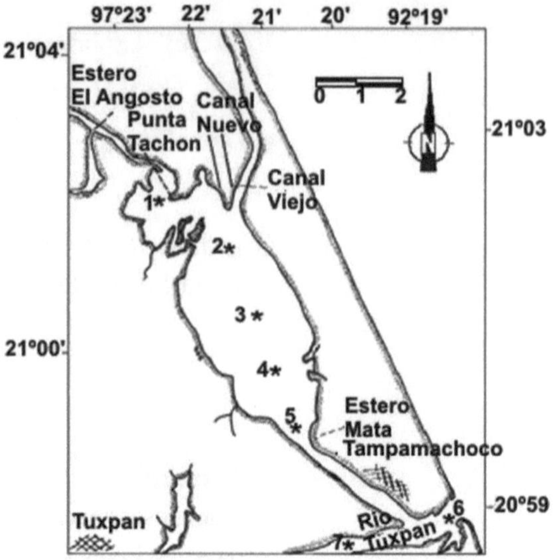

Fig. 2.1 Sampling stations in Laguna Tampamachoco, Veracruz

August, and the depth of the channel between the mouth of the Tuxpan River and the lagoon inlet varies from 6 to 20 m wide. Tide height varies throughout the year with ±0.54 m in March and ±1.09 m in December (CFE 1994). The climate is sub-humid and warm and summer rains are typical (García 1988). Mean annual rainfall totals about 1,900 mm; the driest month is January and the rainiest month is September. The average annual temperature is 25 °C. The lagoon contains several vegetated islands that are populated by *Rhizophora mangle* (red mangrove), *Avicennia germinans* (black mangrove) and patches of *Conocarpus erectus* (button mangrove).

2.3 Materials and Methods

Sampling took place in the morning between 1,000 and 1,100 h during the dry (March) and rainy (August) seasons of 2009, and in the rainy season of 2010 (September). Surface and bottom water samples were collected from seven stations covering the lagoon, the estuary and the river mouth-lagoon inlet (Fig. 2.1). The following parameters were recorded in situ: temperature (°C), depth (m), salinity (psu), dissolved oxygen concentration (mg/L), dissolved oxygen saturation (%), and pH using a YSI 556 MPS sensor. The samples were frozen and later analyzed in the laboratory for concentrations of nitrates, nitrites, ammonium, total nitrogen, orthophosphates and total phosphorus according to Strickland and Parsons (1972). Samples were also analyzed to measure Chemical Oxygen Demand (COD; APHA 2005). To determine nitrogen balance, the Land Ocean Interactions in the Coastal Zone (LOICZ) methods were used and included inputs and outputs between terrestrial and marine materials, as well as Ecological Net Metabolism (MNE).

2.4 Results and Discussion

2.4.1 Temperature

Water temperature in tropical coastal lagoons varies throughout the day between approximately 18 and 32 °C, depending on the season and the geographic location. Sampling in Laguna Tampamachoco took place only during the day therefore the temperature range was small. In March 2009, water temperature varied from 25.94 to 27.63 °C at the surface and 24.40 to 27.41 °C at the bottom. In August 2009 the temperature varied from 26.10 to 29.79 °C at the surface and slightly less at the bottom at most of the stations. However, a difference in temperature between the surface and the bottom of 4.36 °C and 4.32 °C was recorded at stations 6 and 7, respectively. This temperature difference was likely due to stratification of the relatively cool and dense freshwater entering from the Tuxpan River and the warm and less-dense sea water entering with the tide. In September 2010, the temperature difference between the surface and bottom layers was small (<1 °C) at stations 1, 2, and 3. Only surface temperatures were recorded at stations 4, 5 and 6 in September 2010; strong bottom currents inhibited bottom temperature measurements. In March, surface temperatures were slightly lower than those recorded in August. Contreras-Espinosa (1983) in a monthly study carried out in 1979, recorded a wide range of temperatures with a minimum in February (19–23 °C) and a maximum in May (31–32 °C), which included sampling areas in both the estuary and the lagoon. However, an unpublished study carried out in 1985 by CFE (1994) recorded the warmest month in September with a maximum of 33.5 °C and the coldest in February with a minimum of 17 °C, which coincided with the 1979 study. The maximum surface water temperature recorded in the present study was 29.79 °C in August during the rainy season. The temperature difference between the current study and the 1979 study may be an effect of interannual variability or due to the presence of strong meteorological events like the occurrence of La Niña in 2010, which generated low temperatures (NOAA 2013).

2.4.2 Salinity

In the March sampling period, salinity varied between 23.49 and 36.24 psu at the surface and between 32.30 and 36.26 psu at the bottom. This distribution was due to the gradient between the freshwater and seawater inputs resulting in a salt wedge at stations 6 and 7 and a weak salt wedge at stations 4 and 5 (Figs. 2.1 and 2.2). The highest salinities were recorded in the lagoon as a result of high evaporation and shallow depth. In August, and in spite of the rainy season, surface salinity at stations 1–5 was high with values of

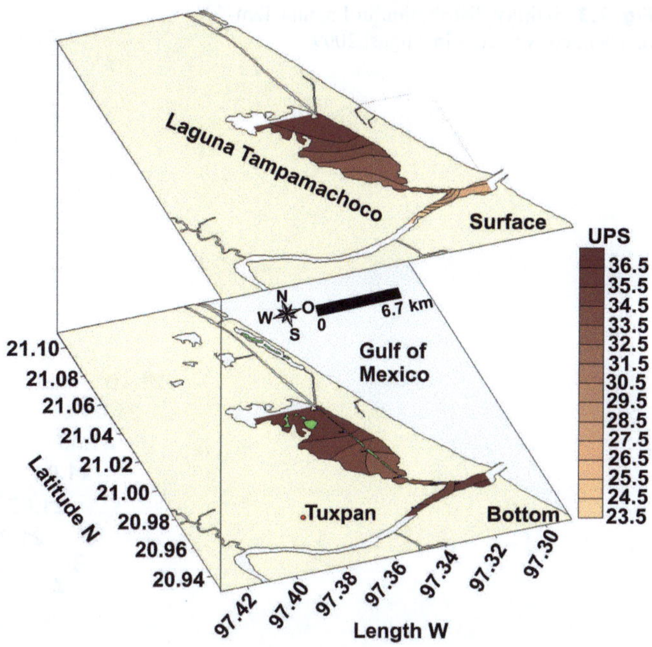

Fig. 2.2 Salinity distribution in Laguna Tampamachoco, Veracruz in March 2009

37.34–37.42 psu and bottom salinity was similar with values of 32.7 to 37.6 psu. In contrast, the estuarine stations (6 and 7) registered salinity values of 28.4 psu at the surface and 32.01 psu at the bottom and 16.80 at the surface and 25.82 psu at the bottom, respectively, in response to the salt wedge located between the river mouth and the lagoon inlet. In September 2010, salinity decreased more than 50 % at all the stations, particularly at the river mouth and the lagoon inlet where the surface water was oligohaline and the bottom water was mesohaline (Fig. 2.3). This marked decrease was in response to heavy rains associated with the presence of a medium to strong La Niña event in 2010. Contreras-Espinosa (1983) recorded similar salinity values more than 30 years ago.

2.4.3 Dissolved Oxygen

The concentration of dissolved oxygen in water varies on a diel cycle due to photosynthetic respiration by primary producers and can be modified by human activities. Dredging and canal construction increase the amount of suspended sediment and organic matter within the water column, which can cause low oxygen levels due to decomposition. Runoff from urban settlements along the margins of lagoons and estuaries, inland agricultural areas, and port activities also increase sediment, organic matter, and nutrients that can cause varying dissolved oxygen. Acceptable oxygen levels were recorded during all three survey periods during the present study, except for sporadic cases in August 2009 when near-

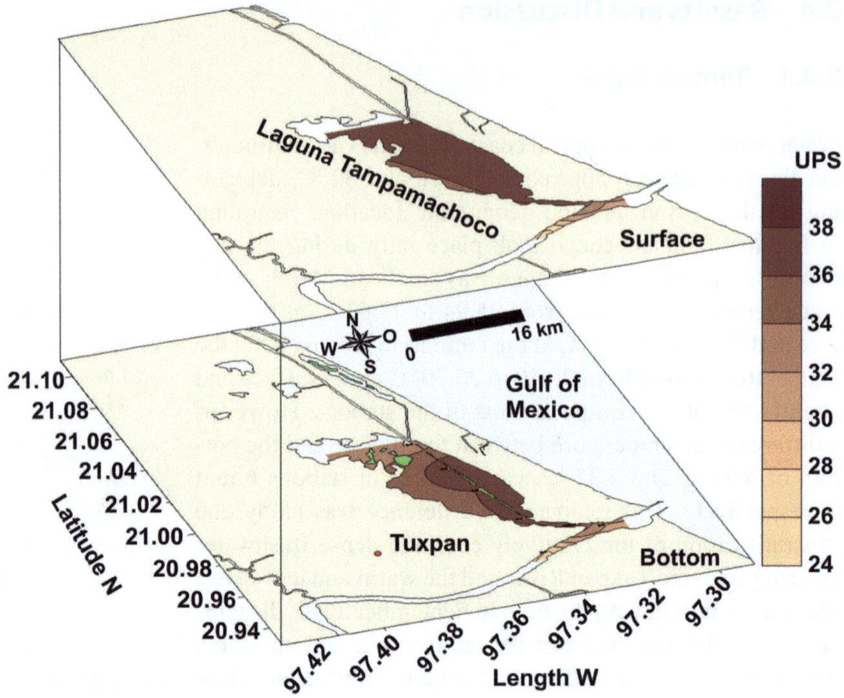

Fig. 2.3 Salinity distribution in Laguna Tampamachoco, Veracruz in August 2009

hypoxic conditions were recorded both in the lagoon and the estuary, and in bottom waters where there was possibly a predominance of organic-matter decomposition (Table 2.1). Despite these sporadic low dissolved oxygen levels, oxygen saturations were acceptable for the survival of benthic organisms. Supersaturation (> 100 %) of dissolved oxygen also occurred due to high photosynthetic activity in some survey periods and water-column locations (Table 2.1).

It is important to mention that lower dissolved oxygen concentrations in August and September in the bottom water coincided with the low-salinity river stations. This may be due to a low tide (see tide prediction table in González 2009) occurring simultaneously with high organic-matter discharge coming from the port and the urban settlements along the river margin. de la Lanza-Espino et al. (1996) characterized this phenomenon both temporally (in dry, rainy and north-wind seasons) and spatially (based on the content of silicates) and defined three characteristic areas within the Tampamachoco lagoon-marine ecosystem with correspondence to a gradient of dissolved oxygen concentrations from high to low: the estuarine-marine, the lagoon proper and the inner region of the lagoon. This distribution of dissolved oxygen is similar to the values recorded in the present study. Contreras-Espinosa (1983) also recorded hypoxic conditions in various seasons in Laguna Tampamachoco, as did the CFE (1994) in its environmental impact study. CFE (1994) documented low concentrations of dissolved oxygen at the lagoon inlet and the river in the same season. These results demonstrate that hypoxia has been a characteristic of this coastal system for more than 30 years, and varies depending on the hydrodynamics, geomorphology, depth, season, and urban activities.

2.4.4 pH

The pH varied from neutral to alkaline among the three sampling periods as a result of the input of river (September) and marine (March and August) waters (Table 2.1). These conditions are normal in coastal riverine-lagoon systems during daylight when photosynthesis is greatest as a result of CO_2 assimilation.

2.4.5 Total Nitrogen, Ammonium, Nitrites and Nitrates

Both nitrates and nitrites were low in the three sampling periods and ammonium varied. Nitrates and nitrites varied from 0–0.7 μM with a slight predominance of nitrites. These low concentrations were similar to values recorded in 1979 by Contreras-Espinosa (1983). High values of ammonium were recorded in March 2009 (12.14–15.71 μM) and September 2010 (20.00–25.71 μM). The approximate 60 % increase in ammonium from March to September was likely caused by increased runoff during the rainy season in September (Fig. 2.4). Contreras-Espinosa (1983) sampled ammonium throughout the year and recorded maximum values in May (at the start of the rainy season) with 52 μM, which corroborate values recorded in the present study during the rainy season. In August 2009, the concentration of ammonium was markedly low throughout the system and varied from 0 to 2.86 μM; this nutrient was particularly low in the Rio Tuxpan. This difference in ammonium could be to normal interannual variability in river discharge (Fig. 2.4).

Table 2.1 Descriptive statistics of physicochemical parameters of the Laguna Tampamachoco, Veracruz, Mexico in March and August 2009 and September 2010

Parameter	Statistic	March 2009		August 2009		September 2010	
		Surface	Bottom	Surface	Bottom	Surface	Bottom
Temperature °C	Max	27.63	27.41	29.79	29.14	27.96	27.42
	Min	25.94	24.4	26.10	23.26	25.45	25.99
	Mean	26.83	25.98	27.74	26.00	26.67	26.61
	SD	0.61	1.19	1.28	1.80	0.86	0.60
Salinity (psu)	Max	36.24	36.55	37.43	37.60	27.39	19.79
	Min	23.49	32.76	16.80	25.84	3.40	1.77
	Mean	31.66	34.83	33.49	34.12	16.14	10.65
	SD	4.32	1.66	7.41	4.31	8.55	7.40
Dissolved Oxygen (mg/L)	Max	10.48	7.90	9.65	6.12	7.58	6.53
	Min	3.38	3.10	5.08	3.15	3.63	5.69
	Mean	6.03	5.28	7.35	4.18	5.63	6.01
	SD	2.34	1.86	1.61	1.27	1.35	0.41
Oxygen Saturation (%)	Max	150.30	119.90	140.40	88.80	100.00	93.50
	Min	51.10	47.60	81.90	43.80	53.60	72.60
	Mean	88.98	77.43	111.67	60.14	75.15	80.20
	SD	31.88	24.43	20.76	20.02	15.64	9.64
Ammonium (μM)	Max	15.71	15.71	11.43	4.29	23.57	35.00
	Min	12.14	12.14	1.43	0.71	19.29	20.00
	Mean	14.02	14.59	3.37	1.94	21.7	25.00
	SD	1.07	1.22	3.60	1.29	1.61	4.77
Total Nitrogen (μM)	Max	37.86	35.71	22.86	11.43	38.57	38.57
	Min	16.43	19.29	5.00	7.14	30.00	30.00
	Mean	26.79	27.24	10.09	8.76	33.93	33.87
	SD	7.35	5.90	5.39	1.58	3.10	3.00
Orthophos-phates (μM)	Max	1.61	2.58	3.55	3.23	6.77	4.19
	Min	0.65	0.97	0.32	0.00	0.65	0.97
	Mean	1.13	1.57	1.74	1.24	2.58	2.39
	SD	0.30	0.54	0.91	1.48	2.02	1.42
Total Phosphorus (μM)	Max	6.13	5.48	6.77	7.74	28.06	24.84
	Min	2.58	2.58	3.23	3.23	9.35	4.19
	Mean	3.63	3.46	4.68	4.80	14.67	13.54
	SD	1.39	1.00	1.10	1.97	7.43	8.45
Chemical Oxygen Demand (mg/L)	Max	2.99	2.99	5.98	5.98	7.62	5.71
	Min	1.63	1.63	5.44	5.44	4.08	3.54
	Mean	2.24	2.25	5.51	5.18	5.31	4.39
	SD	0.65	0.70	0.19	0.20	1.31	0.77

Total nitrogen was highest in September 2010 (range 30.00–38.57 μM) and coincided with the rainy season and the period of maximum runoff. The estuarine sampling areas (stations 6 and 7) and those located at the pier and near the shoreline villages (station 5) were especially high in total nitrogen due to human-waste discharge. Values of total nitrogen were also high in March 2009, although unevenly distributed as evidenced by the large standard deviations (Table 2.1). These concentrations are uncommon in Mexican coastal lagoons and may be relatively high due to the decomposition of organic matter from urban runoff, decomposition of dead organisms and their excreta, suspension of sediments generated by continuous boat traffic, and ammonium inputs from other anthropogenic sources (Campos-Villegas 1996). The high content of ammonium recorded in September 2010 was associated also with the high concentration of total nitrogen due to generation of ammonium from the decomposition of organic matter (i.e., ammonification). Considering data from Contreras-Espinosa (1983), the concentration of ammonium has been high for more than 30 years, with values about 60 % greater in September 2010 in the current study.

Despite the low concentration of nitrates and nitrites and the high concentration of ammonium, a nutrient that is assimilated in photosynthesis by some phytoplanktonic species, the Margalef Index of community diversity for most of the stations studied by Sánchez-Santillán in 1989 (unpublished data)

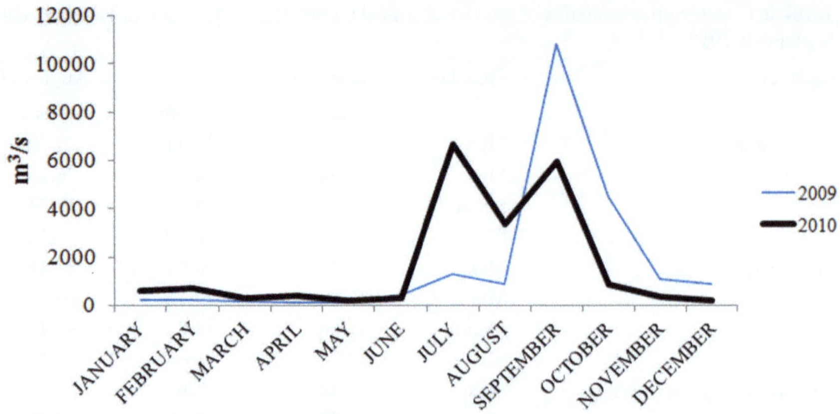

Fig. 2.4 Rio Tuxpan flow in two different years

Table 2.2 Land Ocean Interactions in the Coastal Zone (LOICZ) balance of inorganic nitrogen of Laguna Tampamachoco, Mexico. Acronym abbreviations are: dissolved inorganic nitrogen balance (ΔNID), dissolved inorganic phosphorus balance (ΔPID), and ecological net metabolism (MNE)

ΔNID (mmol/m²day)	ΔPID (mmol/m²day)	MNE	Balance of N
0.11	0.03	−2.9	−0.330
−0.07	−0.02	2.3	0.280
0.24	0.02	−1.6	−0.003

indicated a condition of generalized succession of Laguna Tampamochoco with values from 3 to >5.1. In this trophic state, fast-growing species dominate more static species, which may result in a bloom (high reproductive rate) of the dominant primary producers. The high abundance of primary producers will then die, causing an increase in the organic nitrogen load, which was observed in the current study. Margalef (1975) calculated trophic values of 3.4–5.5 in Laguna de Alvarado (located approximately 1,500 km to the south on the Gulf of Mexico coast), indicating a relatively active phytoplankton community with scarce chlorophyll. Similarly, the physicochemical characteristics and high trophic level of Laguna Tampamachoco are likely related.

2.4.6 Orthophosphates and Total Phosphorus

The orthophosphate content recorded during the three survey periods was within the normal range of values for non-impacted lagoons in Mexico (de la Lanza-Espino 1994). For example, values in September 2010 at the river mouth (station 6; 2.58 µM at the surface and 0.97 µM at the bottom) and in the lagoon near the pier (station 4; 0.97 µM at the surface and 3.87 µM at the bottom) were similar to de la Lanza-Espino (1994). However, the concentration of total phosphorus was high at station 4 (9.35 µM at the surface and 15.48 µM at the bottom) and station 6 (28.06 µM at the surface and 28.84 µM at the bottom) and is indicative of eutrophication. High total phosphorous concentrations are likely due to inputs by the urban settlements, high river discharge and a consequent suspension of sediments due to the rainy

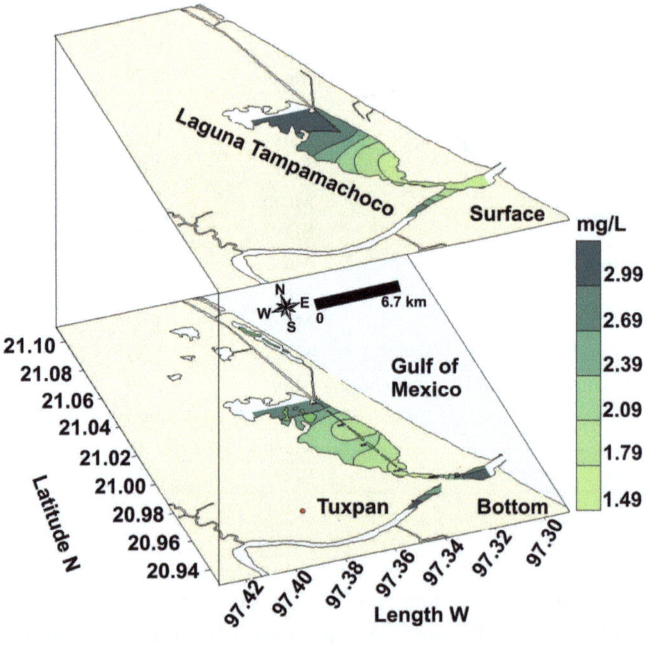

Fig. 2.5 COD distribution in Laguna Tampamachoco, Veracruz in March 2009

season, and the movement of small and large vessels. Total phosphorus in March 2009 and August 2009 was about 20 % lower than September 2010, with levels considered normal for this coastal system (de la Lanza-Espino 1994) (Table 2.1).

In relation to LOICZ nitrogen balance, the Laguna Tampamachoco showed similar export-import processes (Table 2.2) as other coastal lagoons in Mexico (e.g., Laguna El Yucateco in the southern Gulf of Mexico), a process that

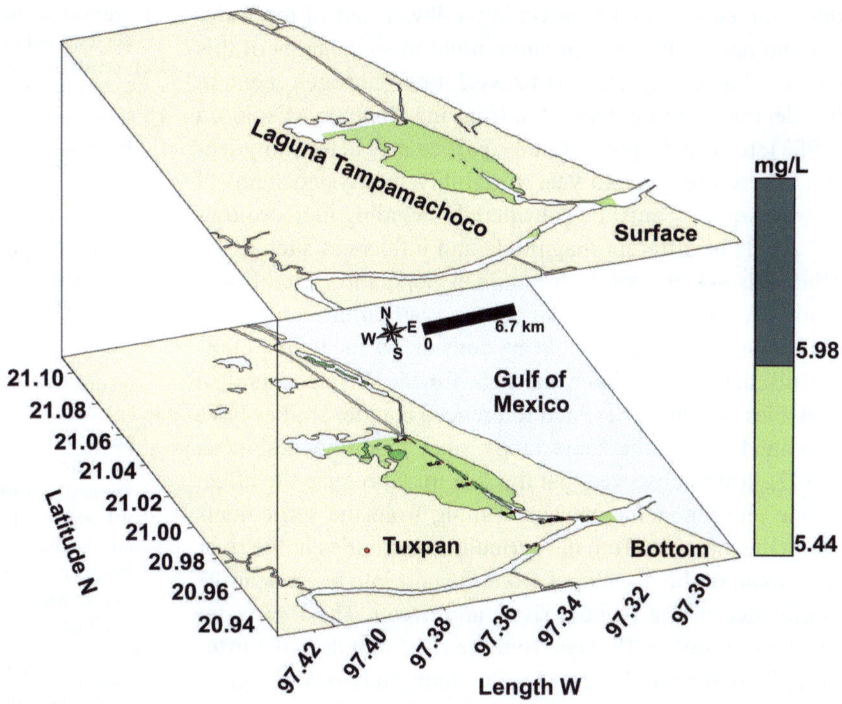

Fig. 2.6 COD distribution in Laguna Tampamachoco, Veracruz in August 2009

depends on tide, meteorological effects, circulation dynamics, differential runoff, and human activities. The Ecological Net Metabolism (MNE) was a state of heterotrophy.

2.4.7 Chemical Oxygen Demand (COD)

In March 2009, COD values varied from 2.99 to 1.63 mg/L of O_2 indicating low concentrations of organic matter and a rapid and efficient remineralization, (Table 2.2, Fig. 2.5). In August, COD was homogeneous and approximately twice the value recorded in March, with no high values and/or an efficient decomposition that used the excess oversaturated dissolved oxygen. Low COD concentrations were recorded at only a few stations and at the bottom of the water column (Fig. 2.6).

2.4.8 Ichthyoplankton

The high concentration of ammonium, total nitrogen and total phosphorus that has caused eutrophication of Laguna Tampamachoco over the past few decades has likely not affected the fish larvae. Adult fishes that are found in Laguna Tampamachoco spawn at sea; eggs and larvae then enter the lagoon during the tidal cycles (Román-Hernández et al. 2006). Ocaña-Luna and Sanchez-Ramirez (2000) estimated the effect of temperature and salinity on the spatial and temporal variation of ichthyoplankton in Laguna Tampamachoco. Fish larvae were relatively abundant at night and were likely feeding during a period of low predation probability; larvae concentration varied from 54 to 307 org/100 m^3 in spring and summer, respectively. In November 1987 and August 1988, fish larvae densities varied between 100 and 153 larvae/100 m^3, respectively, and up to 188 larvae/100 m^3 were recorded in February of 1988 (Ocaña-Luna and Sanchez-Ramirez 2000). The fish larvae densities are within the range recorded by Flores-Coto et al. (1986). Román-Hernández et al. (2006) quantified a maximum average abundance of 1308 larvae/100m^3 in December 2003 and a low abundance of 192 larvae/100 m^3 in April 2004; abundance increased in May to 639 larvae/100 m^3. Based on the current and previous studies, larval densities were likely associated with variations in temperature and salinity. This suggests that the water quality of Laguna Tampamachoco since 1979 has likely not affected ichthyoplankton density. Additionally, Ponce et al. (in press) recently found no contamination of heavy metals and hydrocarbons in oysters and sediments of Laguna Tampamachoco

2.5 Conclusions

Laguna Tampamachoco has three distinct morphologic regions: the estuarine area where the river discharges near the lagoon inlet, the lagoon area proper and the more inner and shallow lagoon area with a complex bathymetry. Since the study by Contreras-Espinosa (1983), several studies have documented seasonal hypoxia in the lower part of the water column and high levels of ammonium in some areas of this riverine-lagoon system. However, total nitrogen content has decreased more than 50% from the Contreras-Espinosa (1983) study to the present study, particularly in the rainy season. Therefore, year to year variability in physicochemical or water quality must be considered, including meteorological events of different magnitude and influences such as La Niña. Also, increases in the human population established along the margins of the lagoon and agricultural activities along the river basin should be considered for their pollutant discharges. However, the data for the physicochemical variables presented here and referenced in other studies have remained within the same range since Contreras-Espinosa (1983). It was observed that the tide might reduce the effect of the anthropogenic sewage coming from the settlements along the margins, from the agricultural activities in the river basin, and of the discharges from the port into areas near the confluence of the lagoon, river, and ocean. The low water residence times of 19 days from the river mouth to the estuarine-lagoon area, 40 days for the main body of the lagoon and 50 days for the whole system including its inner shallow areas, have helped to prevent an increase in the concentrations of ammonium contributed by the adjacent populated areas since Contreras-Espinosa (1983). The low oxygen concentrations at the bottom of the water column are likely compensated by significant photosynthetic activity from an active phytoplankton community. Based on the LOICZ model, nutrients are both exported to and imported from the sea, and the Ecological Net Metabolism (MNE) of Laguna Tampamachoco was calculated to be in a heterotrophic state.

Acknowledgments We thank Comision Federal de Electricidad for permitting the publication of this document, and Dr. A. V. Botello for his invitation to participate in this project, as well as his support during the field sampling

References

APHA, AWWA and WEF (2005) Standard methods for the examination of water and wastewater, 21 st ed. American Public Health Association, Washington, DC.

Calva BLG, Torres-Alvarado R (2000) Distribución de carbohidratos, carbono y nitrógeno orgánico en sedimentos de tres lagunas costeras del Golfo de México. Hidrobiológica 10(2):101–114

Campos-Villegas L (1996) Flujo de nutrientes (N03, N02, NH4) en la interface sedimento-agua en los drenes agroindustriales, e hidrología en la laguna de Ensenada del Pabellón, Sin. Dissertation, Universidad Nacional Autónoma de México

Contreras-Espinosa F (1983) Variaciones en la hidrología y concentraciones de nutrientes del área estuarino-lagunar de Tuxpan Tampamachoco, Veracruz, México. Biótica 8(2):201–213

Contreras-Espinosa F (1985) Comparación hidrológica de tres lagunas costeras del estado de Veracruz. Univ Ciencia 2(3):47–56

Contreras-Espinosa F (2010) Ecosistemas Costeros Mexicanos, una actualización. UAM-Iztapalapa, Mexico

CFE (Comisión Federal de Electricidad) (1994) Manifestación de Impacto Ambiental Modalidad General. Construcción y operación de las Unidades 5 y 6 de la C.T. Adolfo López Mateos. Subdirección de Construcción, Coordinación de Proyectos Termoeléctricos, Mexico

de la Lanza-Espino G (1994) Química. In: de la Lanza Espino G, Cáceres Martínez C (eds) La-gunas Costeras y el Litoral Mexicano. Universidad Autónoma de Baja California Sur, La Paz, pp 127–182

de la Lanza-Espino G, Sánchez-Santillán N, Esquivel A (1996) Análisis temporal y espacial fisicoquímico de una laguna tropical a través del análisis multivariado. Hidrobiológica 8(2):89–96

de la Lanza-Espino G, Flores-Baca B, Hernández-Pulido S (2012) La calidad del agua en la la-guna Tampamachoco, Veracruz, México. In: Ramírez-León H, Navarro-Pineda JM, Barrios-Piña H (eds) Dinámica Ambiental de Ecosistemas Acuáticos Costeros Elementos y Ejemplos Prácticos de Diagnóstico. Instituto Politécnico Nacional, Sección de Estudios de Posgrado e Investigación del Instituto Politécnico Nacional, Asociación Mexicana de Infraestructura Portuaria, Marina y Costera AC, Sociedad de Egresados de Ingeniería Civil del Instituto Politécnico Nacional, Mexico, pp 127–182

Flores-Coto C, Ducoing V, Zavala-García F et al (1986) Efecto de la marea en el paso de las larvas de algunas especies de la Familia Clupeidae (Pisces), en la boca del Carmen, laguna de Términos, Campeche, México. An Inst Cienc del Mar y Limnol UNAM 14(1):53–68

García E (1988) Modificaciones al sistema de clasificación climática de Köppen para adaptarlo a las condiciones de la República Mexicana. Offset Larios, México

González JI (2009) Predicción de mareas de México. http://oceanografia.cicese.mx/predmar. Accessed 22 Nov 2009

Margalef R (1975) Fitoplancton invernal de la laguna costera de Alvarado (México). An Inst Bot Cavanilles 32(2):381–387

NOAA (2013) Cold and warm episodes by season. http://www.cpc.ncep.noaa.gov/products/analysis_monitoring/ensostuff/ensoyears.shtml. Accessed 12 March 2013

Ocaña-Luna A, Sánchez-Ramírez M (2000) Diversidad del ictioplancton en las lagunas Madre y Almagre, Tamaulipas, y laguna de Tampamachoco, Veracruz. Instituto Politécnico Nacional, Escuela Nacional de Ciencias Biológicas, Mexico

Ponce G, Villanueva S, Botello AV et al (2013) Distribución espacial de contaminantes en sedimentos de la laguna de Tampamachoco, Veracruz, México. Procesos Geoquímicos de la Superficie en América Latina, Red Iberoamericana de Física y Química Ambiental (SiFyQA), Mexico (in press)

Román-Hernández U, Valdez-Zenil J, Zavala-García F (2006) Composición y abundancia del ictioplancton durante la temporada de estiaje en la laguna de Tampamachoco, Veracruz, México. Revista UDO Agrícola 6(1):138–149

Sánchez-Santillán N, de la Lanza-Espino G (2012) Cambio o variabilidad climática: un problema semático con graves consecuencias. In: Ramírez-León H, Navarro-Pineda JM, Barrios-Piña H (eds) Dinámica Ambiental de Ecosistemas Acuáticos Costeros Elementos y Ejemplos Prácticos de Diagnóstico. Instituto Politécnico Nacional, Sección de Estudios de Posgrado e Investigación del Instituto Politécnico Nacional, Asociación Mexicana de Infraestructura Portuaria, Marina y Costera AC, Sociedad de Egresados de Ingeniería Civil del Instituto Politécnico Nacional, Mexico, pp 49–66

Strickland JDH, Parsons TR (1972) A practical handbook of seawater analysis. Fisheries Research Board of Canada, Ottawa

Mercury in Fish, Crustaceans and Mollusks from Estuarine Areas in the Pacific Ocean and Gulf of Mexico Under Varying Human Impact

3

Jorge R. Ruelas-Inzunza, Ofelia Escobar-Sánchez and Federico Páez-Osuna

Abstract

Mercury (Hg) is the most dangerous trace element present in the edible parts of fishes and invertebrates. With the aim of having a general view on Hg occurrence in commercially exploited biota (fish and invertebrates) from selected estuarine systems of Mexico, we compiled information related to Hg levels in fish (elasmobranchs and teleosts), shrimps, clams, mussels and oysters from impacted estuarine areas and other coastal ecosystems in the Pacific Ocean and the Gulf of Mexico. Levels of Hg in the Asiatic clam *Corbicula fluminea* (a freshwater species) were relatively low (<0.32 μg g^{-1}) in comparison to individuals collected in moderate or severely impacted sites. In the case of marine mollusks (*Crassostrea corteziensis* and *Mytella strigata*) Hg concentrations were comparable to those from low or moderately contaminated sites. In shrimps, Hg values were low (<0.72 μg g^{-1}) and consistently higher in hepatopancreas tissue than in muscle. Rays had lower Hg levels (<0.4 μg g^{-1} wet weight) than sharks (<2.0 μg g^{-1} wet weight). Teleost fish have been studied more thoroughly than other groups; Hg levels in muscle tissue varied by two orders of magnitude (from 0.02 to 1.58 μg g^{-1} dry weight). Among studied organisms, fish are known as the main pathway of Hg entrance to humans. It is necessary to generate information of the rates of consumption of fish, especially of predator species. Considering legal limits of Hg and methyl Hg (1.0 and 0.5 μg g^{-1} wet weight, respectively) in edible portion of fish in Mexico, at present there is risk to the human population for the consumption of the scalloped hammerhead shark *Sphyrna lewini*.

Keywords

Mercury · Fish · Mollusks · Crustaceans · Health risk

3.1 Introduction

Mercury (Hg) is an element with an atomic weight of 200.59 g/mole and a boiling point of 356.6 °C; it is a dense, silvery-white liquid at room temperature (Hunter 1975). Hg is distributed worldwide and it is mobilized naturally through the earth's crust, atmosphere, oceans, and life forms. Although Hg occurs naturally in the environment, it is concentrated in geographical belts. In fact, significant Hg deposits belong to one of the two Tertiary or Quarternary orogenic and volcanic belts: the Circum-Pacific and the Mediterranean-Himalayan belt (WGMF 1980). For example, it can be found in

J. R. Ruelas-Inzunza (✉) · O. Escobar-Sánchez
Laboratorio de Metales Pesados, Instituto Tecnológico de Mazatlán, Calle Corsario 1, No 203, Colonia Urías, C.P., 82040 Mazatlán, Sin., México
e-mail: ruelas@ola.icmyl.unam.mx

O. Escobar-Sánchez
e-mail: escobars.ofelia@gmail.com

F. Páez-Osuna
Instituto de Ciencias del Mar y Limnología, Universidad Nacional Autónoma de México, Av. Joel Montes Camarena s/n, C.P., 82040 Mazatlán, Sinaloa, México
e-mail: paezos@ola.icmyl.unam.mx

F. Amezcua, B. Bellgraph (eds.), *Fisheries Management of Mexican and Central American Estuaries*, Estuaries of the World, DOI 10.1007/978-94-017-8917-2_3,

natural deposits, such as the Hg bed under the Mediterranean Sea, which holds some of the richest reserves of mercury in the world (Bacci 1989). The main ore of mercury is the red sulphide cinnabar (HgS), which has been mined throughout the world in places such as: Spain, Italy, Yugoslavia, Russia, China, Japan, Mexico, California, and British Columbia (Hunter 1975). Mercury is naturally emitted into the air as a result of off-gassing from the earth's surface and from volcanoes. Mercuric vapor can remain in the atmosphere for significant amounts of time and travel long distances before it cycles back to the earth in rainwater. After Hg is released into the environment in inorganic form, it is methylated by bacteria in water and converted to an organic form, usually methylmercury (Rasmussen et al. 2005). Anaerobic microorganisms like sulfate-reducing bacteria are the main producers of methylmercury (Parks et al. 2013). Such transformation enhances the entrance of Hg to the food chain, eventually resulting in biomagnification. Although Hg intake in humans occurs through contaminated food, drink, or air, exposure to organic mercury is almost exclusively a result of consumption of fish and shellfish (Gunderson 1995; NRC 2000). In this sense, Storelli et al. (2002) estimated that fish consumption accounts for 80–90 % of the total exposure to mercury.

An estuary is defined as a semi-enclosed coastal body of water that is either permanently or periodically open to the sea and within which seawater is diluted with water derived from land drainage (Kennish 2001). Estuaries have elevated levels of biological productivity and play important ecological roles (Clark 1996). On the other hand, areas surrounding estuaries have become more densely populated in the last few decades and this has resulted in a wide range of environmental threats: damming of rivers, pollution, urbanization, industrial development, construction of flood-protection devices, mariculture, and recreation (Lindeboom 2002). Estuarine environments are highly dynamic and very sensitive to anthropogenic discharges since they function as sinks for fine-grain sediments that are usually associated with contaminants such as trace metals (Rosales-Hoz et al. 2003). In this chapter, we report Hg levels in biota from estuarine systems of selected areas in Mexico with different degrees of human impact. Data were compiled from published documents related to Hg levels in invertebrates (shrimps, clams and oysters) and vertebrates (elasmobranchs and teleosts) in estuarine systems and other coastal ecosystems of the Pacific Ocean and the Gulf of Mexico.

3.2 Environmental Impact in Studied Areas

The estuarine areas where biota was collected are located in states of the Pacific coast and in the Gulf of Mexico (Table 3.1). Human impacts vary from extreme in Veracruz (specifically in Coatzacoalcos estuary; Ortíz-Lozano et al. 2005) to severe in some parts of Sinaloa state (Baluarte estuary; Table 3.1).

Table 3.1 Degrees of human impact in coastal states of Mexico

State	Human impact
Sonora	Severe (fisheries and aquaculture, industrial, harbors, urban development, wastewater discharges, mining)
Sinaloa	Severe (aquaculture, fisheries, agriculture, industrial, tourism and recreation, harbors, mining and wastewater discharges)
Baja California	Severe (urban development, industrial, garbage presence, wastewater discharges)
Veracruz	Extreme (industrial, harbors, urban development, garbage presence, habitat destruction, wastewater discharges)

Scale for impacts: 0 = not present; 1 = light; 2 = moderate; 3 = intense; 4 = severe; 5 = extreme (Ortíz-Lozano et al. 2005)

Wastewater discharges are the agents of human impact that exist in all of the study areas. Other relevant activities that affect estuarine environments considered in this study were industrial, agriculture, mining, and aquaculture (Ortíz-Lozano et al. 2005). Several of the activities mentioned in Table 3.1 emit Hg to the environment; however, estimations of the relative contributions for every source are not complete.

3.3 Mercury Sources of Mexico

In 1999, it was reported that the main sources of annual Hg emissions to the atmosphere were: mining and refining of gold (11.27 t; equivalent to 36.0 % of the total), mining and refining of Hg (9.666 t; 30.8 %), chloralkali plant processes (4.902 t; 15.7 %), copper smelting (1.543 t; 4.9 %), residential combustion of wood (1.168 t; 3.7 %), carboelectric plants (0.7855 t; 2.5 %), and oil refining (0.680 t; 2.2 %; Acosta y Asociados 2001). Other Hg emission sources (e.g., thermoelectrical plants, lead and zinc smelting, fluorescent lamps and dental amalgams) accounted for 0.9413 t (3.0 %; Acosta y Asociados 2001). Considering other environmental compartments as well as Hg contained in wastes and products, total releases are considered in Table 3.2.

Four of the six states included in Table 3.2 correspond to the locations of the studied estuarine areas; for the purpose of contrasting results, the states with the highest (Durango) and lowest (Baja California Sur) Hg releases were also listed. The decreasing order of total releases of Hg in the studied areas was Sonora > Veracruz > Sinaloa > Baja California. It is worth mentioning that Hg levels in certain areas do not correspond to local releases; in this sense, in some sites it has been found that mine wastes can more greatly affect the surrounding areas of the mines than the mining operation itself (Fernández-Caliani et al. 2009); this is because of the allocation of Hg through mine drainage and its transportation in the atmosphere (Chopin and Alloway 2007).

Table 3.2 Total releases of mercury (Hg, tons) in selected states of Mexico during 2004

	Air	Water	Soil	Waste	Product	Total	National rank
Durango	0.68	0.13	53.34	2.53	0.09	56.77	1st
Sonora	1.90	0.17	38.95	5.71	2.03	48.73	2nd
Veracruz	4.21	0.48	5.53	14.11	1.30	25.63	5th
Sinaloa	0.88	0.21	4.24	4.35	0.15	9.84	18th
Baja California	1.02	0.22	0.97	4.65	0.19	7.05	21st
Baja California Sur	0.16	0.05	0.17	0.85	0.03	1.26	32nd (last)

The issue of Hg supply to the Mexican environment through anthropogenic activities is not recent. Between 1540 and 1850, about 45,000 t of Hg were sent from Spain for the extraction of gold and silver (de la Peña-Sobarzo 2003); as a consequence, there still exist many places considered "hot spots" (i.e. soils with Hg concentrations above 10 ppb).

3.4 Benthic Invertebrates

The dietary importance of benthic invertebrates to numerous species of fish, birds and mammals underscores their importance in the trophic transfer of mercury and their potential significance as biological indicators (Wiener et al. 2007). Benthic invertebrates such as oysters, mussels, clams and shrimps are also currently consumed by humans, providing a direct pathway for human exposure to mercury. Of all the benthic invertebrate groups, mollusks have been the group mostly employed in coastal biomonitoring (Rainbow and Phillips 1993; Zhou et al. 2008). In the last 15 years (1997–2012), 364 documents related to the topic of biomonitoring have been published regarding the Mexican coasts; fishes and mollusks have been the most studied.

There are difficulties associated with comparing metal levels in mollusks of different species and from different geographical regions (NAS 1980; Rainbow 2002; Osuna-Martínez et al. 2010). Mercury levels among species of mollusks from distinct regions of the world vary depending on metabolism (regulation and accumulation rates), feeding habits, and ecological conditions. Additionally, mercury levels may vary depending on sampling season, size (age), and gonadal maturation. Consequently, comparison of mercury levels among mollusks must be made with caution and data should only be used in preliminary analyses. In this context, Osuna-Martínez et al. (2010) found that Hg concentration in the soft tissue of *Crassostrea corteziensis* (Y) and *C. gigas* (X) was positively correlated (linear equation: $Y = 0.616X + 0.084$) when both species of oysters were exposed in the field to the same Hg concentration. Therefore, this equation may be used when comparing Hg levels between different sites involving these two species.

Mercury content of aquatic mollusks varies throughout the world and can be classified based on mollusk habitats including freshwater, mangrove, and lagoon/estuarine environments (Table 3.3). The only bivalve species collected in freshwater environments of Mexico for Hg studies is the Asiatic clam *Corbicula fluminea*. Mercury levels were relatively low (<0.32 μg g^{-1}) in comparison to specimens from moderate or severely impacted regions. South Virginia (USA) and the Ebro River (Spain) contained clams with mean Hg concentrations of 1.89 and 2.30 μg g^{-1}, respectively (Neufeld 2010; Faria et al. 2010). These results suggest that the upper region (Hidalgotitlán) of Coatzacoalcos estuary, Mexico, and Cerro Prieto geothermal field in Baja California, Mexico are relatively un-impacted by Hg.

Mollusk species collected from mangroves in northwestern Mexico are *C. corteziensis* and the tropical mussel *Mytella strigata*. Mercury concentrations of these species varied from 0.30 to 0.56 μg Hg/g in Tobari lagoon (Sonora) and from 0.032 to 0.145 μg g^{-1} in Mazatlán harbor (Urías lagoon, Sinaloa). Mercury levels of bivalves from northwest Mexico corresponded to low or moderately contaminated sites as compared to mangrove oyster *C. rhizophorae* from other regions (Table 3.3; Vaisman et al. 2005; Olivares-Rieumont et al. 2012).

Two oyster species, *C. gigas* and *C. corteziensis*, were collected in estuarine or lagoon environments (oyster farms) of the study area (Table 3.3). Additionally, various wild clam species were examined from these lagoon systems. The levels of Hg were similar among the sites in northwest Mexico and varied from 0.08 to 0.93 μg g^{-1} for *C. gigas*, and 0.18–0.56 μg g^{-1} for *C. corteziensis*. These ranges are comparable to levels measured in *C. virginica* in estuarine systems of the Gulf of Mexico (Reimer and Reimer 1975; Aguilar et al. 2012; Apeti et al. 2012); *C.gigas* in a Moroccan coastal lagoon (Maanan, 2008); and *Perna perna* in Ghana, west Africa (Joiris et al. 2000; Otchere et al. 2003). The higher Hg levels found in the aforementioned regions are considered to have moderate or clearly impacted sites, similar to northwest Mexico.

The Hg content of marine crustaceans is variable throughout Mexico and other parts of the world (Table 3.4). Reimer and Reimer (1975), the first study to document Hg content of shrimp collected from local markets in Mexico, found comparable Hg levels in the muscle of three species: *Penaeus setiferus* from Veracruz, *Farfantepenaeus californiensis* from Mazatlán, and *Litopenaeus stylirostris* from Topolobampo and Guaymas. Ruelas-Inzunza et al. (2004) found similar or

Table 3.3 Mercury concentrations ($\mu g\ g^{-1}$ dry weight) in mollusks from different coastal areas throughout the world

Species	Common name	Area	Hg	Reference
		Freshwater environments		
Corbicula fluminea	Asian clam	Cerro Prieto Geothermal field, (Mexico)	0.11 (0.01–0.32)	Gutierrez-Galindo et al. (1988)
Corbicula fluminea	Asian clam	Coatzacoalcos River, (Mexico)	0.09 ± 0.008	Ruelas-Inzunza et al. (2009)
Corbicula fluminea	Asian clam	North River, Virginia (USA)[a]	0.12 ± 0.01	Neufeld (2010)
Corbicula fluminea	Asian clam	South River, Virginia (USA)[c]	1.89 ± 0.11	Neufeld (2010)
Corbicula fluminea	Asian clam	Ebro River (Spain)[c]	2.30 ± 0.49 (1.8–3.0)	Faria et al. (2010)
		Mangrove environments		
Crassostrea corteziensis	Cortez oyster	Tobari lagoon, (Mexico)	0.43 ± 0.07 (0.30–0.56)	Jara-Marini et al. (2013)
Crassostrea corteziensis	Cortez oyster	Urías lagoon, (Mexico)	0.056 ± 0.017 (0.032–0.078)	Jara-Marini et al. (2008)
Mytella strigata	Mussel	Urías lagoon, (Mexico)	0.067 ± 0.035 (0.034–0.145)	Jara-Marini et al. (2008)
Crassostrea gasar	Mangrove oyster	Sine-Saloum estuary, (Senegal)[a]	0.063 ± 0.003 (0.060–0.103)	Bodin et al. (2013)
Crassostrea gasar	Mangrove oyster	(Ghana)[b]	0.155 ± 0.060 (0.09–0.34)	Otchere et al. (2003)
Crassostrea rhizophorae	Mangrove oyster	Sepetiva Bay (Brazil)[b]	(0.015–0.023)	Kehrig et al. (2006)
Crassostrea rhizophorae	Mangrove oyster	Jaguaribe estuary,, (Brazil)[a]	0.052 ± 0.024 (0.022–0.123)	Vaisman et al. (2005)
Crassostrea rhizophorae	Mangrove oyster	Pacoti estuary, (Brazil)[a]	0.045 ± 0.019 (0.021–0.065)	Vaisman et al. (2005)
Crassostrea rhizophorae	Mangrove oyster	Cocó estuary, (Brazil)[b]	0.084 ± 0.024 (0.039–0.116)	Vaisman et al. (2005)
Crassostrea rhizophorae	Mangrove oyster	Ceará estuary, (Brazil)[c]	0.154 ± 0.060 (0.056–0.300)	Vaisman et al. (2005)
Crassostrea rhizophorae	Mangrove oyster	Santa Cruz River (Brazil)[c]	(0.270–2.210)	Vaisman et al. (2005)
Crassostrea rhizophorae	Mangrove oyster	Sagua la Grande River (Cuba)[c]	0.570 (0.190–0.690)	Olivares-Rieumont et al. (2012)
		Estuarine or lagoon environments		
Crassostrea gigas (cultured)	Giant oyster	Tobari lagoon, (Mexico)	0.40 ± 0.13 (0.24–0.77)	Jara-Marini et al. (2013)
Anadara tuberculosa	Pustulose ark	Tobari lagoon, (Mexico)	0.21 ± 0.07 (0.05–0.31)	Jara-Marini et al. (2013)
Chione fluctifraga	Smooth venus	Tobari lagoon, (Mexico)	0.28 ± 0.10 (0.12–0.38)	Jara-Marini et al. (2013)
Chione gnidia	Gnidia venus	Tobari lagoon, (Mexico)	0.51 ± 0.19 (0.26–0.80)	Jara-Marini et al. (2013)
Crassostrea gigas (cultured)	Giant oyster	Coastal lagoons, SE Gulf of California, (Mexico)	0.427 ± 0.348 (0.08–0.93)	Osuna-Martínez et al. (2010)
Crassostrea corteziensis (cultured)	Cortez oyster	Coastal lagoons, SE Gulf of California, (Mexico)	0.370 ± 0.269 (0.18–0.56)	Osuna-Martínez et al. (2010)
Crassostrea virginica	American oyster	Boca de Atasta, (Mexico)	0.15 ± 0.10[d] (0.05–0.30)	Reimer and Reimer (1975)
Crassostrea virginica	American oyster	Términos lagoon, (Mexico)	0.73 (0.20–2.00)	Aguilar et al. (2012)
Crassostrea virginica	American oyster	Tamiahua lagoon, (Mexico)	0.10 ± 0.05[d] (0.05–0.30)	Reimer and Reimer (1975)
Polymesoda caroliniana	Carolina marshclam	Coatzacoalcos estuary, (Mexico)	0.142 ± 0.045 (0.105–0.225)	Ruelas-Inzunza et al. (2009)
Crassostrea virginica	American oyster	N Gulf of Mexico (USA)	(0.03–0.500)	Apeti et al. (2012)
Crassostrea gigas	Giant oyster	Oualidia lagoon, Morocco[b,c]	0.08–0.84	Maanan (2008)
Perna perna	Brown mussel	Ghana[b]	0.29 (0.01–0.76)	Joiris et al. (2000)
Perna perna	Brown mussel	Ghana[b]	0.334 ± 0.200 (0.19–0.84)	Otchere et al. (2003)
Anadara senilis	Bloody cockle	Ghana[b]	0.254 ± 0.185 (0.10–0.86)	Otchere et al. (2003)

[a] Not impacted area
[b] Minor or moderately impacted
[c] Impacted from chloro-alkali plants and/or industrial effluents
[d] Calculated with a humidity of 80 % in soft tissue

Table 3.4 Concentrations of mercury (µg g^{-1} dry weight) in shrimp sampled from coastal areas throughout the world

Species	Common name	Area	Tissue	Hg content	Reference
		Mexico			
Farfantepenaeus brevirostris	Crystal shrimp	AEP lagoon, SE Gulf of California	Hepatopancreas	0.35±0.07	Ruelas-Inzunza et al. (2004)
			Muscle	0.21±0.07	Ruelas-Inzunza et al. (2004)
Farfantepenaeus californiensis	Brown shrimp	AEP lagoon, SE Gulf of California	Hepatopancreas	0.62±0.11	Ruelas-Inzunza et al. (2004)
			Muscle	0.13±0.08	Ruelas-Inzunza et al. (2004)
Litopenaeus stylirostris	Blue shrimp	AEP lagoon, SE Gulf of California	Hepatopancreas	0.57±0.01	Ruelas-Inzunza et al. (2004)
			Muscle	0.30±0.036	Ruelas-Inzunza et al. (2004)
Litopenaeus vannamei	Whiteleg shrimp	AEP lagoon, SE Gulf of California	Hepatopancreas	0.72±0.07	Ruelas-Inzunza et al. (2004)
			Muscle	0.20±0.01	Ruelas-Inzunza et al. (2004)
Xiphopenaeus kroyery	Seabob	AEP lagoon, SE Gulf of California	Hepatopancreas	0.27±0.04	Ruelas-Inzunza et al. (2004)
			Muscle	0.13±0.04	Ruelas-Inzunza et al. (2004)
Litopenaeus vannamei	Whiteleg shrimp	Mazatlán harbor, SE Gulf of California	Whole body	0.04±0.01	Jara-Marini et al. (2012)
Farfantepenaeus californiensis	Brown shrimp	Mazatlán harbor, SE Gulf of California	Whole body	0.039±0.006	Jara-Marini et al. (2012)
Penaeus setiferus	White shrimp	Veracruz, W Gulf of Mexico	Muscle[a]	0.16±0.08	Reimer and Reimer (1975)
Farfantepenaeus californiensis	Brown shrimp	Mazatlán, SE Gulf of California	Muscle[a]	0.48±0.40	Reimer and Reimer (1975)
Litopenaeus stylirostris	Blue shrimp	Topolobampo, SE Gulf of California	Muscle[a]	0.20±0.20	Reimer and Reimer (1975)
Litopenaeus stylirostris	Blue shrimp	Guaymas, E Gulf of California	Muscle[a]	0.36±0.28	Reimer and Reimer (1975)
		International			
Crangon crangon	Common shrimp	Limfjord, Denmark	Muscle	0.09±0.03	Riisgard and Famme (1986)
Penaeus sp	Shrimp	Malaysia	Muscle	0.36±0.13	Rahman et al. (1997)
Penaeus semisulcatus	Green tiger prawn	Gulf of Arabia	Whole body	0.013±0.007	Al-Saleh and Al-Doush (2002)
Penaeus semisulcatus	Green tiger prawn	Northern Persian Gulf	Muscle	0.19±0.05	Elahi et al. (2007)
Litopenaeus stylirostris	Blue shrimp	New Caledonia	Muscle	0.20±0.06	Chouvelon et al. (2009)
Penaeus monodon	Giant tiger prawn	Mekong River Delta, S Vietnam	Muscle	0.06±0.04	Tu et al. (2008)
Penaeus monodon	Giant tiger prawn	Mekong River Delta, S Vietnam	Exoskeleton	<0.05	Tu et al. (2008)
Penaeus monodon	Giant tiger prawn	Mekong River Delta, S Vietnam	Hepatopancreas	0.07±0.02	Tu et al. (2008)
Metapenaeus ensis	Middle prawn	Guangdong Province, S China	Whole body	0.012	Li et al. (2013)
Penaeus japonicus	Kuruma prawn	Guangdong Province, S China	Whole body	0.017	Li et al. (2013)
Penaeus monodon	Giant tiger prawn	Guangdong Province, S China	Whole body	0.017	Li et al. (2013)

AEP Altata-Ensenada del Pabellón

[a] Obtained from fish market and fishermen; calculated with a humidity of 75%

lower Hg content in muscle of five shrimp species (*F. brevirostris, F. californiensis, L. stylirostris, L. vannamei, Xiphopenaeus kroyery*) collected in Altata-Ensenada del Pabellón (AEP) lagoon. This work demonstrated that hepatopancreas contained higher Hg levels than muscle. In a more recent study in Urías lagoon (Mazatlan harbor), Jara-Marini et al. (2012) found the lowest Hg concentrations measured from whole-body samples of *L. vannamei* and *F. californiensis*.

The elevated Hg levels in hepatopancreas tissue is probably related to the biological functions performed by this organ (e.g., metabolism of xenobiotics, digestion of food, storing of lipids and carbohydrates, and synthesis of enzymes and proteins; Manisseri and Menon 1995; Ruelas-Inzunza et al. 2013). Values reported in muscle of shrimp from AEP lagoon were higher than the Hg content found in *P. semisulcatus* from the Gulf of Arabia (Al-Saleh and Al-Doush 2002); *P. monodon* from the Mekong River Delta in south Vietnam (Tu et al. 2008); and *Metapenaeus ensis*, *P. japonicus* and *P. monodon* from Guangdong Province in southern China (Li et al. 2013). However, Hg levels in muscle of shrimp from AEP lagoon were lower than the values reported by Rahman et al. (1997) in *Penaeus* spp. from Malaysia. Chouvelon et al. (2009) reported 0.20 $\mu g\ g^{-1}$ of Hg in muscle of *L. stylirostris* from New Caledonia. Ruelas-Inzunza et al. (2004) found a level of 0.30 $\mu g\ g^{-1}$ in the same species from the AEP lagoon (northwest Mexico), indicating a similar level of contamination.

3.5 Elasmobranchs

There are about 970 species of elasmobranchs (sharks and rays) in the world that live in a broad range of marine habitats varying from the deep ocean to shallow coastal waters, including estuaries (Nelson 2006). Even though sharks and rays are considered to be primarily oceanic species, they are commonly found in estuarine waters. As with other fish, the estuary is a nursery site for elasmobranchs (Simpfendorfer et al. 2005) where batoid fishes (rays) occur more frequently than sharks.

Rays and juvenile sharks use the shallow and protected water of estuaries to escape from their potential predators and to feed on an abundance of prey. Elasmobranch fishes are among the top predators in the marine environment and play an important role in the transfer of energy within marine ecosystems; they regulate the size and dynamics of prey populations through predation (Cortés 1999; Wheterbee and Cortés 2004). In the estuaries, elasmobranchs are also considered to be top predators, which makes them susceptible to accumulation of contaminants like mercury.

More importantly, Hg can be assimilated by marine organisms and consequently transferred to the upper trophic levels, which can eventually lead to adverse effects on humans due to the consumption of contaminated food (Wang 2002). Contaminants have increased pressure on coastal and estuarine ecosystems over the past decades because of enhanced human activities in these areas. The input of toxic chemicals into estuarine areas from various sources can result in deleterious effects on wildlife habitats, degradation of the ecosystem, and possible poisoning of humans (Moreno et al. 1984; Morton and Blackmore 2001; Ip et al. 2004; Pan and Wang 2012). The human population is often the ultimate receptacle of anthropogenic pollutants that may magnify the concentration of Hg in the food chain.

Within the elasmobranchs, the bull shark (*Carcharhinus leucas*) is perhaps the most notorious species for migrating into estuarine ecosystems (Ortega et al. 2009). Mercury concentrations reported for this shark in Mexican estuaries (Altata-Ensenada del Pabellón, Sinaloa) ranged from 0.06 $\mu g\ g^{-1}$ wet weight in muscle to 0.18 $\mu g\ g^{-1}$ wet weight in liver (Ruelas-Inzunza and Páez-Osuna 2005). In the same area, the scalloped hammerhead shark (*Sphyrna lewini*) showed more contrasting Hg values of 0.03 $\mu g\ g^{-1}$ wet weight in liver and 1.45 $\mu g\ g^{-1}$ in muscle (Table 3.5). The estuarine area of Altata-Ensenada del Pabellon is considered to be severely impacted by humans; habitat degradation has resulted from agriculture, aquaculture, and industrial activities (Ortíz-Lozano et al. 2005; Ruelas-Inzunza and Páez-Osuna 2005). However, the variation of Hg levels in sharks collected in this estuary could be due to differences in shark diet (Monteiro et al. 1996). In Florida estuaries, the highest Hg level in muscle tissue of bull shark (0.97 $\mu g\ g^{-1}$ wet weight) was higher than in specimens from Altata-Ensenada del Pabellón (0.06 $\mu g\ g^{-1}$ wet weight). *S. lewini* from Florida (Adams et al. 2003) had Hg concentrations (1.25 $\mu g\ g^{-1}$ wet weight) that were comparable to values reported in Altata-Ensenada del Pabellón (1.45 $\mu g\ g^{-1}$ wet weight; Table 3.5). Contrastingly, elevated concentrations were found in Cape Canaveral estuary (Florida) relative to other estuarine areas (i.e. Mexican estuaries and other estuarine areas in Florida), where Hueter et al. (1995) found Hg concentrations in muscle tissue of 1.27 $\mu g\ g^{-1}$ wet weight for *C. leucas* and 1.99 $\mu g\ g^{-1}$ wet weight in *C. limbatus*.

Total Hg concentrations in muscle of batoids were consistently low ($<0.4\ \mu g\ g^{-1}$ wet weight) and less than sharks. This difference is likely due to the higher trophic status of sharks. The trophic dynamics of estuaries tend to be complex and the concentration of Hg is magnified in upper trophic levels of the food web (Day et al. 1989). Sharks are primarily carnivores and their diet consists mostly of fish (i.e., proportion by biomass and number; Wetherbee and Cortés 2004), crabs, shrimp, and other elasmobranchs. Contrastingly, batoids feed on organisms from lower trophic levels including crustaceans and mollusks; they rarely prey on fishes.

Table 3.5 Mercury concentrations ($\mu g\ g^{-1}$ wet weight) in muscle of elasmobranchs (sharks and rays) from estuaries worldwide

Estuarine Site	Species	Common name	Hg	Reference
		Sharks		
Altata-Ensenada del Pabellón México	*Carcharhinus leucas*	Bull shark	0.06	Ruelas-Inzunza and Páez-Osuna (2005)
Altata-Ensenada del Pabellón México	*Carcharhinus leucas*	Bull shark	0.18[a]	Ruelas-Inzunza and Páez-Osuna (2005)
Altata-Ensenada del Pabellón México	*Sphyrna lewini*	Scalloped hammerhead shark	1.45	Ruelas-Inzunza and Páez-Osuna (2005)
Altata-Ensenada del Pabellón Mexico	*Sphyrna lewini*	Scalloped hammerhead shark	0.03[a]	Ruelas-Inzunza and Páez-Osuna (2005)
Cape Canaveral, Florida	*Carcharhinus leucas*	Bull shark	1.27	Hueter et al. (1995)
Cape Canaveral, Florida	*Carcharhinus limbatus*	Blacktip shark	1.99	Hueter et al. (1995)
Cape Canaveral, Florida	*Carcharhinus plumbeus*	Sandbar shark	0.86	Hueter et al. (1995)
Indian River Lagoon, Florida	*Carcharhinus leucas*	Bull shark	0.77	Adams and McMichael (1999)
Charlotte Harbor, Florida	*Carcharhinus leucas*	Bull shark	0.97	Adams et al. (2003)
Indian River Lagoon, Florida	*Carcharhinus leucas*	Bull shark	0.78	Adams et al. (2003)
Tampa Bay	*Carcharhinus leucas*	Bull shark	0.66	Adams et al. (2003)
Charlotte Harbor, Florida	*Negaprion brevirostris*	Lemon shark	0.70	Adams et al. (2003)
Charlotte Harbor, Florida	*Carcharhinus limbatus*	Blacktip shark	0.79	Adams et al. (2003)
Tampa Bay	*Carcharhinus limbatus*	Blacktip shark	0.54	Adams et al. (2003)
Volusia County, Florida	*Rhizoprionodon terraenovae*	Atlantic sharpnose shark	0.57	Adams et al. (2003)
Indian River Lagoon, Florida	*Sphyrna lewini*	Scalloped hammerhead shark	0.44	Adams et al. (2003)
Tampa Bay	*Sphyrna lewini*	Scalloped hammerhead shark	1.25	Adams et al. (2003)
Charlotte Harbor, Florida	*Sphyrna tiburo*	Bonnethead shark	0.34	Adams et al. (2003)
Choctawhatchee Bay	*Sphyrna tiburo*	Bonnethead shark	0.58	Adams et al. (2003)
Indian River Lagoon, Florida	*Sphyrna tiburo*	Bonnethead shark	0.39	Adams et al. (2003)
Tampa Bay	*Sphyrna tiburo*	Bonnethead shark	0.59	Adams et al. (2003)
Cape Canaveral, Florida	*Negaprion brevirostris*	Lemon shark	0.34	Nam et al. (2011)
Tampa Bay	*Negaprion brevirostris*	Lemon shark	0.18	Adams et al. (2003)
Bahía Blanca Bay, Argentina	*Mustelus schmitti*	Gatuzo shark	0.85	Marcovecchio et al. (1986)
		Rays		
Tampa Bay	*Dasyatis americana*	Southern stingray	0.17	Adams et al. (2003)
Charlotte Harbor, Florida	*Dasyatis sabina*	Atlantic stingray	0.25	Adams et al. (2003)
Choctawhatchee Bay	*Dasyatis sabina*	Atlantic stingray	0.25	Adams et al. (2003)
Indian River Lagoon, Florida	*Dasyatis sabina*	Atlantic stingray	0.16	Adams et al. (2003)
Tampa Bay	*Dasyatis sabina*	Atlantic stingray	0.33	Adams et al. (2003)
Tampa Bay	*Dasyatis say*	Bluntnose stingray	0.2	Adams et al. (2003)
Charlotte Harbor, Florida	*Dasyatis say*	Bluntnose stingray	0.02	Adams et al. (2003)
Indian River Lagoon, Florida	*Dasyatis say*	Bluntnose stingray	0.07	Adams et al. (2003)
Indian River Lagoon, Florida	*Gymnura micrura*	Smooth butterfly ray	0.15	Adams et al. (2003)
Volusia County, Florida	*Gymnura micrura*	Smooth butterfly ray	0.07	Adams et al. (2003)
Indian River Lagoon, Florida	*Myliobatis freminvillei*	Bullnose ray	0.12	Adams et al. (2003)

[a] Liver

Use of sediments as a food source by benthic organisms is common (Hoffman et al. 1995). Consequently, mercury in sediments can be transferred up the trophic web of the ecosystem (Díaz-Jaramillo et al. 2013). Even in pristine areas, Hg can be magnified within food webs. This magnification is more severe in estuaries that contain high levels of Hg (due to anthropogenic activities) in low trophic levels; trophic transfer then elevates concentrations in top predators.

Estuaries have the greatest food availability of any ecoregion in the world (Haedrich and Hall 1976) and provide important habitat for elasmobranchs. Some shark and ray species have the osmoregulatory ability to tolerate abrupt estuarine changes in salinity and oxygen concentrations over seasonal cycles (Ortega et al. 2009). Due to this use of variable habitats, high levels of Hg in estuarine areas may contribute to increased physiological stress in these species (Saiz-Salinas and González-Oreja 2000).

Table 3.6 Average mercury concentrations (μg g^{-1} dry weight) in muscle tissue of teleost fish from selected estuarine areas in Mexico

Site	Species (common name)	Hg	Reference
Altata-Ensenada del Pabellón	*Caulolatilus princeps* (ocean whitefish)	0.57	Ruelas-Inzunza et al. (2008)
	Mugil cephalus (striped mullet)	0.13	Ruelas-Inzunza and Páez-Osuna (2005)
	Cynoscion xanthulus (orangemouth corvine)	0.11	Ruelas-Inzunza and Páez-Osuna (2005)
	Lutjanus colorado (colorado snapper)	0.89	Ruelas-Inzunza and Páez-Osuna (2005)
	Galeichthys peruvianus (catfish)	1.58	Ruelas-Inzunza and Páez-Osuna (2005)
Guaymas	*Mugil cephalus* (gray mullet)	0.03	Reimer and Reimer (1975)
	Opisthonema libertate (Pacific thread herring)	0.21	Ruelas-Inzunza and Páez-Osuna (2005)
	Cathorops fuerthii (catfish)	0.46	Ruelas-Inzunza and Páez-Osuna (2005)
	Mugil cephalus (striped mullet)	0.02	Ruelas-Inzunza and Páez-Osuna (2005)
	Seriola lalandi (yellowtail)	0.76	Ruelas-Inzunza and Páez-Osuna (2005)
Urías	*Mugil cephalus* (gray mullet)	0.06	Reimer and Reimer (1975)
	Anisotremus interruptus (burrito grunt)	0.12	Reimer and Reimer (1975)
	Lutjanus colorado (colorado snapper)	0.18	Ruelas-Inzunza et al. (2011a)
Baluarte estuary	*Mugil curema* (white mullet)	0.449	Ruelas-Inzunza et al. (2011b)
	Agonostomus monticola (mountain mullet)	0.389	Ruelas-Inzunza et al. (2011b)
	Guavina guavina (guavina)	0.386	Ruelas-Inzunza et al. (2011b)
	Oreochromis sp. (red tilapia)	0.351	Ruelas-Inzunza et al. (2011b)
	Gobiesox fluviatilis (mountain clingfish)	0.27	Ruelas-Inzunza et al. (2011b)
	Oreochromis urolepis (Wami tilapia)	0.274	Ruelas-Inzunza et al. (2011b)
Coatzacoalcos estuary	*Centropomus* sp. (snook)	0.27	Reimer and Reimer (1975)
	Oreochromis sp. (tilapia)	0.054	Ruelas-Inzunza et al. (2009)
	Gobiomorus polylepis (finescale sleeper)	0.492	Ruelas-Inzunza et al. (2009)
	Gerres cinereus (yellowfin mojarra)	0.135	Ruelas-Inzunza et al. (2009)
	Centropomus viridis (white snook)	0.612	Ruelas-Inzunza et al. (2009)
	Lepisosteus osseus (longnose gar)	0.141	Ruelas-Inzunza et al. (2009)

3.6 Teleosts

Levels of Hg in teleosts varied from 1.58 μg g^{-1} in the catfish *Galeichthys peruviaus* from Altata-Ensenada del Pabellón lagoon to 0.02 μg g^{-1} in the striped mullet *Mugil cephalus* from Guaymas (Table 3.6). Due to the common reporting of Hg concentrations in fish muscle tissue from all cited works, it was used to compare between different sites within Mexico. Mean Hg concentrations (μg g^{-1} dry weight) in ichthyofauna was highest in Altata-Ensenada del Pabellón (0.656) followed by Baluarte estuary (0.353), Guaymas (0.296), Coatzacoalcos estuary (0.284), and Urías (0.12).

Although Sinaloa is ranked the 18th highest state in Mexico for Hg inputs to the environment (Table 3.2), fish had variable Hg levels among the sites studied (Altata-Ensenada del Pabellón, Baluarte estuary and Urías; Table 3.6). Mercury inputs from human activities around Altata-Ensenada del Pabellón include waste effluents from intensive agriculture where Hg compounds are used as fungicides (Boening 2000); urban sewage from the cities of Culiacán and Navolato (population around 1 million) also contribute Hg to this system. In addition, seasonal operation (from December to June) of a sugar-cane processing industry supplies effluents containing Hg to the Altata-Ensenada del Pabellón system (Páez-Osuna et al. 1998). Baluarte estuary is impacted by Hg inputs from polymetallic deposits (Clark et al. 1982). In past decades, Pb and Au were extracted within the watershed of this estuary (DIFOCUR 1991) and currently there is a renewed interest for mineral extraction; these mining activities will likely increase the input of Hg into Baluarte estuary. The Urías lagoon is impacted by urban activities including the processing of fishery products, the existence of tuna and shrimp fleets, a thermoelectric plant and three shrimp farms (Páez-Osuna 2007). Mercury levels in fish of Guaymas estuary were not elevated despite the state of Sonora's high Hg emissions (2nd in Mexico) that is primarily due to the mining of Au and Cu, one of the main industrial activities. However, the coastal area of Guaymas estuary receives very little freshwater input and thus very little Hg is transported to the estuary from inland areas. Lastly, Coatzacoalcos estuary has industrial facilities that manufacture petrochemical products; such activities supply significant inputs of trace elements to the environment, including Hg (Croudace and Cundy 1995). Nevertheless, seasonal rains (from June to September) and winds (from October to February) as well as high flows of freshwater have mitigated the impact of Hg on the atmospheric and aquatic media (Ruelas-Inzunza et al. 2009).

The order of total releases of Hg in the compared states (Table 3.2) and the degree of anthropogenic impact (Table 3.1) in the studied areas does not match the average Hg levels in fish. Hg supply to the studied sites is through natural and human processes, another issue of importance is the feeding habit of specimens; i.e, fish of carnivorous habits (*G. peruvianus* and *Lutjanus colorado*) had more elevated Hg levels than non-carnivorous species. More research is necessary in order to elucidate pathways of Hg sources, mobility and destinations in the environment.

3.7 Conclusions

In Mexico, studies related to the occurrence of Hg in biota from aquatic environments are scarce. Published information on this topic includes mollusks, crustaceans, elasmobranchs and teleost fish. Mercury levels of concern were registered in elasmobranchs (sharks) and carnivorous fish. Total releases of Hg in the states where fish were collected did not correspond to Hg concentrations in fish.

Levels of Hg in freshwater mollusks (*C. fluminea*) from Mexico were lower than corresponding values in specimens from the USA and Spain. In the case of mollusk species (*C. corteziensis* and *M. strigata*) from mangrove environments, Hg levels in individuals from northwest Mexico corresponded to low or moderately contaminated sites. For estuarine or lagoon environments (oyster farms), Hg concentrations in *C. gigas* and *C. corteziensis* and other wild species were comparable to individuals from the Gulf of Mexico, the Moroccan coast and Ghana.

Regarding crustaceans, most of the published information is related to shrimps. Mercury concentration was consistently higher in hepatopancreas tissue than in muscle. Penaeid shrimps from Altata-Ensenada del Pabellón (Sinaloa) had Hg levels comparable or lower than individuals from Veracruz (Veracruz) and Mazatlán (Sinaloa).

In elasmobranchs, previous studies on Hg content corresponded to sharks and rays. In the case of sharks, levels of Hg in *C. leucas* from Altata-Ensenada del Pabelón were lower than in specimens from Florida. The hammerhead shark *S. lewini* had comparable Hg concentrations in specimens from Altata-Ensenada del Pabellón and Florida. Batoids (rays) had low levels (<0.4 µg g^{-1} wet weight) of Hg, which were lower than sharks.

Teleost fish have been the most widely studied group of aquatic animals regarding Hg bioaccumulation. Among species, mercury levels in muscle tissue were variable (by two orders of magnitude). Mercury concentration of teleosts was the highest in Altata-Ensenada del Pabellón, followed by Baluarte estuary, Guaymas, Coatzacoalcos, and Urías.

Given the importance of mollusks, crustaceans and fish for human consumption, more studies are necessary about these fishing resources. Fish are known to be the main pathway of Hg entrance to humans but information related to the Hg concentration of most of the exploited fish species, as well as to rates of fish consumption in Mexico is scarce. It is relevant to generate this sort of information, especially for predator species in the upper trophic levels of the ecosystem, which are most likely to bioaccumulate Hg. Published information on methyl Hg in fish collected in Mexico is also very scarce. Given the legal limits of Hg and methyl Hg (1.0 and 0.5 µg g^{-1} wet weight, respectively) in edible portion of fish in Mexico, the species that represents the highest risk to humans is the scalloped hammerhead shark (*S. lewini*). Fisheries management programs should consider the risk to humans of consuming fish and other marine resources contaminated with Hg to preclude or limit their consumption. Further, management programs that decrease the integration of Hg into the food web, and additional research that addresses the methods of integration are needed to decrease the negative impacts of Hg contamination to these important human food supplies.

References

Acosta y Asociados (2001) Inventario preliminar de emisiones atmosféricas de mercurio en México. Acosta y Asociados, Agua Prieta

Adams DH, McMichael RH Jr (1999) Mercury levels in four species of sharks from the Atlantic coast of Florida. Fish Bull 97:372–379

Adams DH, McMichael RH Jr, Henderson GE (2003) Mercury levels in marine and estuarine fishes of Florida 1989–2001. Florida Marine Research Institute, St. Petersburg

Aguilar CA, Montalvo C, Rodríguez L et al (2012) American oyster (*Crassostrea virginica*) and sediments as a coastal zone pollution monitor by heavy metals. Int J Environ Sci Tech 9(4):579–586

Al-Saleh I, Al-Doush I (2002) Mercury content in shrimp and fish species from the Gulf coast of Saudi Arabia. Bull Environ Contam Toxicol 68:576–583

Apeti DA, Lauenstein GG, Evans DW (2012) Recent status of total mercury and methyl mercury in the coastal waters of the northern Gulf of Mexico using oysters and sediments from NOAA´s mussel watch program. Mar Pollut Bull 64(11):2399–2408

Bacci E (1989) Mercury in the Mediterranean. Mar Pollut Bull 20:59–63

Bodin N, N'Gom-Ká R, Ká S et al (2013) Assessment of trace metals contamination in mangrove ecosystems from Senegal, West Africa. Chemosphere 90(2):150–157

Boening DW (2000) Ecological effects, transport, and fate of mercury: a general review. Chemosphere 40:1335–1351

Chopin EIB, Alloway BJ (2007) Distribution and mobility of trace elements in soils and vegetation around the mining and smelting areas of Tharsis, Ritinto and Huelva, Iberian Pyrite Belt, SW Spain. Water Air Soil Pollut 182:245–261

Chouvelon T, Warnau M, Churlaud C et al (2009) Hg concentrations and related risk assessment in coral reef crustaceans, Mollusks and fish from New Caledonia. Environ Pollut 157:331–340

Clark JR (1996) Coastal zone management handbook. CRC Marine Science Series, Boca Raton

Clark KF, Foster CT, Damon PE (1982) Cenozoic mineral deposits and subduction-related magmatic arcs in Mexico. Geol Soc Am Bull 93:533–544
Cortés E (1999) Standardized diet composition and trophic levels of sharks. ICES J Mar Sci 56:707–715
Croudace IW, Cundy A (1995) Heavy metal and hydrocarbon pollution in recent sediments from Southampton water, southern England: a geochemical and isotopic study. Environ Sci Technol 29:1288–1296
Day JW Jr, Hall CAS, Kemp WM et al (1989) Estuarine Ecology. Wiley, USA
de la Peña-Sobarzo P (2003) Focos rojos de mercurio en América del Norte. El Faro III(31):8–9
Díaz-Jaramillo M, Muñoz C, Rudolph I et al (2013) Seasonal mercury concentrations and $\delta^{15}N$ and $\delta^{13}C$ values of benthic macroinvertebrates and sediments from a historically polluted estuary in south central Chile. Sci Total Environ 442:198–206
Dirección de Investigación y Fomento de la Cultura Regional (DIFOCUR) (1991) Minería y espacio económico de la región minera de El Rosario en el siglo XVIII. DIFOCUR, Mexico
Elahi M, Esmaili-Sari A, Bahramifar N (2007) Total mercury levels in selected tissues of some marine crustaceans from Persian Gulf, Iran: variations related to length, weight and sex. Bull Environ Contam Toxicol 88:60–64
Faria M, López MA, Díez S et al (2010) Are native naiads more tolerant to pollution than exotic freshwater bivalve species? An hypothesis tested using physiological responses of three species transplanted to mercury contaminated sites in the Ebro River (NE, Spain). Chemosphere 81:1218–1226
Fernández-Caliani JC, Barba-Brioso C, González I et al (2009) Heavy metal pollution in soils around the abandoned mine sites of the Iberian Pyrite Belt (southwest Spain). Water Air Soil Pollut 200:211–226
Gunderson EL (1995) Dietary intakes of pesticides, selected elements, and other chemicals: FDA total diet study, June 1984-April 1986. J AOAC Int 78:910–921
Gutiérrez-Galindo EA, Flores-Muñoz G, Aguilar-Flores A (1988) Mercury in freshwater fish and clams from the Cerro Prieto geothernal field of Baja California, Mexico. Bull Environ Contam Toxicol 41:201–207
Haedrich RL, Hall CAS (1976) Fishes and estuaries. Oceanus 19(5):55–63
Hoffman DJ, Rattner BA, Burton GA, Cairns J (1995) Handbook of ecotoxicology. Lewis Publishers, USA
Hueter RE, Fong WG, Henderson G et al (1995) Methylmercury concentration in shark muscle by species, size and distribution of sharks in Florida Coastal waters. Water Air Soil Pollut 80:893–899
Hunter D (1975) The diseases of occupations. Hodder and Stoughton, London
Ip CCM, Li XD, Zhang G et al (2004) Over one hundred years of trace metal fluxes in the sediment of the Pearl River Estuary, South China. Environ Pollut 132:157–172
Jara-Marini ME, Soto-Jiménez MF, Páez-Osuna F (2008) Trace metals accumulation patterns in a mangrove lagoon ecosystem, Mazatlán Harbor, southeast Gulf of California. J Environ Sci Health A Tox Hazard Subst Environ Eng 43:995–1005
Jara-Marini ME, Soto-Jiménez MF, Páez-Osuna F (2012) Mercury transfer in a subtropical coastal lagoon food web (SE Gulf of California) under two contrasting climatic conditions. Environ Toxicol 27(9):526–536
Jara-Marini ME, Tapia-Alcaraz JN, Dumer-Gutiérrez JA et al (2013) Comparative bioaccumulation of trace metals using six filter feeder organisms in a coastal lagoon ecosystem (of the central-east Gulf of California). Environ Monit Assess 185:1071–1085
Joiris CR, Holsbeek L, Otchere FA (2000) Mercury in the Bivalves *Crassostrea tulipa* and *Perna perna* from Ghana. Mar Pollut Bull 40:457–460
Kennish MJ (ed) (2001) Practical handbook of marine science, 3rd edn. CRC Press, Boca Raton
Kehrig H, Costa M, Moreira I et al (2006) Total and methyl mercury in different species of mollusks from two estuaries in Rio de Janeiro State. J Braz Chem Soc 17:1409–1418
Li P, Feng X, Chan HM et al (2013) Mercury in the seafood and human exposure in coastal area of Guangdong province, South China. Environ Toxicol Chem 32:541–547
Lindeboom H (2002) The coastal zone: an ecosystem under pressure. In: Field GF, Hempel G, Sumerhayes CP (eds) Oceans 2020 Science, trends, and the challenge of sustainability. Island Press, Washington DC, pp 49–84
Maanan M (2008) Heavy metal concentrations in marine Mollusks from the Moroccan coastal region. Environ Pollut 153:176–183
Manisseri MK, Menon MN (1995) Copper-induce damage to hepatopancreas of the penaeid shrimp *Metapenaeus dobsoni* an ultrastructural study. Dis Aquat Org 22:51–57
Marcovecchio JE, Moreno VJ, Pérez A (1986) Bio-magnification of total mercury in Bahia Blanca estuary shark. Mar Pollut Bull 17(6):276–278
Monteiro LR, Costa V, Furness RW et al (1996) Mercury concentrations in prey fish indicate enhanced bioaccumulation in mesopelagic environments. Mar Ecol Prog Ser 141:21–25
Moreno VJ, Pérez A, Bastida RO et al (1984) Distribución de mercurio total en los tejidos de un delfín nariz de botella (*Tursiops gephyreus* Lahille, 1908) de la provincia de Buenos Aires (Argentina). Rev Inv Des Pesq 4:93–102
Morton B, Blackmore G (2001) South China Sea. Mar Pollut Bull 42(12):1236–1263
Nam DH, Adams DH, Reyier EA et al (2011) Mercury and selenium levels in lemon sharks (*Negaprion brevirostris*) in relation to a harmful red tide event. Environ Monit Assess 176:549–559
NAS (1980) The International Mussel Watch. National Academy of Sciences, Washington, DC
Nelson JS (2006) Fishes of the world. Wiley, New York
Neufeld DSG (2010) Mercury accumulation in caged *Corbicula*: rate of uptake and seasonal variation. Environ Monit Assess 168:385–396
NRC (2000) Toxicological Effects of Methylmercury. National Academy Press, Washington DC
Olivares-Rieumont S, Lima L, Rivero S et al (2012) Mercury Levels in sediments and Mangrove Oysters, *Crassostrea rizophorae*, from the north coast of Villa Clara, Cuba. Bull Environ Contam Toxicol 88:589–593
Ortega LA, Heupel MR, Van Beynen P et al (2009) Movement patterns and water quality preferences of juvenile bull sharks (*Carcharhinus leucas*) in a Florida estuary. Environ Biol Fishes 84:361–373
Ortíz-Lozano L, Granados-Barba A, Solís-Weiss V et al (2005) Environmental evaluation and development problems of the Mexican coastal zone. Ocean Coast Manag 48:161–176
Osuna-Martínez CC, Páez-Osuna F, Alonso-Rodríguez R (2010) Mercury in cultured oysters (*Crassostrea gigas* Thunberg, 1793 and *C. corteziensis* Hertlein, 1951) from four coastal lagoons of the SE Gulf of California, Mexico. Bull Environ Contam Toxicol 85:339–343
Otchere FA, Joiris CR, Holsbeek L (2003) Mercury in the bivalves *Anadara* (*Senilia*) *senilis*, *Perna perna* and *Crassostrea tulipa* from Ghana. Sci Total Environ 304:369–375
Páez-Osuna F (ed) (2007) La contaminación por nitrógeno y fósforo en Sinaloa: flujos, fuentes, efectos y opciones de manejo. UNAM, Mexico
Páez-Osuna F, Bojórquez-Leyva H, Green-Ruiz C (1998) Total carbohydrates: organic carbon in lagoon sediments as an indicator of organic effluents from agriculture and sugar-cane industry. Environ Pollut 102:321–326
Pan K, Wang WX (2012) Trace metal contamination in estuarine and coastal environments in China. Sci Total Environ 421:3–16

Parks JM, Johs A, Podar M et al (2013) The genetic basis for bacterial mercury methylation. Science 339(6125):1332–1335
Rahman SA, Wood AK, Sarmani S, Majid AA (1997) Determination of mercury and organic mercury contents in Malaysian seafood. J Radioanal Nucl Chem 217:53–56
Rainbow PS (2002) Trace metal concentrations in aquatic invertebrates: why and so what? Environ Pollut 120:497–507
Rainbow PS, Phillips DJH (1993) Cosmopolitan biomonitors of trace metals. Mar Pollut Bull 26:593–601
Rasmussen RS, Nettleton J, Morrissey MT (2005) A Review of mercury in seafood: special focus on tuna. J Aquat Food Prod Technol 14(4):71–100
Reimer AA, Reimer RD (1975) Total mercury in some fish and shellfish along the Mexican coast. Bull Environ Contam Toxicol 14(1):105–111
Riisgard HG, Famme P (1986) Accumulation of inorganic and organic mercury in shrimp, *Crangon crangon*. Mar Pollut Bull 17:255–257
Rosales-Hoz L, Cundy AB, Bahena-Manjarrez JL (2003) Heavy metals in sediment cores from a tropical estuary affected by anthropogenic dicharges: Coatzacoalcos estuary, Mexico. Estuar Coast Shelf Sci 56:1–10
Ruelas-Inzunza J, Páez-Osuna F (2005) Mercury in fish and shark tissues from two coastal lagoons in the Gulf of California, Mexico. Bull Environ Contam Toxicol 74:294–300
Ruelas-Inzunza J, García-Rosales SB, Páez-Osuna F (2004) Distribution of mercury in adult penaeid shrimps from Altata-Ensenada del Pabellón lagoon (SE Gulf of California). Chemosphere 57:1657–1661
Ruelas-Inzunza J, Meza-López G, Páez-Osuna F (2008) Mercury in fish that are of dietary importance from the coasts of Sinaloa (SE Gulf of California). J Food Compos Anal 21:211–218
Ruelas-Inzunza J, Páez-Osuna F, Zamora-Arellano N et al (2009) Mercury in biota and surficial sediments from Coatzacoalcos estuary, Gulf of Mexico: distribution and seasonal variation. Water Air Soil Pollut 197:165–174
Ruelas-Inzunza J, Hernández-Osuna J, Páez-Osuna F (2011a) Total and organic mercury in ten fish species for human consumption from the Mexican Pacific. Bull Environ Contam Toxicol 86:679–683
Ruelas-Inzunza J, Green-Ruiz C, Zavala-Nevárez M et al (2011b) Biomonitoring of Cd, Cr, Hg and Pb in the Baluarte river basin associated to a mining area (NW Mexico). Sci Total Environ 409:3527–3536
Ruelas-Inzunza J, Frías-Espericueta MG, Páez-Osuna F (2013) Mercury in the atmospheric and coastal environments of Mexico. Rev Environ Contam Toxicol 226:65–99
Saiz-Salinas JI, González-Oreja JA (2000) Stress in estuarine communities: lessons from the highly-impacted Bilbao estuary (Spain). J Aquat Ecosyst Stress Recovery 7:43–55
Simpfendorfer CA, Freitas GG, Wiley RR et al (2005) Distribution and habitat partitioning of immature bull sharks (*Carcharhinus leucas*) in a Southwest Florida Estuary. Estuar 28:78–85
Storelli MM, Stuffler RG, Marcotrigiano GO (2002) Total and methylmercury residues in tuna-fish from the Mediterranean sea. Food Addit Contam 19:715–720
Tu NPC, Ha NN, Ikemoto T et al (2008) Regional variations in trace element concentrations in tissues of black tiger shrimp *Penaeus monodon* (Decapoda: Penaeidae) from South Vietnam. Mar Pollut Bull 57:858–866
Vaisman AG, Marins RV, Lacerda LD (2005) Characterization of the mangrove oyster, *Crassostrea rhizophorae*, as a biomonitor for mercury in tropical estuarine systems, Northeast Brazil. Bull Environ Contam Toxicol 74:582–588
Wang WX (2002) Interactions of trace metals and different marine food chains. Mar Ecol Prog Ser 243:295–309
WGMF (1980) Report on mercury in fish and fish products. Working Group Mercury in Fish, Department of Primary Industry. Australian Government Publishing Service, Canberra
Wheterbee B, Cortés E (2004) Food consumption and feeding habits. In: Carrier JF, Musik JA, Heithaus M (eds) Biology of sharks and their relatives. CRC Press, Boca Raton
Wiener JG, Bodaly RA, Brown SS et al (2007) Monitoring and evaluating trends in methylmercury accumulation in aquatic biota. In: Harris R, Krabbenhoft DP, Mason R et al (eds) Ecosystem responses to mercury contamination indicators of change. CRC Press, Pensacola, pp 87–122
Zhou Q, Zhang J, Fu J et al (2008) Biomonitoring: an appealing tool for assessment of metal pollution in the aquatic ecosystem. Anal Chim Acta 606:135–150

Establishing Management Strategies for Boca de Camichín Estuarine System Using Nutrient Carrying Capacity

4

Omar Calvario-Martínez, Carlos H. Lechuga-Devéze and Miguel A. Sánchez-Rodríguez

Abstract

Coastal lagoons are considered one of the most important ecosystems on the planet because they provide the necessary conditions for the biological cycle of a large number of species. However, in recent years, water quality in these systems has deteriorated because of population growth and human activities, which has led to increased inputs of organic matter and nutrients such as nitrogen and phosphorus. In order to determine the nutrient carrying capacity of coastal lagoons in tropical areas, a study was carried out at the estuarine system of Boca de Camichín, Nayarit using a model developed by the International Land Ocean Interactions in the Coastal Zone This work was conducted during three climatic seasons (hot-dry, rainy and cold-dry) between April 2008 and February 2009. Results obtained show that the system is clearly heterotrophic, which is caused by considerable production of organic matter. High productivity is maintained due to hydrodynamic movement, the penetration of sea water through a deep channel as well as the influence of a large river (San Pedro River). In all of the studied seasons, mass balances indicate that different concentrations of phosphate (19–90 tP/month) and nitrogen (90–162 tN/month) accumulate in the interior of the estuary. These results show that the Boca de Camichín system may be very close to its carrying capacity; therefore potential human activities that generate new organic matter that are discharged to Boca de Camichín lagoon should be controlled. Our results also show that it is essential to determine the nutrient carrying capacity of these types of systems in order to take proper management actions that ensure resource sustainability.

Keywords

Nutrient carrying capacity · Coastal lagoons · LOICZ · Nayarit · Marismas Nacionales

O. Calvario-Martínez (✉) · M. A. Sánchez-Rodríguez ·
C. H. Lechuga-Devéze
Av Sabalo-Cerritos s/n, Centro de Investigación
en Alimentación y Desarrollo, CP 82100 Mazatlán, Sinaloa, México
e-mail: ocalvario@ciad.mx

M. A. Sánchez-Rodríguez
e-mail: msanchez@ciad.mx

C. H. Lechuga-Devéze
e-mail: clechuga@cibnor.mx

4.1 Introduction

A coastal lagoon is defined as a depression in the coastal zone below the mean high tide elevation that has permanent or ephemeral connection with the sea and is protected by a terrestrial barrier (Lankford 1977). In general, coastal lagoons are shallow and lie on an axis parallel to the coast. These systems are considered amongst the most important ecosystems on Earth because they provide habitat for a large number of species. Coastal lagoons are home to a large diversity of vertebrates from different taxonomic levels: 50 Orders, 445 Families and approximately 22,000 species (Ray 1991; Toledo 2003).

F. Amezcua, B. Bellgraph (eds.), *Fisheries Management of Mexican and Central American Estuaries*, Estuaries of the World, DOI 10.1007/978-94-017-8917-2_4, © Springer Science+Business Media Dordrecht 2014

Areas of high productivity serve as reproduction and breeding grounds for different fishery resources, both coastal and marine (Alonso-Rodríguez and Páez-Osuna 2001). Coastal lagoons contain a large diversity of habitats that are created by the presence of: 1) sandy bottoms between coastal waters and the marine ecosystem, 2) muddy inland waters, 3) mangroves 4) aquatic sea grasses, and 5) haloclines (i.e. a drastic change in the vertical salinity distribution), among others.

The total surface area of coastal lagoons in Mexico is approximately 12,555 km^2 (Cárdenas 1969). Some lagoons are very productive and while others are potentially productive; many are in precarious states of exploitation whereas others are irrationally exploited. However, all coastal lagoons have the potential to be contaminated in the near future because of human impacts on the natural environment (Bechtel and Copeland 1971; Foyn 1969).

The quality of coastal waters in many regions of the world has deteriorated in recent years due to an increase in population and human activities in coastal regions (NRC 2000). The contributions of organic matter and nutrients such as nitrogen and phosphorus have increased due to industrial contributions, commercial or anthropogenic drains and shrimp-farm effluents (Páez-Osuna et al. 1997). All of these changes have an impact on the ecological carrying capacity of the system, which is defined as "the stocking or farm density above which unacceptable ecological impacts begin to manifest. From a practical standpoint, this process begins with the definition of components of interest (e.g., species, habitats) and acceptable levels of change for each of these" (Inglis et al. 2000; Mckindsey et al. 2006).

The use of biogeochemical models that include meteorological, dissolved nutrients and water flow data from coastal systems and adjacent oceanic areas can be used to understand the ability of these systems to transform, accumulate or export organic elements. Modeling studies contribute actively to the Land-Ocean Interactions in the Coastal Zone (LOICZ) international program; models are also used to investigate strategies for the sustainable use of coastal ecosystems.

Based on the utility of these models, the objective of this study was to calculate the nutrient carrying capacity of the estuarine system of Boca de Camichín, Nayarit and to assess its ability to transform, accumulate or export organic elements. Ultimately, these data will help to define the management strategies for sustainable use of the Boca de Camichín ecosystem.

4.2 Description of the Study Area

This study was undertaken in the Boca de Camichín estuarine system, which is located along the central coast of the State of Nayarit (Pacific coast of Mexico) between 21° 43' 26" N to 21° 45' 41" N and 105° 29' 2" W to 105° 30' 22" W (Fig. 4.1). Boca de Camichín has an approximate area of 2.07 km^2. This system is at the southern limit of the estuarine complex of Marismas Nacionales (i.e., National Flood plains), which is considered to be the most extensive mangrove region of the American Pacific. Boca de Camichín is located also at the mouth of the San Pedro River—together they form one of the largest flood plains and estuaries in the region, inhabited by large quantities of shrimp and other marine species such as oyster, sea bass, and red snapper. Several fishing grounds and fishing communities are located in the estuaries of Palmar de Cuautla, Puerta de Palapares and Boca de Camichín, with the latter involved in oyster farming as well (INFDMGEN 2005).

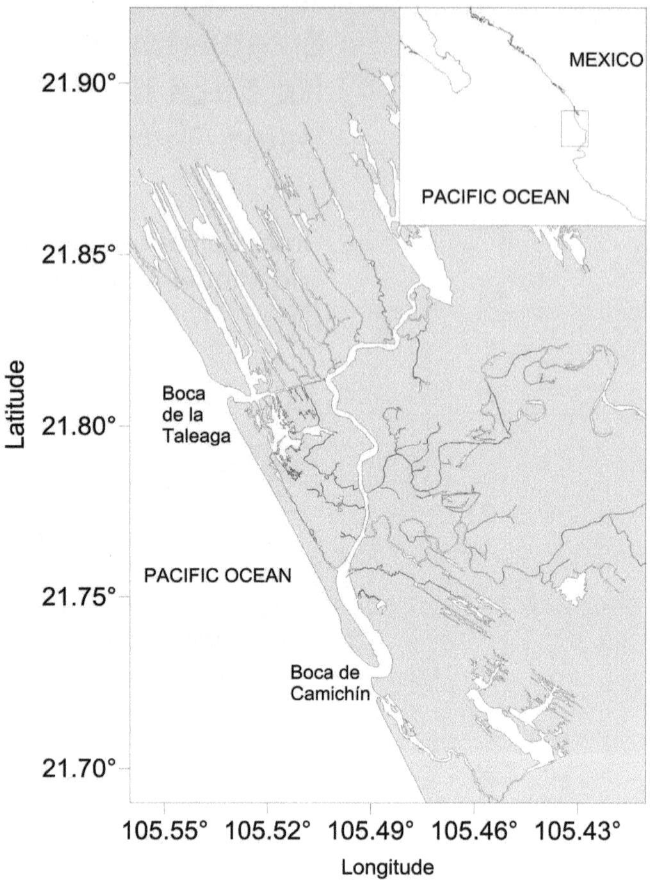

Fig. 4.1 The estuary system of Boca de Camichín, Nayarit, México

The San Pedro River is one of the most important rivers in Nayarit. In the State of Durango it is called La Sauceda River; however, in Nayarit it is more formally known as the compounded name San Pedro-Mezquital. The San Pedro River watershed drains an area that is 15.6% of the size of Nayarit and the main river flows through the center of the basin in a north-south direction. In the southern portion of the basin, the river changes its course and flows west until it discharges into the Pacific Ocean. Near its mouth, the San

Table 4.1 Size (ha) and percentage by municipality of the San Pedro River sub-basin. (Universidad Autónoma de Nayarit 2004)

Hydrographic region 11	Basin	Sub-basin	Municipality	Area (ha)	%
Presidio -San Pedro 946,805 ha	Río San Pedro 397,125 ha	Río San Pedro. 307,780.09 ha	Acaponeta	44,037.5	14.3
			Huajicori	26,127.6	8.5
			El Nayar	79,451.4	25.8
			Rosamorada	43,188.8	14.0
			Ruiz	43,466.3	14.1
			Santiago Ixcuintla	55,469.5	18.0
			Tuxpan	16,039.1	5.2
			Total	307,780.2	100

Table 4.2 Median precipitation and temperature in the state of Nayarit, Mexico (Comisión Nacional del Agua 2009)

Months	Median precipitation (mm)	Median temperature (°C)
	Period (1941–2005)	Period (1980–2004)
January	18.8	20.4
February	9.8	20.8
March	4.5	22.0
April	4.0	23.8
May	7.4	25.6
June	136.2	27.1
July	280.5	26.4
August	277.2	26.3
September	222.6	26.2
October	76.0	25.5
November	15.3	23.4
December	16.3	21.1
Total	1068.7	24.1

Pedro River does not have a well-defined channel and is part of the coastal lagoon and estuary (INEGI 1999).

Considering the size of the San Pedro River basin (~4000 km^2 in the State of Nayarit alone), it is classified as a large basin according to criteria by Singh (1995) (Table 4.1). The San Pedro River sub-basin has a size of about 3010 km^2 and is also considered to be a large basin according to Singh 1995 (i.e., because it is more than 1000 km^2), although it is officially classified by the Mexican National Institute of Statistics and Geography (INEGI) as a sub-basin.

The coastal plain of Mexcaltitán covers an area of 7,392 ha and includes the subsystems of Grande de Mexcaltitan, Los Patos, Las Gallinas, and the estuarine systems of Boca de Camichín, Toro Mocho and Las Conchas (Carta Nacional Pesquera 2004). According to the classification by Lankford (1977), this system is considered a Type II-C coastal plain depression.

The climate is considered hot and dry with rainfall occurring between June and October; the highest rainfall occurs in July, August and September. The hottest months are between May and October and June and July are the warmest as shown in Table 4.2.

4.3 Methods

In order to accomplish the objectives, laboratory and field activities were performed along with the development of a nutrient carrying capacity model for the Boca de Camichín system.

4.3.1 Laboratory and Field Activities

The National Water Commission (CONAGUA) identifies the months of June through October as "humid" months, while the "dry" months occur from November through May. During the five humid months, 93 % of the annual rainfall occurs, while only 7 % of precipitation occurs in the remaining 7 months (Comisión Nacional del Agua 2009). In order to describe the seasonal variability of precipitation and its influence on the nutrient carrying capacity of Boca de Camichín, water samples were collected at three times under different atmospheric conditions during the year. According to historical data presented by CONAGUA, the month of October was selected for sampling during the rainy season, February for sampling in the cold-dry season and April for sampling during the hot-dry season (Fig. 4.2).

During each of the three sampling periods, water samples were collected and analyzed to determine a series of quantitative physical-chemical parameters. Sampling was conducted in the estuarine system of Boca de Camichín at a network of seven stations within the estuary proper (stations 3–9), two stations in the open sea near the entrance to the estuary (stations 1 and 2) and four locations upstream of the estuary (stations 10–13). The station at the mouth of the Ensenada River (station 14) is also called Boca de la Talega. All sampling stations are shown in Fig. 4.3. Sampling was carried out on the 19th of April, 2008 for the hot-dry period, the 5th of October, 2008 for the rainy season and the 2nd February, 2009 for the cold-dry period.

At each of the sampling stations, the in situ temperature was determined with a mercury thermometer by partial immersion following the established recommendations of the Mexican norm, NMX-AA-007-SCFI-2000. Salinity was measured with a refractometer and pH was quantified with a Corning potentiometer (model 313) according to the method

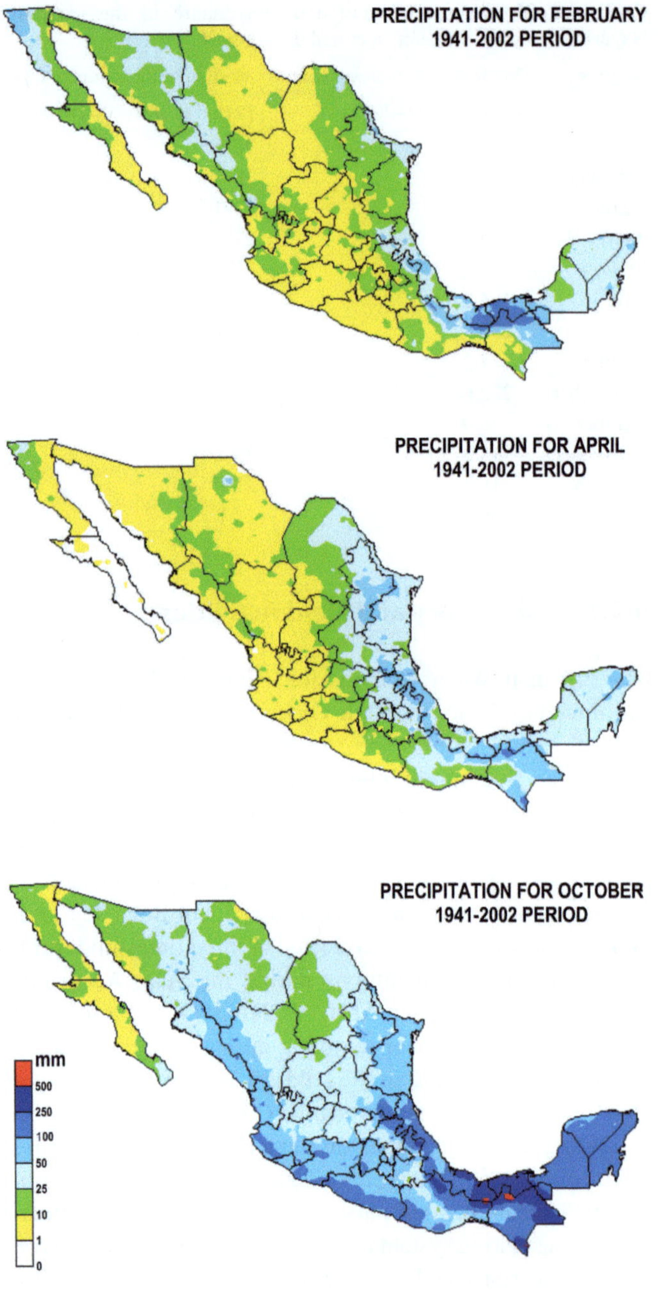

Fig. 4.2 Historical precipitation throughout México during the months of February, April and October. (Reproduced from Comisión Nacional del Agua, 2009)

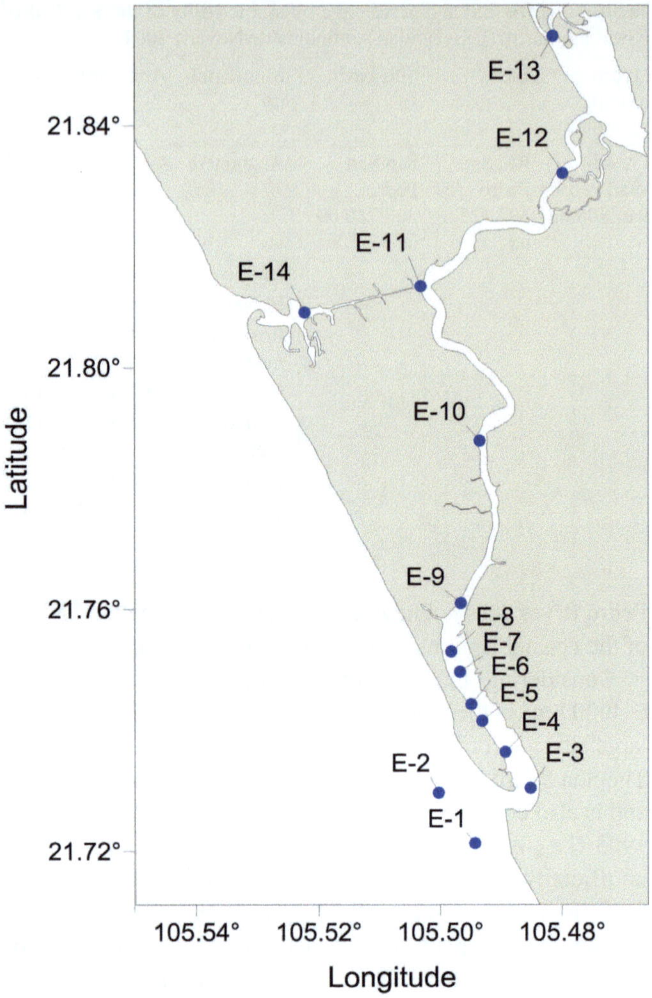

Fig. 4.3 Sampled stations in the Boca de Camichín estuary system

described in the Mexican norm NMX-AA-008-SCFI-2000. Turbidity was determined with a Secchi disk and dissolved oxygen concentrations were measured with a YSI oximeter (model 59). The concentration of nitrates, nitrites, ammonium, orthophosphates and chlorophyll a were evaluated from surface water collected with a polyethylene bottle according to the methods described by Parsons et al. (1984). Total soluble phosphorus and total particulate phosphorus were evaluated according to Valderrama (1981) and organic matter was determined by loss on ignition as described in the Mexican norm NMX-AA-34-SCFI-2001.

The bathymetry of Boca de Camichín system was obtained by measuring the bottom depth with a Garmin GPS-map 5215 ecosonde. Several depth measurements were made by following a zig-zag track that started at station 3 and finished at station 9. The average bottom depth, together with the surface area enclosed between station 3 and 9 was used to calculate the volume of the system.

4.3.2 Carrying Capacity of the Estuarine System

In order to determine the carrying capacity of the estuarine system, a model developed by the LOICZ group was used (Gordon et al. 1996). The law of the conservation of mass is used as the fundamental principle of the model. Therefore, it was necessary to obtain or generate data on all inputs and outputs of the estuarine system including both its volume and chemical composition. The result was a model that stoichiometrically bound the water-salt-nutrients balance.

The nutrients of particular interest in the model are nitrogen and phosphorus. The model provides information about the hydraulic balance of the system, saline balance, nutrient flows and stoichiometry. The model incorporates the volume and flow of nitrogen and phosphorus obtained empirically.

The hydraulic balance includes all inputs and outputs of water to the Boca de Camichín system. There must be flow inputs or outputs to compensate for water losses or gains, respectively, in the system. Assuming that the change in water volume over time in the Boca de Camichín system is constant, the net water outflow from the system was estimated by the difference, or residual flow (V_r), using the equation:

$$V_r = -V_q - V_p - V_g - V_o + V_e \tag{4.1}$$

Where V_q is the volume of river inputs, V_p is precipitation, V_g is groundwater, V_o is urban effluents and V_e is evaporation.

Different water inputs to the estuary have different levels of salinity (including groundwater) and mix with the estuarine waters. The average salinity of the inputs and the estuary itself is defined as residual salinity (S_r). When multiplied by the residual flow (S_r * V_r), the volume of water required to reach equilibrium with the ocean is the product. Since the true equilibrium cannot be reached, the Volume of Actual Mix (V_x) is calculated by:

$$V_x = \frac{-V_r \times S_r}{Sal._{ocean} - Sal._{lagoon}} \tag{4.2}$$

From this mixture volume (V_x), combined with the volume of the estuary, the time for total water renewal (i.e., turnover) can be calculated with the equation:

$$T = \frac{V_{lagoon}}{V_x + V_r} \tag{4.3}$$

The balance of non-conservative material (e.g., N and P) is based on the exchange flow established for saline balance and the concentrations of N and P that exchange by different contributors. The residual flow of element Y is explained by:

$$\frac{V_r(Y_{lagoon} - Y_{ocean})}{2} \tag{4.4}$$

While the flow-mix is:

$$V_x(Y_{lagoon} - Y_{ocean}) \tag{4.5}$$

The results of these equations are the amounts of element Y, per surface unit and unit of time, in excess or lacking in the system. For example, if phosphorus flows are positive it is presumed that there is excess production of dissolved inorganic phosphorus (DIP), probably derived from effective oxidation of organic matter in the system. The stoichiometric link between the balance of non-conservative elements is based on the conclusions for N:P by Redfield (16:1) (Redfield 1934). If the values for DIP are positive and indicate a net oxidation of organic matter in the system, it would be expected that the flow of Dissolved Inorganic Nitrogen (DIN) is equal to 16 * DIP. Most of the reactions that involve the transfer of non-conservative flows of nitrogen are dominated by the gaseous phase of nitrogen. Therefore, if the result of the previous operation is positive, the system behaves as a denitrificant (i.e., it has an excess of nitrogen that will be released to the atmosphere). If the result is negative, the system consumes all produced nitrogen.

The difference between the production of organic carbon (*p*) and respiration (*r*) provides information on the net metabolism of the ecosystem based on the conclusions for C: P by Redfield (106:1) (Redfield 1934):

$$(p - r) = DIP \times (C:P)part \tag{4.6}$$

Where (C:P)part is the carbon to phosphorus ratio in the particulate organic matter that is being recycled in the system. If $(p-r)$ is positive, the system clearly indicates autotrophic behavior. The magnitudes of the volume and flow of each component and element of this model allow us to define the processes in the estuarine system in order to evaluate their contributions, whether natural or anthropogenic, and their trophic tendencies.

Three main considerations were taken into account when using this model. Firstly, the model needs to define the boundaries (see previous section for surface area definition) and they must represent the system as closely as possible. If larger surface values are used, the bottom depth of the system will introduce greater errors in the results generated by the model. Secondly, the average total water volume for the San Pedro River (220,611 m^3/day) was calculated using information available between the years 1988–1999 as reported by IMTA (2010). Thirdly, the Boca de Camichín estuarine system represents the only main water discharge of the San Pedro River towards the Pacific Ocean as shown in the water simulations discharge program developed by INEGI (2012).

4.4 Results

4.4.1 Water Quality Parameters

4.4.1.1 Hot-dry season

During the sampling period of April 2008 the values of sea surface temperature (SST) showed less homogeneity than

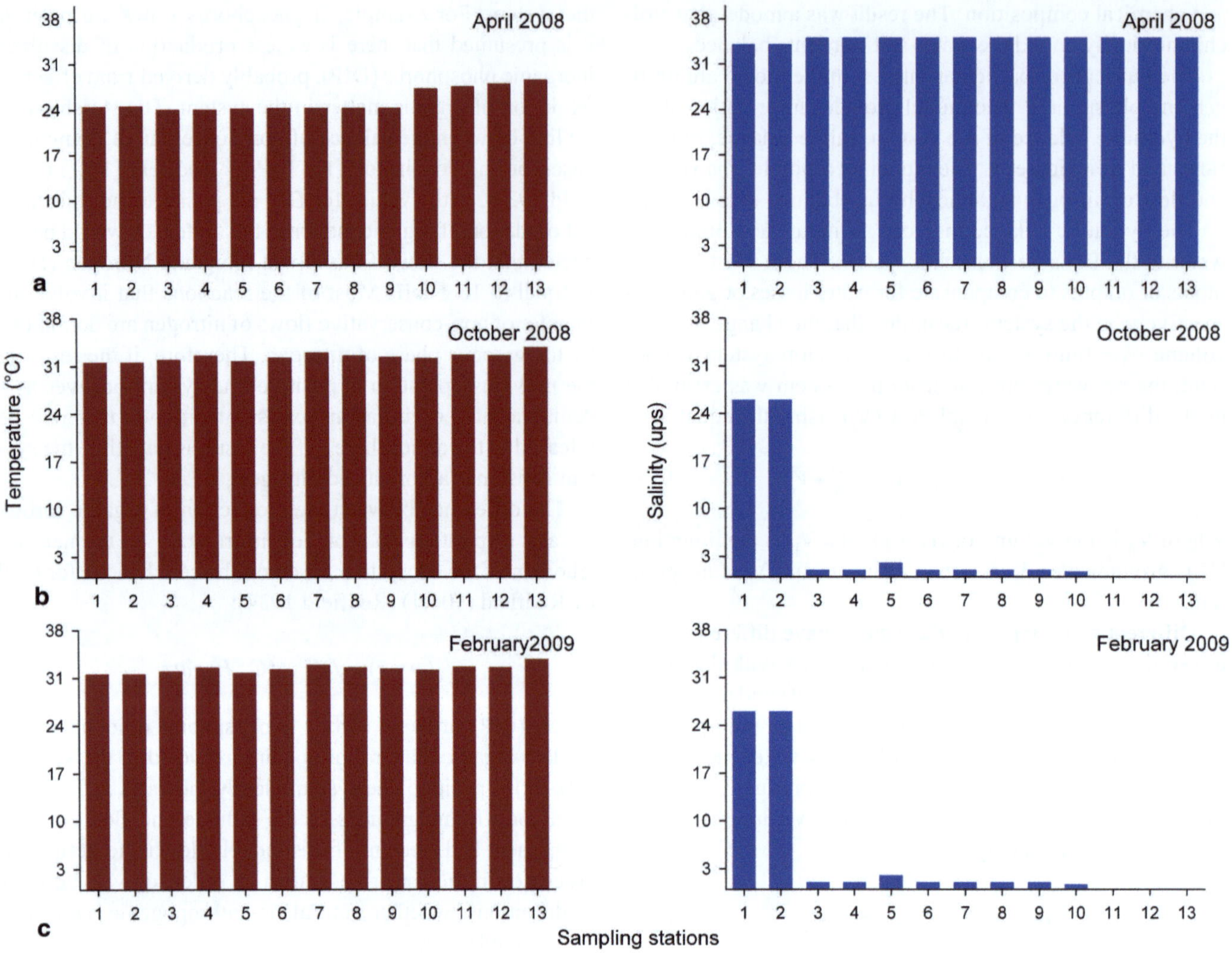

Fig. 4.4 Distribution of temperature and salinity at 13 sampling stations in the Boca de Camichín estuary system during the hot-dry season (April 2008), rainy season (October 2008), and cold-dry season (February 2009)

during October 2008 (rainy season) and February 2009 (cold-dry season). Maximum temperature values in the upstream area were above 28.5 °C. The minimum temperatures occurred in the estuarine system (24.2 °C) and near the entrance of the study area adjacent to the ocean (24.4–24.8 °C). The spatial distribution of salinity did not show a marked change; however, salinity values that clearly indicated marine conditions (range from 34–35 ups) (Fig. 4.4).

The values for pH during the hot-dry season varied between 7.46 and 7.88. Spatially, the maximum pH value was obtained farthest downstream in the estuary system, adjacent to the ocean, and the minimum was measured at the station farthest upstream (Fig. 4.5). The maximum value of pH coincided with the maximum dissolved oxygen (8.15 mg/L) measured at station 3 (Fig. 4.5). These conditions, corroborated by the maximum values of chlorophyll a (13.8 ± 0.7 mg/m^3), suggest the presence of a phytoplankton patch in the area.

The distribution of total phosphorous was relatively homogeneous at all of the sampling stations in April 2008 and varied between 0.070 ± 0.003 and 0.097 ± 0.027 mg/L (Fig. 4.6). Orthophosphate concentrations were higher near the open sea (near the entrance of the estuary); at the upstream stations, orthophosphates were between 0.095 ± 0.000 and 0.115 ± 0.005 mg/L (Fig. 4.7).

Concentrations of nitrogen nutrients, especially nitrates and ammonia, were highest at the upstream stations (0.017 and 0.023 mg/L for nitrates and 0.034 ± 0.005–0.046 ± 0.004 mg/L for ammonia). Nitrites did not vary significantly among the stations with values ranging between undetected and 0.005 mg/L (Fig. 4.7).

4.4.1.2 Rainy season

The spatial distribution of SST during October 2008 was homogeneous and ranged between 31.5 and 33.7 °C (Fig. 4.4). A similar distribution was observed during the hot-dry season.

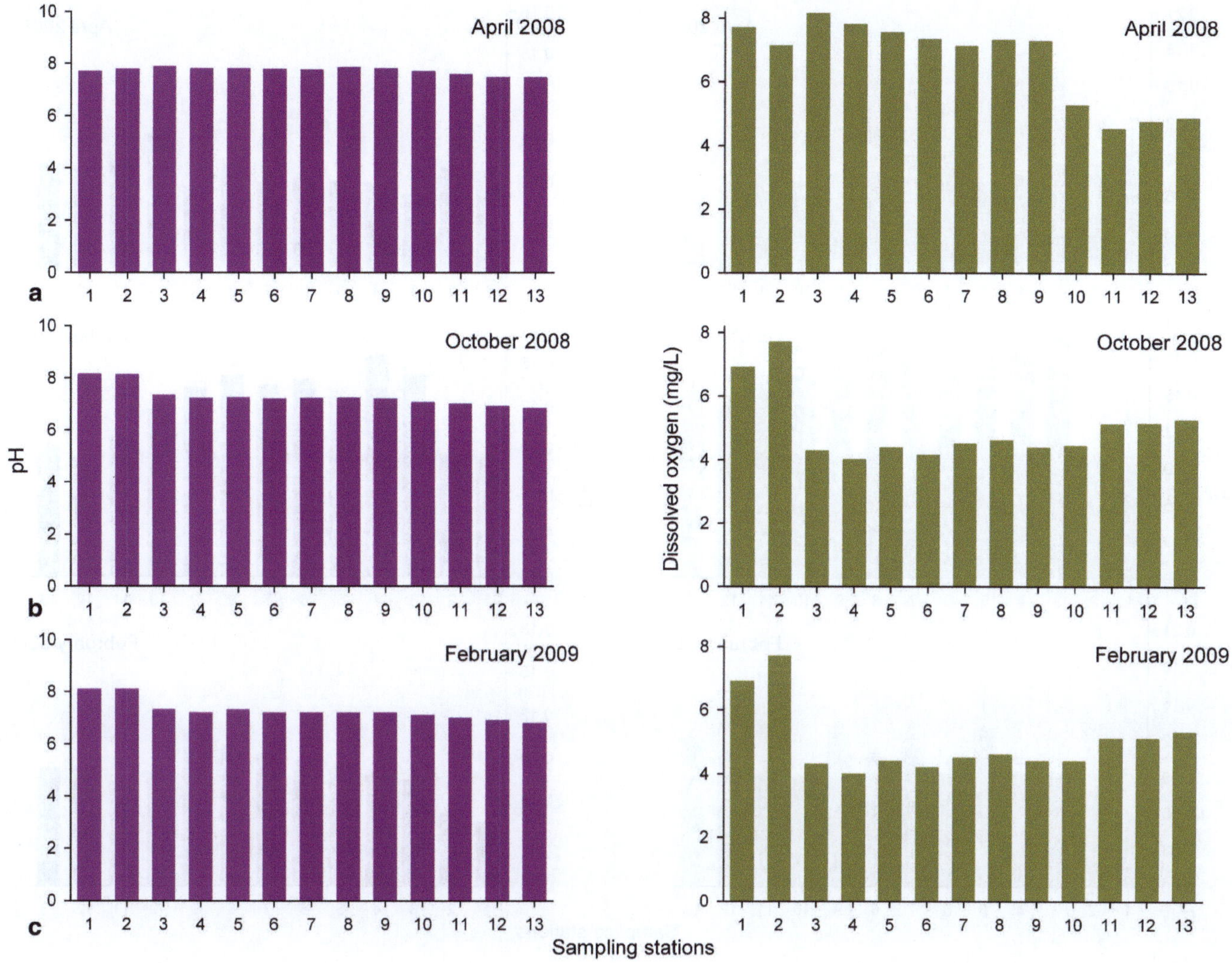

Fig. 4.5 Distribution of pH and dissolved oxygen at 13 sampling stations in the Boca de Camichín estuary system during the hot-dry season (April 2008), rainy season (October 2008) and cold-dry season (February 2009)

In the case of salinity, maximum values were found near the open sea, close to the entrance of the estuary (26 ups) and at the mouth of the Ensenada River (station 14, 11 ups). Minimum salinity was observed in the estuary and farther upstream (values ranging from 0–1 ups), clearly showing the influence of freshwater from the San Pedro River (Fig. 4.4).

The minimum values of pH were observed in the widest part of the river (stations 12 and 13), with values ranging from 6.85–6.92. The maximum pH was observed near the open sea (stations 1 and 2), with values of 8.13 and 8.14, respectively (Fig. 4.5). Dissolved oxygen varied from 4.01–7.70 mg/L, with maximum values in the open sea area and minimum values inside the estuary (Fig. 4.5).

The maximum concentrations of total phosphorus were observed inside the estuary, with values ranging from 0.184–0.201 mg/L. In the open sea, values ranged from 0.095±0.005–0.098±0.005 mg/L (Fig. 4.6). The values for orthophosphates showed a similar pattern to those observed for total phosphorus, with maximum values in the estuary ranging from 0.124±0.006–0.139±0.005 mg/L and lower values in the open sea and upstream areas with values of 0.056±0.005–0.119±0.002 mg/L, respectively (Fig. 4.7).

Nitrogen nutrients were at the highest concentrations inside the estuary, with values of 0.008±0.005–0.062±0.007 mg/L for ammonia and 0.013±0.001–0.016±0.000 mg/L for nitrites. Maximum nitrate concentrations were registered near the interface of the open sea and estuary, with values ranging from 0.057 and 0.076 mg/L (Fig. 4.7).

4.4.1.3 Cold-dry Season

During February 2009, SST was homogeneous and slightly higher temperatures were measured at the upstream stations (Fig. 4.4). Minimum temperatures were located within the estuary near the ocean interface (Fig. 4.4). Salinity showed considerable spatial variation with the highest salinities occurring at the ocean-estuary interface (36 ups) and decreas-

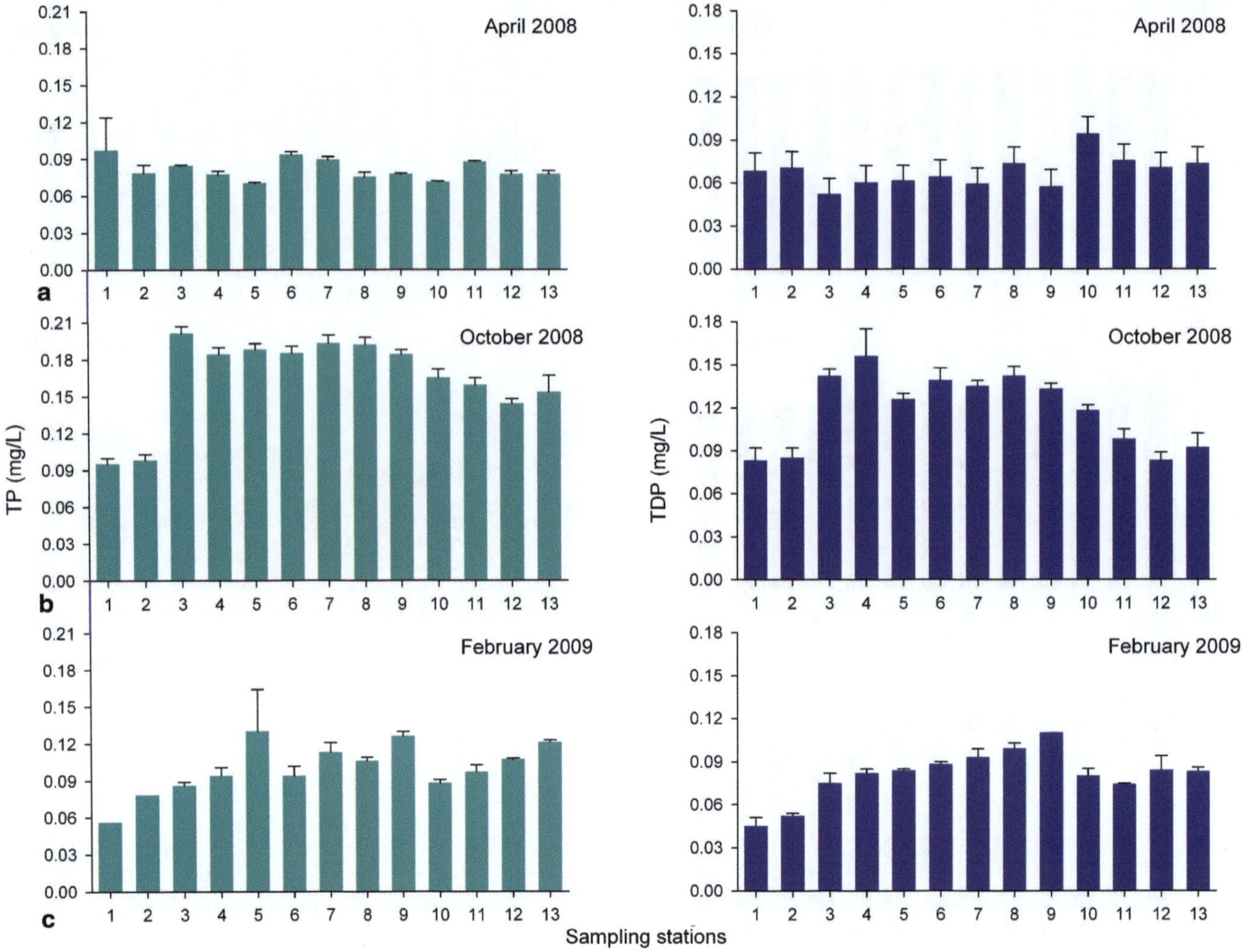

Fig. 4.6 Distribution of total phosphorous (TP) and total dissolved phosphorous (TDP) at 13 sampling stations in the Boca de Camichín estuary system during the hot-dry season (April 2008), rainy season (October 2008) and cold-dry season (February 2009)

ing values of salinity occurring farther upstream (8 ups; Fig. 4.4).

The values of pH during the cold-dry season were between 7.60 and 7.90 (Fig. 4.5). Spatially, maximum pH was recorded adjacent to the entrance of the estuary nearest the ocean and pH decreased in an upstream direction (Fig. 4.5). Values for dissolved oxygen ranged from 7.6 and 13.2 mg/L, with the highest values at the stations farthest downstream in the estuary (Fig. 4.5). Oxygen distribution was not related to the concentration of chlorophyll a which varied between 1.0±0.2 and 4.8±2.0 mg/m^3 (Fig. 4.5). It is possible that the variation in oxygen concentration is related more strongly to physical factors than biological factors.

Maximum concentrations of total phosphorus were observed in the estuary with values up to 0.126±0.004 mg/L; the lowest values were observed in the open sea area and varied from 0.056±0.000–0.078±0.000 mg/L (Fig. 4.6). Values for orthophosphates showed a similar pattern to that observed for total phosphorous with maximum values in the estuary of 0.097±0.002 mg/L and lower values in the open sea varying from 0.024±0.001–0.039±0.003 mg/L (Fig. 4.7).

Nitrogen concentrations were highest in the estuary with values of 0.075±0.005–0.131±0.005 mg/L for ammonia and 0.061–0.114 mg/L for nitrates. Nitrite concentrations were relatively low at all sampling stations with values ranging from undetected to 0.007±0.001 mg/L (Fig. 4.7).

4.4.2 Carrying Capacity of the Boca De Camichín System

The San Pedro River is the main tributary in the estuarine system with a basin size of 26,480 km^2; the Santiago River is the most influential tributary of the southern part of the coastal plain (Carta Nacional Pesquera 2004). The calculated estuarine system of Boca de Camichín covers an area of

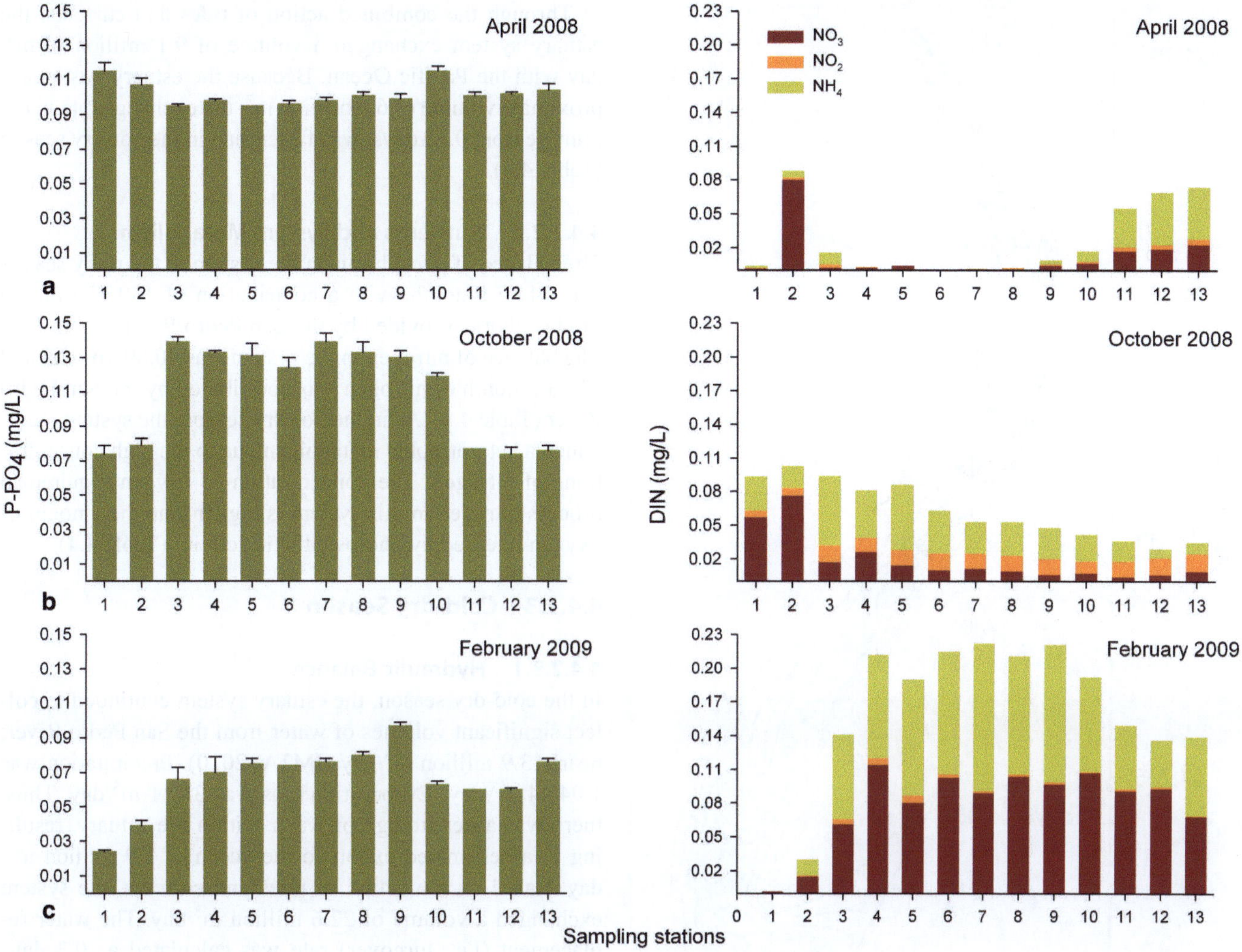

Fig. 4.7 Distribution of orthophosphates (P-PO4) and dissolved inorganic nitrogen (DIN) at 13 sampling stations in the Boca de Camichín estuary system during the hot-dry season (April 2008), rainy season (October 2008) and cold-dry season (February 2009)

2.11 km² and has an average depth of 2.9 m (Fig. 4.8). The total volume of the system has an estimated area of 5.9 million m³.

Precipitation was 5.31 mm/month during the hot-dry season with an average salinity of 35.00 ups. For the rainy season, precipitation was 198.66 mm/month, which decreased salinity to 1.14 ups. In the cold-dry season, precipitation was 15.05 mm/month and salinity was 27.54 ups.

4.4.2.1 Hot-dry Season

4.4.2.1.1 Hydraulic balance

A total of 366.4 m³/day of precipitation was estimated for the Boca de Camichín estuary during the hot-dry season based on measured precipitation of 5.31 mm/month. Furthermore, the San Pedro River supplied 220,611 m³/day to the estuary (IMTA, 2010). The Boca de Camichín estuary lost a total of 387.1 m³/day through evaporation. Total inputs to the system exceeded losses so the calculated discharge to the ocean was 220,590 m³/day. Through the action of tides and currents, the system exchanged a volume of 15.3 million m³/day with the Pacific Ocean. From these input and output volumes, a total water replacement of 0.4 days was calculated for the whole system (Table 4.3).

4.4.2.1.2 Nutrients and System Metabolism

The balance of phosphorus in the system in the hot-dry season was 19.11 t/month, with 2.14 t/month of phosphorus coming from the San Pedro River (Table 4.3). The balance of nitrogen in the system was 90.86 t/month, with a contribution of 1.57 t/month of nitrogen coming from the San Pedro River (Table 4.3). Due to the requirement of high concentrations of oxygen to process the amounts of nitrogen found in the system, the oxygen released from photosynthetic activity was insufficient. Thus, the Boca de Camichín estuary is

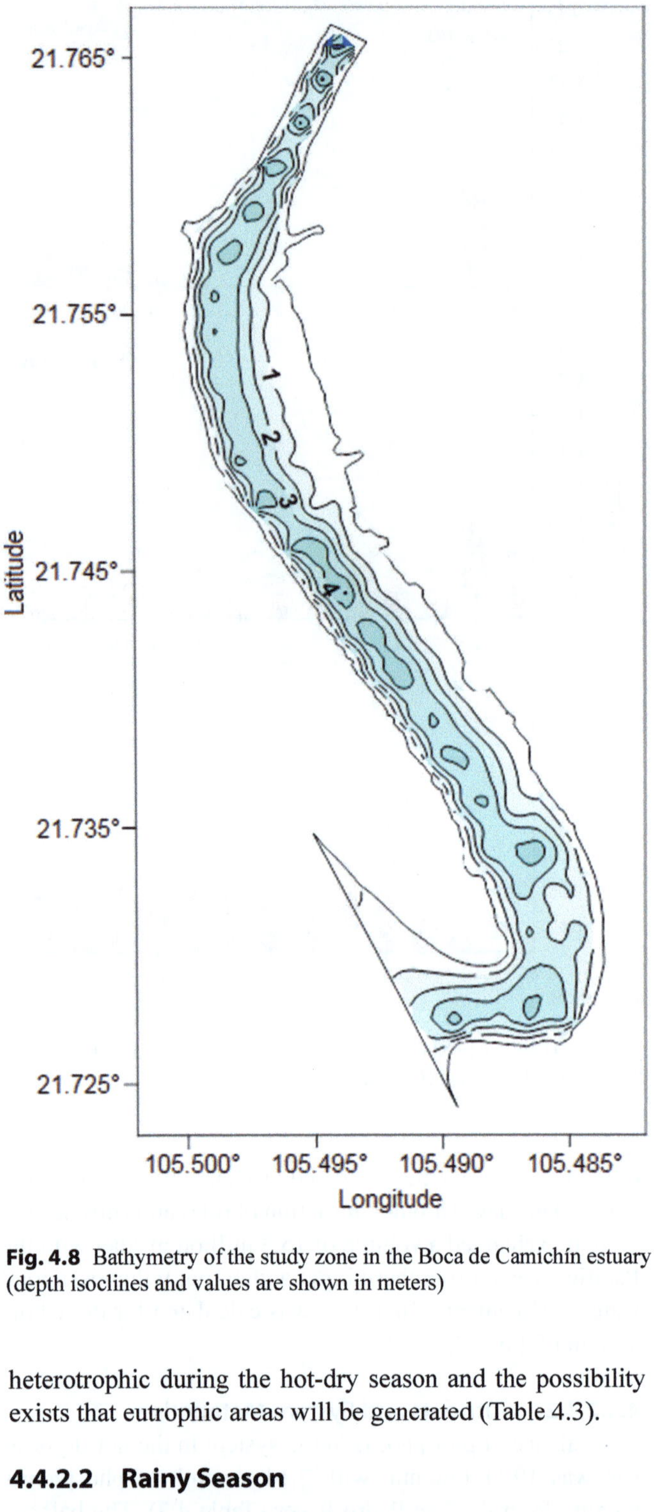

Fig. 4.8 Bathymetry of the study zone in the Boca de Camichín estuary (depth isoclines and values are shown in meters)

heterotrophic during the hot-dry season and the possibility exists that eutrophic areas will be generated (Table 4.3).

4.4.2.2 Rainy Season

4.4.2.2.1 Hydraulic Balance

The estuarine system collected 16.3 million m^3/day of water from the San Pedro River in the rainy season (IMTA 2010) and 13,733.9 m^3/day of precipitation. Evaporation losses were 387.1 m^3/day. As in the hot-dry season, contributions to the Boca de Camichín estuary exceeded losses in the rainy season. Consequently, there was a net export of water of about 16.3 million m^3/day (Table 4.4).

Through the combined action of tides and currents, the estuary system exchanged a volume of 9.1 million of m^3/day with the Pacific Ocean. Because the estuary has an approximate volume of 6.0 million m^3, the exchange rate in the rainy season, 0.82 days, was faster than in the hot-dry season (Table 4.4).

4.4.2.2.2 Nutrients and System Metabolism

The balance of phosphorus in the system in the rainy season was 74.35 t/month, with a contribution of 130.45 t/month of phosphorus provided by the San Pedro River (Table 4.4). The balance of nitrogen in the system was 20.69 t/month and 23.53 t/month of nitrogen was contributed by the San Pedro River (Table 4.4). As in the hot-dry season, the system maintained a heterotrophic metabolism due to the high concentrations of nitrogen. The concentration of oxygen required to process nitrogen in this system is higher than the amount of oxygen released by photosynthetic activity (Table 4.4).

4.4.2.3 Cold-dry Season

4.4.2.3.1 Hydraulic Balance

In the cold-dry season, the estuary system continued to collect significant volumes of water from the San Pedro River; nearly 3.9 million m^3/day (IMTA, 2010). Precipitation was 1,040.4 m^3/day. Evaporation loss was 387.1 m^3/day. Thus, there was a net storage of water within the estuary, resulting in an estimated export to the ocean of 3.9 million m^3/day. Based on the action of tides and currents, the system exchanged a volume of 22.6 million m^3/day. The water replacement (i.e., turnover) rate was calculated as 0.3 days (Table 4.5).

4.4.2.3.2 Nutrients and System Metabolism

The balance of phosphorus in the system was 90.62 t/month, with a contribution of 23.16 t/month of phosphorus contributed by the San Pedro River (Table 4.5). The balance of nitrogen in the system was 162.78 t/month, while the contribution of nitrogen coming from the San Pedro River was 24.21 t/month (Table 4.5). As in the hot-dry and rainy seasons, the system again maintained a heterotrophic metabolism during the cold-dry season due to the high concentrations of nitrogen. In order to process the nitrogen, higher concentrations of oxygen are required than are produced during photosynthesis (Table 4.5).

4.5 Discussion

The Boca de Camichín estuary appears to be dominated by heterotrophic processes caused by a large production of organic matter. These heterotrophic processes imply that a higher amount of dissolved oxygen is consumed than that produced by photosynthetic activity. However, clear levels

Table 4.3 Carrying capacity in the Boca de Camichin estuary system, hot-dry season

General	m	m^2	m^3	km^2
Surface		2,073,980		2.11
Average depth	2.9			
Volume			5,993,802	
Water fluxes		April 2008		Units
Precipitation (V_p)		0.0002		m/day
Evaporation (V_e)		0.0002		m/day
Aquaculture effluents (V_q)		0		m^3/day
Agriculture effluents (V_o)		0		m^3/day
Rivers (V_c)		220,611		m^3/day
Water quality		April 2008		Units
System				
Salinity		35.00		ups
Nitrates (+nitrites+ammonium)		0.357		mmol/m^3
Orthophosphates		3.146		mmol/m^3
Ocean				
Salinity		34.00		ups
Nitrates (+nitrites+ammonium)		3.570		mmol/m^3
Orthophosphates		3.584		mmol/m^3
Aquaculture effluents				
Salinity		0.00		ups
Nitrates (+nitrites+ammonium)		0.000		mmol/m^3
Orthophosphates		0.000		mmol/m^3
Agriculture effluents				
Salinity		0.00		ups
Nitrates (+nitrites+ammonium)		0.000		mmol/m^3
Orthophosphates		0.000		mmol/m^3
Rivers				
Salinity		35.00		ups
Nitrates (+nitrites+ammonium)		3.840		mmol/m^3
Orthophosphates		3.410		mmol/m^3
Other effluents				
Salinity		0.0		ups
Nitrates (+nitrites+ammonium)		0.000		mmol/m^3
Orthophosphates		0.00		mmol/m^3
Meteorological data		April 2008		Units
Precipitation		0.177		mm/day
Evaporation		0.233		mm/day
Hydraulical balance		April 2008		
V_p=pp (m/month) * Sup (m^2)		366.4		m^3/day
V_e=ev (m/month) * Sup (m^2)		387.1		m^3/day
V_q=aquaculture effluents		0.0		m^3/day
V_o=agriculture effluents		0.0		m^3/day
V_c=river		220,611.00		m^3/day
$V_r=-(V_p)-(-V_e)-(V_q)-(V_o)-(V_c)$		−220,590.3		m^3/day
Salinity balance				
Sr=(Sext+Sist)/2		34.50		ups
V_r*S_r		−7,610,363.98		ups/m^3/day
$V_x*(S_{ext}\text{-}S_{int})$		15,331,748.98		ups/m^3/day
$V_x=(V_q*S_q+V_o*S_o+V_c*S_c-V_rS_r)/(S_{ext}-S_{syst})$		−15,331,748.98		m^3/day
Exchange rate				
$T=V_{syst}/V_x+V_r$		0.4		day
Nutrient fluxes				
$V_r*DIP_r=V_r*(DIP_{ext}+DIP_{syst})/2$		−742,286		mmol/day
$V_x*(DIP_{ext}\text{-}DIP_{syst})$		−6,715,306		mmol/day

Table 4.3 (continued)

General	m	m²	m³	km²
V_oP_o (aquaculture effluent)		0		mmol/day
V_qP_q (aquaculture)		0		mmol/day
V_cP_c		751,401		
$\Delta DIP_t = -V_o{*}DIP_o - V_q{*}DIP_q - V_r{*}DIP_r - V_x{*}(DIP_{ext} - DIP_{syst})$		6,706,191		mmol/day
$\Delta DIP = (-DIP_r - DIP_x)/Sup_{syst}$		3.233		mmol/m²/day
$V_r{*}DIN_r = V_r{*}(DIN_{ext} + DIN_{syst})/2$		−433,129		mmol/day
$V_x{*}(DIN_{ext}\text{-}DIN_{syst})$		−49,260,909		mmol/day
V_oN_o (aquaculture effluent)		0		mmol/day
V_qN_q (aquaculture)		0		mmol/day
V_cN_c		846,484		
$\Delta DIN_t = -V_o{*}DIN_o - V_q{*}DIN_q - V_r{*}DIN_r - V_x{*}DIN_{ext} - DIN_{syst})$		48,847,554		mmol/day
$\Delta DIN = (-DIN_r - DIN_x)/Sup_{syst}$		23.553		mmol/m²/day
Stoichiometric relationships				
$\Delta DIN_{exp} = (N:P){*}\ \Delta DIP_t$		107,299,059		mmol/day
$\Delta DIN_{obs} = \Delta DIN_t$		48,847,554		mmol/day
$(nfix\text{-}denit)t = \Delta DIN_{obs} - \Delta DIN_{exp}$		−58,451,505		mmol/day
(nfix-denit)/sup $_{syst}$		−28.183		mmol/m²/day
Net metabolism				
$(p\text{-}r) = -\Delta DIP_t{*}(C:P)$		−710,856,268		mmolC/day
$(p\text{-}r) = -\Delta DIP_t{*}(C:P)/sup_{syst}$		−342.7		mmolC/m²/day
PP_{syst}		No data available		mmolC/m²/day

Table 4.4 Carrying capacity in the Boca de Camichin estuary system, rainy season

General	m	m²	m³	km²
Surface		2,073,980		2.11
Average depth	2.9			
Volume			5,993,802	
Water fluxes		October 2008		Units
Precipitation (V_p)		0.0066		m/day
Evaporation (V_e)		0.0002		m/day
Aquaculture effluents (V_q)		0		m³/day
Agriculture effluents (V_o)		0		m³/day
Rivers (V_c)		16,387,802		m³/day
Others (V_v)		0		m³/day
Water quality		October 2008		Units
System				
Salinity		1.14		ups
Nitrates (+nitrites+ammonium)		1.484		mmol/m³
Orthophosphates		4.285		mmol/m³
Ocean				
Salinity		26.00		ups
Nitrates (+nitrites+ammonium)		2.130		mmol/m³
Orthophosphates		2.47		mmol/m³
Aquaculture effluents				
Salinity		0.00		ups
Nitrates (+nitrites+ammonium)		0.000		mmol/m³
Orthophosphates		0.000		mmol/m³
Agriculture effluents				
Salinity		0.00		ups
Nitrates (+nitrites+ammonium)		0.000		mmol/m³
Orthophosphates		0.000		mmol/m³
Rivers				
Salinity		0.18		ups
Nitrates (+nitrites+ammonium)		0.77		mmol/m³

Table 4.4 (continued)

General	m	m^2	m^3	km^2
Orthophosphates		2.79		$mmol/m^3$
Meteorological data		October 2008		Units
Precipitation		6.622		mm/day
Evaporation		0.233		mm/day
Hydraulical balance		October 2008		
V_p=pp (m/month) * Sup (m^2)		13,733.9		m^3/day
V_e=ev (m/month) * Sup (m^2)		387.1		m^3/day
V_q=aquaculture effluents		0.00		m^3/day
V_o=agriculture effluents		0.00		m^3/day
V_c=river		16,387,802.00		m^3/day
$V_r=-(V_p)-(-V_e)-(V_q)-(V_o)-(V_c)$		16,401,148.8		m^3/day
Salinity balance				
$S_r=(S_{ext}+S_{int})/2$		13.57		ups
V_r*S_r		−222,563,588.57		ups/m^3/day
$V_x*(S_{ext}\text{-}S_{int})$		225,513,392,93		ups/m^3/day
$V_x=(V_q*S_q+V_o*S_o+V_c*S_c-V_rS_r)/(S_{ext}-S_{syst})$		9,071,335.19		m^3/day
Exchange rate				
$T=V_{syst}/V_x+V_r$		0.82		day
Nutrient fluxes				
$V_r*DIP_r=V_r*(DIP_{ext}+DIP_{syst})/2$		−55,394,880		mmol/day
$V_x*(DIP_{ext}\text{-}DIP_{syst})$		−16,464,473		mmol/day
V_oP_o (agriculture effluents)		0		mmol/day
V_qP_q (aquaculture)		0		mmol/day
V_cP_c		45,771,131		
$\Delta DIP_t=-V_o*DIP_o-V_q*DIP_q-V_r*DIP_r-V_x*(DIP_{ext}-DIP_{syst})$		26,088,222		mmol/day
$\Delta DIP=(-DIP_r-DIP_x)/Sup_{syst}$		12.579		mmol/m^2/day
$V_r*DIN_r=V_r*(DIN_{ext}+DIN_{syst})/2$		−29,636,876		mmol/day
$V_x*(DIN_{ext}\text{-}DIN_{syst})$		5,860,083		mmol/day
V_oN_o (agriculture effluents)		0		mmol/day
V_qN_q (aquaculture)		0		mmol/day
V_cN_c		12,651,383		
$\Delta DIN_t=-V_o*DIN_o-V_q*DIN_q-V_r*DIN_r-V_x*DIN_{ext}-DIN_{syst})$		11,125,410		mmol/day
$\Delta DIN=(-DIN_r-DIN_x)/Sup_{syst}$		5.364		mmol/m^2/day
Stoichiometric relationships				
$\Delta DIN_{exp}=(N:P)*\Delta DIP_t$		417,411,557		mmol/day
$\Delta DIN_{obs}=\Delta DIN_t$		11,125,410		mmol/day
(nfix-denit)t=$\Delta DIN_{obs}\text{-}\Delta DIN_{exp}$		−406,286,147		mmol/day
(nfix-denit)/sup $_{syst}$		−195,897		mmol/m^2/day
Net metabolism				
$(p-r)=-\Delta DIPt*(C:P)$		−2,765,351,564		mmolC/day
$(p-r)=-\Delta DIPt*(C:P)/sup_{syst}$		−1333.4		mmolC/m^2/day
PP_{syst}		No data available		mmolC/m^2/day

of eutrophication have not occurred, mainly because a large dynamic exists that allows water replacement in a very short time (about half a day). The maintenance of this dynamic is essential to maintain productivity levels in the system. Any infrastructure created on the banks of the estuary could create areas of low dynamism with the potential of generating eutrophic sites.

In all the seasons studied, the mass-balance indicates differences in the concentrations of phosphates (19–90 tP/month) and nitrogen (90–162 t N/month). In order to process these amounts of N and P, it is essential to maintain the hydrodynamics of water exchange, so that the supply of water is assured. Maintaining hydrodynamics is not sufficient in itself to allow complete oxidation of excess N and P; nonetheless, it is vital for the health of the ecosystem.

Table 4.5 Carrying capacity in the Boca de Camichin estuarine system during the cold-dry season

General	m	m^2	m^3	km^2
Surface		2,073,980		2.11
Average depth	2.9			
Volume			5,993,802	
Water fluxes		February 2009		Units
Precipitation (V_p)		0.0005		m/day
Evaporation (V_e)		0.0002		m/day
Aquaculture effluents (V_q)		0		m^3/day
Agriculture effluents (V_o)		0		m^3/day
Rivers (V_c)		3,888,300		m^3/day
Others (V_v)		0		m^3/day
Water quality		February 2009		Units
System				
Salinity		27.54		ups
Nitrates (+nitrites+ammonium)		4.385		mmol/m^3
Orthophosphates		2.481		mmol/m^3
Ocean				
Salinity		35.85		ups
Nitrates (+nitrites+ammonium)		0.348		mmol/m^3
Orthophosphates		1.017		mmol/m^3
Aquaculture effluents				
Salinity		0.00		ups
Nitrates (+nitrites+ammonium)		0.000		mmol/m^3
Orthophosphates		0.000		mmol/m^3
Agriculture effluents				
Salinity		0.00		ups
Nitrates (+nitrites+ammonium)		0.000		mmol/m^3
Orthophosphates		0.000		mmol/m^3
Rivers				
Salinity		16.65		ups
Nitrates (+nitrites+ammonium)		3.35		mmol/m^3
Orthophosphates		2.09		mmol/m^3
Meteorological data		February 2009		Units
Precipitation		0.502		mm/day
Evaporation		0.233		mm/day
Hydraulical balance		February 2009		
V_p=pp (m/month) * Sup (m^2)		1,040.4		m^3/day
V_e=ev (m/month) * Sup (m^2)		387.1		m^3/day
V_q=aquaculture effluents		0		m^3/day
V_o=agriculture effluents		0		m^3/day
V_c=river		3,888,300.00		m^3/day
$V_r=-(V_p)-(-V_e)-(V_q)-(V_o)-(V_c)$		−3,888,953.3		m^3/day
Salinity balance				
$S_r=(S_{ext}+S_{int})/2$		31.70		ups
V_r*Sr		−123,260,374.96		ups/m^3/day
$V_x*(S_{ext}-S_{int})$		188,000,569.96		ups/m^3/day
$V_x=(V_q*S_q+V_o*S_o+V_c*S_c-V_rS_r)/(S_{ext}-S_{syst})$		22,623,413.95		m^3/day
Exchange rate				
$T=V_{syst}/V_x+V_r$		0.3		day
Nutrient fluxes				
$V_r*DIP_r=V_r*(DIP_{ext}+DIP_{syst})/2$		−6,801,779		mmol/day
$V_x*(DIP_{ext}-DIP_{syst})$		−33,120,678		mmol/day
V_oP_o (agriculture effluents)		0		mmol/day
V_qP_q (aquaculture)		0		mmol/day
V_cP_c		8,126,547		

Table 4.5 (continued)

General	m	m^2	m^3	km^2
$\Delta DIP_t = -V_o{*}DIP_o - V_q{*}DIP_q - V_r{*}DIP_r - V_x{*}(DIP_{ext} - DIP_{syst})$		31,795,910		mmol/day
$\Delta DIP = (-DIP_r - DIP_x)/Sup_{syst}$		15.331		mmol/m²/day
$V_r{*}DIN_r = V_r{*}(DIN_{ext} + DIN_{syst})/2$		−9,203,208		mmol/day
$V_x{*}(DIN_{ext} - DIN_{syst})$		−91,330,722		mmol/day
V_oN_o (agriculture effluents)		0		mmol/day
V_qN_q (aquaculture)		0		mmol/day
V_cN_c		13,018,028		
$\Delta DIN_t = -V_o{*}DIN_o - V_q{*}DIN_q - V_r{*}DIN_r - V_x{*}DIN_{ext} - DIN_{syst})$		87,515,902		mmol/day
$\Delta DIN = (-DIN_r - DIN_x)/Sup_{syst}$		42.197		mmol/m²/day
Stoichiometric relationships				
$\Delta DIN_{exp} = (N{:}P){*}\ \Delta DIP_t$		508,734,566		mmol/day
$\Delta DIN_{obs} = \Delta DIN_t$		87,515,902		mmol/day
$(nfix\text{-}denit)t = \Delta DIN_{obs}\text{-}\Delta DIN_{exp}$		−421,218,664		mmol/day
$(nfix\text{-}denit)/sup_{syst}$		−203,097		mmol/m²/day
Net metabolism				
$(p-r) = -\Delta DIP_t{*}(C{:}P)$		−3,370,366,498		mmolC/day
$(p-r) = -\Delta DIP_t{*}(C{:}P)/sup_{syst}$		−1,625.1		mmolC/m²/day
PP_{syst}		No data available		mmolC/m²/day

During the dry-hot season, the Boca de Camichín system is dominated by the presence of sea water, which causes high salinity values throughout the system (except near the San Pedro River). However, during the rainy season the river is a dominant force and is able to influence salinity, even at the mouth. An intermediate process is observed during the cold-dry period, where the flow of the San Pedro River reduces the salinity of the system. These hydrodynamics should be maintained to avoid eutrophic conditions. The great hydrodynamic influence of the area is fostered by the depth of the main channel (between 3 and 6 m), which is maintained by the strong estuary currents.

The information obtained in this study indicates that the carrying capacity of the estuarine system of Boca de Camichín may be nearing its maximum nutrient threshold. A higher contribution of organic matter (from urban development, industrial or aquatic) may lead to greater use of the available dissolved oxygen, which could lead to eutrophication of the system. For example, activities such as oyster farming should be performed under conditions that ensure a low input of organic matter to the system. The input of organic matter at areas farther upstream should also be considered.

Finally, the Boca de Camichín estuary system is clearly heterotrophic, maintaining high productivity conditions due to the imposed hydrodynamics such as penetration by the sea, the presence of a deep channel that allows a fast turnover rate of the estuary, and by the influence of the San Pedro River that aids the flushing of nutrients. The system is likely to be near its nutrient carrying capacity, so activities generating potential new organic matter input to this system should be monitored and regulated.

Acknowledgments This work was carried out thanks to financial support provided by PD/A CRSP, University of Hawaii and the Universidad Autónoma de Sinaloa.

References

Alonso-Rodríguez R, Páez-Osuna F (2001) La presencia del fitoplancton en los estanques de cultivo de camarón: composición, abundancia y variación. In: Páez Osuna F (ed) Camaronicultura y el Medio Ambiente. Instituto de Ciencias del Mar y Limnología, Universidad Autónoma de Sinaloa. Colegio de Sinaloa y Programa Universitario de Alimentos, Mexico, pp 137–155

Bechtel TJ, Copeland BJ (1971) Contribution in marine science: fish species diversity indices as indicators of pollution in Galveston Bay, vol 15. Inst Marine Science, University of Texas, pp 103–132

Cárdenas FM (1969) Pesquerías en las lagunas litorales de México. In: Agustín Ayala Castañares y Fred B (ed) Lagunas Costeras, un simposio. UNAM-UNESCO, Mexico, pp 645–652

Carta Nacional Pesquera (2004) Acuerdo mediante el cual se aprueba la actualización de la Carta Nacional Pesquera y su Anexo. Diario Oficial de la Federación, 15 de marzo 2004. Segunda, Tercera, Cuarta y Quinta Sección, Secretaría de Agricultura, Ganadería, Desarrollo Rural, Pesca y Alimentación, Mexico

Comisión Nacional del Agua (2009) Servicio Meteorológico Nacional: Temperatura y Precipitación. http://smn.cna.gob.mx/. Accessed 4 Feb 2009

Foyn E (1969) Waste disposal and pollution in coastal lagoons. In: Ayala-Castañares A, Fred B (eds) Lagunas Costeras, un simposio. UNAM-UNESCO, Mexico, pp 281–290

Gordon DC Jr, Boudreau PR, Mann KH, et al (1996) LOICZ Biogeochemical modelling guidelines. LOICZ Reports and Studies 5, LOICZ, Texel

IMTA (2010) Banco Nacional de Datos de Aguas Superficiales (BANDAS) http://www.imta.gob.mx/index.php?Itemid=145&option=com_wrapper&view= wrapper. Accessed 12 Jan 2013
INEGI (1999) Síntesis de información geográfica del estado de Nayarit. INEGI, Mexico
INEGI (2012) Simulador de flujos de agua de cuencas hidrográficas (SIATL). http://antares.inegi.org.mx/analisis/red_hidro/SIATL/#. Accessed 12 Jan 2013
INFDMGEN, Gobierno del Estado de Nayarit (2005) Enciclopedia de los Municipios de México. Registrado en http://www.e-local.gob.mx/work/templates/enciclo/nayarit/mpios/18015a.htm. Accesed 9 Feb 2009
Inglis GJ, Hayden BJ, Ross AH (2000) An overview of factors affecting the carrying capacity of coastal embayments for mussel culture. National Institute of Water and Atmospheric Research, Auckland
Lankford RR (1977) Coastal lagoons of Mexico: their origin and classification. In: Wiley M (ed) Estuarine processes, vol 2. Academic Press Inc., New York, pp 182–215
McKindsey CW, Thetmeyer H, Landry T, et al (2006) Review of recent carrying capacity models for bivalve culture and recommendations for research and management. Aquaculture 261(2):451–462
NRC (2000) Clean coastal waters: understanding the effects of nutrient pollution. National Research Council, Washington, D.C.
Páez-Osuna F, Guerrero-Galvan SR, Ruiz-Fernández AC, et al (1997) Fluxes and balances of nutrients in a semi-intensive shrimp farm in North-Western Mexico. Mar Pollut Bull 34:290–297
Parsons TR, Maita Y, Lalli CM (1984) A manual of chemical and biological methods for seawater analysis. Pergamon Press, Oxford
Ray CG (1991) Coastal zone biodiversity patterns. BioScience 41:490–498
Redfield AC (1934) On the proportions of organic derivations in sea water and their relation to the composition of plankton. University Press of Liverpool, Liverpool
Toledo A (2003) Ríos, costas, mares: Hacia un análisis integrado de las regiones hidrológicas de México. Secretaria de Medio Ambiente y Recursos Naturales, Instituto Nacional de Ecología, El Colegio de Michoacán, Mexico
Singh VP (1995) Watershed Modeling. In: Singh VP (ed) Computer models of watershed hydrology. Water Resources Publications, Colorado, pp 1–22
Universidad Autónoma de Nayarit (2004) Ordenamiento territorial del estado de Nayarit. CUVEDES-UAN, Tepic
Valderrama JC (1981) The simultaneous analysis of total nitrogen and total phosphorus in natural waters. Mar Chem 10:109–122

Part II
Ecological Considerations

Trophic Relationships within a Subtropical Estuarine Food Web from the Southeast Gulf of California through Analysis of Stable Isotopes of Carbon and Nitrogen

5

Martín E. Jara-Marini, Federico Páez-Osuna and Martín F. Soto-Jiménez

Abstract

We identified the sources of carbon supporting an estuarine food web in the Southeast Gulf of California. The trophic food web in the Estero de Urías Lagoon (EUL) was studied through the carbon and nitrogen isotopes in the potential food sources (plankton, macroalgae, plants) and organisms including filter-feeding mollusks, crustaceans, fishes and seabirds. The isotopic composition of sediment suspended organic matter (SSOM) and suspended particulate organic matter (SPOM) showed that there are diverse organic matter sources in EUL. The greater inputs of mangrove to detritus were reflected in their similar $\delta^{13}C$ values with respect to SSOM and SPOM. The $\delta^{13}C$ data suggest a direct transfer of C from SSOM and zooplankton to filter-feeders organisms and to a lesser degree from SPOM and phytoplankton. The isotopic composition of the different groups of organisms showed the complexity of the food web. However, there was a continuous gradient of ^{15}N-enrichment from SSOM and SPOM to seabird with intermediate values for filter-feeders and crustaceans. The $\delta^{15}N$ values in the EUL food web were consistent with 5 trophic levels. Fishes were strongly dependent on macrobenthos and pelagic derived nutrition. Cormorants occupied the highest trophic level and its major diet contributors were fishes. The studied food web was not segregated by time because isotopic trends were similar between dry and wet seasons.

Keywords

Coastal ecosystems · Food sources · C and N isotopes · Trophic levels · Gulf of California

M. E. Jara-Marini (✉)
Departamento de Ciencia de los Alimentos, Centro de Investigación en Alimentación y Desarrollo, A.C., km 0.6 Carretera a la Victoria, Ejido La Victoria, 83304 Hermosillo, SON, México
e-mail: mjara@ciad.mx

Programa de Posgrado en Ciencias del Mar y Limnología, Instituto de Ciencias del Mar y Limnología, Unidad Académica Mazatlán, Universidad Nacional Autónoma de, México, Av. Joel Montes Camarena s/n, 82040 Mazatlán, Sinaloa, México

F. Páez-Osuna · M. F. Soto-Jiménez
Instituto de Ciencias del Mar y Limnología, Universidad Nacional Autónoma de México, Av. Joel Montes Camarena s/n, 82040 Mazatlán, Sinaloa, México
e-mail: paezos@ola.icmyl.unam.mx

M. F. Soto-Jiménez
e-mail: martin@ola.icmyl.unam.mx

5.1 Introduction

Coastal lagoons are considered as one of the most productive aquatic ecosystem types due to high levels of primary production, large reserves of organic matter and habitat diversity. Additionally, lagoons offer optimal niches for numerous aquatic species, which utilize these areas as refuge and/or breeding grounds (Flores-Verdugo et al. 1996; Vega-Cendejas and Arreguín-Sánchez 2001; Zetina-Rejón et al. 2003). Along the Mexican coast these ecosystems include a great variety of habitats that include mangrove forest, salt marshes, intertidal pools, swamps, freshwater inner lagoons, brackish and sea water systems.

F. Amezcua, B. Bellgraph (eds.), *Fisheries Management of Mexican and Central American Estuaries*, Estuaries of the World, DOI 10.1007/978-94-017-8917-2_5, © Springer Science+Business Media Dordrecht 2014

Several authors have emphasized the importance of estuaries for marine fisheries. A large part of fish landings around the world consists of species that spend at least part of their lives in estuarine waters (Pauly 1988; Pauly and Yañez-Arancibia 1994; Barletta et al. 1998). In the southeast Gulf of California, the estuarine systems are exploited by small-scale fisheries that use a diverse suite of fishing gears (e.g., gill nets, seine nets, trawl nets, cast nets, hook and line, traps among others), to catch shrimp, sharks, rays, finfish, snails, swimming crabs, and other species. The catch from small-scale fisheries in Mexico is about 108,000 t per year (Ramírez-Rodríguez 2009), and a large proportion of that capture comes from estuarine systems.

The establishment of increasing human populations near lagoons has resulted in significant degradation and loss of coastal wetlands (Lapointe et al. 2005; Carlier et al. 2008; Serrano-Grijalva et al. 2011). For example, during past decades mangrove forests worldwide have experienced an enormous loss (Wilkie and Fortuna 2004). In México the highest loss occurred during the last two decades mostly associated with shrimp aquaculture expansion (De La Lanza Espino et al. 1994; Páez-Osuna et al. 2003). Other human activity with high impacts on coastal ecosystems in Mexico is the agriculture, livestock and urbanism (Soto-Jiménez et al. 2003; Piñón-Gimate et al. 2009). Discharges of organic matter from multiple untreated wastes are common also in most coastal ecosystems. Changes in environmental conditions generally provoke a variety of biological processes or responses and consequently, the degradation of ecosystems may occur (Alonso-Rodríguez et al. 2000; Lapointe et al. 2005; Piñón-Gimate et al. 2009).

Nutrient-rich waste discharges alter the metabolism of organisms and produce changes in species composition and diversity and, modify the trophic structure and function of food webs in these ecosystems (Persson et al. 2001; Serrano-Grijalva et al. 2011). This, in turn, stimulates bacterial activity and may result in benthic oxygen depletion, causing long-term changes in the structure of benthic assemblages (Iken et al 2001; Carlier et al 2008). Drastic changes in density and/or biodiversity produce alterations in top-down and bottom-up regulating forces in food webs, and may affect faunal communities (Polis et al. 1997). In this sense, fisheries can be affected by the reduction in fish stock abundance and changes in species composition by a reduction in fish diversity (Persson et al. 2001; Kon et al. 2009). Further, fish removed from the ecosystem, which play an important role in linking the coastal and marine ecosystems, can affect the nutrient budget (Maranger et al. 2008).

Stable isotopes of carbon and nitrogen can help to elucidate the complex task of the conversion of organic matter as it passes through each trophic link. This is one of the most important ways of understanding food web relationships in ecosystems (Minigawa and Wada 1984; Peterson and Fry 1987; Hobson and Welch 1992; Choy et al. 2008). Moreover, these techniques allow a quick assessment of the origin of organic matter assimilated by organisms and a broad overview of the mean trophic structure with limited sampling effort (DeNiro and Epstein 1981; Minigawa and Wada 1984; Peterson and Fry 1987).

The main goal of this study was to assess the trophic origins and pathways within a subtropical coastal lagoon of the southeast Gulf of California. Our main objectives were: (1) to identify the food sources supporting the dominant species in the trophic network; and (2) to estimate the total number of trophic levels with-in both the wet and dry seasons. We hypothesized that there would be differences in consumer isotopic signatures and expected a more marine-derived signature during the dry season and more terrestrial-derived signature during the rainy season.

5.2 Methods

5.2.1 Study Area

The Estero de Urías Lagoon (EUL) is located along the coast of the southeastern Gulf of California (Fig. 5.1). Considering its ecological importance, the lagoon has been included among the Priority Zones determined by the National Commission for the Knowledge and Use of Biodiversity of Mexico (CONABIO). The Estero de Urías is a shallow lagoon of approximately 16 km^2 which has a free and permanent exchange with the sea. The portion proximal to the sea (~3 km) functions as a navigation channel for the shipping terminal of Mazatlan Harbor. In the opposite extreme of the lagoon, there are a group of tidal channels (1–4 m deep) bordered by mangrove forest (*Rhizophora mangle*, *Laguncularia racemosa* and *Avicennia germinans*).

The climate of the region is tropical sub-humid with a monthly average temperature ranging from 19.7 °C in February to 28.0 °C in August. Typically, the Estero de Urías lagoon has an estuarine type regime during the wet season (July-October, salinities from 5.0 to 28.0 ‰), and an anti-estuarine behavior during the dry season (November- June, salinities from 35.0 to 45.0 ‰; Ochoa-Izaguirre et al. 2002). The development of Mazatlan city has profoundly affected the system and many contamination problems have been reported (Páez-Osuna et al. 1997; Soto-Jiménez and Páez-Osuna 2001; Green-Ruiz and Páez-Osuna 2006; Jara-Marini et al. 2009) due to effluents coming from the food industry, the shrimp fishing fleet, a thermoelectric plant, shrimp aquaculture farms, treated and untreated urban wastes and runoff.

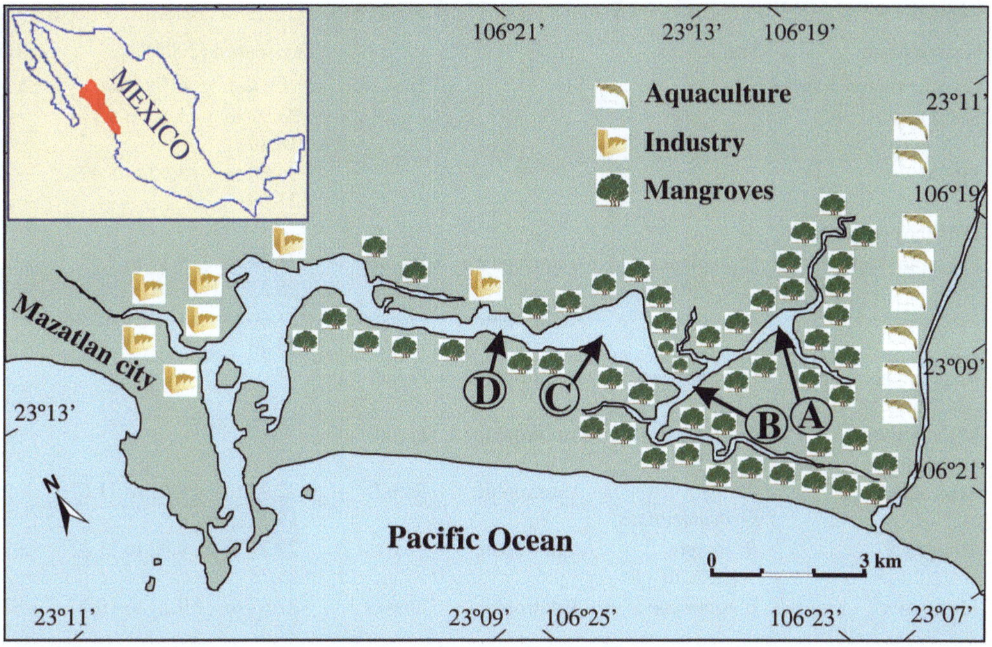

Fig. 5.1 Localization of the sampling stations (A, B, C, and D) in the Estero de Urías lagoon system

5.2.2 Sample Collection and Preparation

Sampling was performed at four sites in the Estero de Urías lagoon during the dry (April 2006) and rainy (September 2006) seasons (Fig. 5.1). Twenty water samples (1 L at 1.0 m depth from the water surface) from sites were collected in each sea-son, using a peristaltic pump equipped with a plastic line (Solinst model 410) and were placed in low density polyethylene bottles. For stable isotope analysis, suspended particulate organic matter (SPOM) was obtained by filtration on pre-combusted (400 °C for 4 h) Whatman GF/F glass fiber membranes. Subsequently, the membranes were placed in a glass chamber vaporized with concentrated HCl for 3 h in order to remove carbonates, dried at 40 °C for 4 h and kept at 4 °C until analysis. Also, 36 surface sediment samples (2.5 cm from the sediment surface) were collected at lagoon sites during two seasons using a plastic tube (ID = 7 cm), in order to characterize the surface sediment organic matter (SSOM).

Primary producers and primary to tertiary consumers representative of the food web of the lagoon system were collected and divided according to their feeding habits (Table 5.1): autotrophic (phytoplankton mixed, macroalgae and magrove), plankton consumer (zooplankton mixed, copepod and juvenile flathead mullet), filter-feeders (barnacle, oyster and mussel), omnivores (polychaete, snail, shrimp and crab), detritivores (adult flathead mullet), and carnivores (fishes: mojarra, snapper and grunt; bird: cormorant). A total of 218 samples of aquatic organisms were collected. Macroalgae, mangrove leaves (*R. mangle* and *L. racemosa*), polychaete (*Streblospio benedicti*) and snails (*Littoraria aberrans*) were collected by hand, rinsed with water to remove particulate material and placed in HCl-washed plastic bags. Oysters (*Crassostrea cortezienzes*), mussels (*Mytela strigata*) and barnacles (*Fistulobalanus dentivarians*) were collected from mangrove roots by using a stainless-steel knife at low tide. Shrimps (*Litopenaeus vannamei* and *Farfantepenaeus californiensis*), crab (*Callinectes arcuatus*) and fishes (adult and juvenile flathead mullet *Mugil cephalus*, yellow fin mojarra *Gerres cinereus*, snapper *Latjanus argentiventris* and white grunt *Haemulopsis leuciscus*) were collected using a gill net, with a mesh size of 8.9 cm. The net was left to drift during 30-minute periods at every site. Cormorant (*Phalacrocorax brasilianus*) samples were obtained from a local hunt club. Surface sediment and biological samples were transported to the laboratory, freeze-dried at −48 °C and 32×10^{-3} mbar (Labconco freeze-dry system) for 72 h and kept at 4 °C until analysis. Subsequently, the samples were placed in a glass chamber vaporized with concentrated HCl for 3 h in order to remove carbonates, dried at 40 °C for 4 h, powdered and kept at 4 °C until analysis. All organisms were collected during both seasons.

5.2.3 Stable Isotope Analysis

Powdered samples were weighed according to sample type (1.0 mg for fauna; 4–5 mg for flora; and 20 mg dry weight for sediment and particulate matter on filter) and put into tin cups before stable isotope analysis. Stable isotopes analyses were performed in the Stable Isotope Facility of the Department of Plant Science of the University of California at Davis, CA, USA. Isotope composition was performed using a Europa Scientific ANCA-NT 20-20 Stable Isotope Analyzer with an ANCA-NT Solid/Liquid Preparation Module (Europa Scientific, Crewe, UK). Analytical precision (standard

Table 5.1 $\delta^{13}C$ and $\delta^{15}N$ (range of values) of the organic matter sources and groups of organisms collected in Urías lagoon system during two seasons

Name/Group	Species			Dry season (2006)		Rainy season (2006)		
Organic matter source		FH	Tissue	$\delta^{13}C$ (‰)	$\delta^{15}N$ (‰)	$\delta^{13}C$ (‰)	$\delta^{15}N$ (‰)	n
SSOM				−25.97 to −18.40	5.20 to 9.17	−26.71 to −19.69	4.52 to 6.50	36
SPOM				−26.80 to −21.68	3.92 to 5.47	−26.42 to −22.89	2.12 to 8.55	40
Primary producer								
Phytoplankton	Phytoplankton	Autotrophic	Whole	−21.31 to −18.52	5.17 to 7.07	−21.52 to −19.29	5.44 to 7.10	12
Macroalgae	*Ulva* sp.	Autotrophic	Fronds	−24.60 to −10.86	11.02 to13.86	-	-	5
Macroalgae	*Gracilaria* sp.	Autotrophic	Fronds	-	-	−26.19 to −15.37	8.39 to 13.43	5
Macroalgae	*C. linum*	Autotrophic	Fronds	-	-	−27.09 to −25.58	8.18 to 9.83	2
Macroalgae	*C. serticularioides*	Autotrophic	Fronds	−22.05 to −19.92	8.01 to 11.67	−25.40 to −23.35	10.48 to 10.57	6
Mangrove	*R. mangle*	Autotrophic	Leaves	−29.21 to −25.73	9.02 to 11.32	−27.49 to −26.10	9.62 to 9.97	6
Mangrove	*L. racemosa*	Autotrophic	Leaves	−27.62 to −27.47	9.21 to 10.17	−27.40 to −26.80	9.72 to 11.22	4
Primary consumer								
Zooplakton	Zooplakton	Planktivorous consumer	Whole	−21.75 to −19.03	7.34 to 10.86	−25.17 to −17.24	7.76 to 8.20	8
Zooplakton	Copepods	Planktivorous	Whole	-	-	−20.10 to −19.40	9.32 to 9.44	2
Oyster	*C. cortezienzes*	Filter feeding	Soft	−26.60 to −23.80	7.70 to 9.50	−24.90 to −18.60	6.60 to 10.70	30
Mussel	*M. strigata*	Filter feeding	Soft	−27.79 to −24.52	7.84to 9.99	−24.06 to −18.25	7.60 to 9.52	30
Barnacles	*F. dentivarians*	Filter feeding	Soft	−22.72 to −22.08	9.01 to 9.31	−21.77 to −20.68	9.34 to 9.53	6
Snail	*L. aberrans*	Omnivorous	Soft	−25.14 to −21.44	7.05 to 9.32	−24.38 to −22.55	7.76 to 8.68	20
Flathead mullet-juvenile	*M. cephalus*	Planktivorous	Muscle	−22.05 to −20.47	10.88 to 11.17	−25.14 to −21.44	7.05 to 9.32	4
Polychaete	*S. benedicti*	Omnivorous	Whole	−22.73 to −21.15	12.46 to 13.59	−20.72 to −20.23	13.25 to 13.29	6
Shrimp	*L. vannamei*	Omnivorous	Muscle	−17.28 to −15.67	12.86 to 14.10	−19.53 to −16.47	13.52 to 14.82	16
Shrimp	*F. californiensis*	Omnivorous	Muscle	−15.28 to −13.21	12.23 to 13.91	-	-	4
Secondary consumers								
Flathead mullet-adult	*M. cephalus*	Detritivore	Muscle	−16.13 to −14.52	16.00 to16.21	−21.70 to −15.81	13.65 to 16.93	8
Crab	*C. arcuatus*	Omnivorous	Muscle	−17.46 to −15.76	14.60 to 15.62	−18.69 to −14.90	14.86 to 15.10	16
Yellow fin mojarra	*G. cinereus*	Carnivorous	Muscle	−17.26 to −15.31	15.99 to 17.40	−15.93 to −14.88	14.72 to 17.02	8
Snapper	*L. argentiventris*	Carnivorous	Muscle	−18.35 to −15.08	16.47 to 18.01	−16.02 to −15.38	16.51 to 17.76	8
White grunt	*H. leuciscus*	Carnivorous	Muscle	−15.23 to −14.92	15.58 to 16.01	−15.92 to −13.94	13.90 to 15.78	6
Tertiary consumers								
Cormorant	*P. brasilianus*	Carnivorous	Muscle	−21.73 to −21.64	17.35 to 18.81	−21.78 to −21.57	17.03 to 18.70	6

SSOM sediment suspended organic matter, *SPOM* suspended particulate organic matter, *n* number of samples, *FH* feeding habits

deviation, n=5) was 0.2‰ for both nitrogen and carbon, as estimated from standards (peach leaves NIST-1547 and bovine liver NIST-1577b, from National Institute of Standard Technology) analyzed every 12 samples. The isotopic composition $\delta^{13}C$ or $\delta^{15}N$) was expressed as the relative difference between isotopic ratios in the sample and their corresponding standard (Vienna Pee Dee Belemnite for carbon and atmospheric N2 for nitrogen):

$$\delta^{13}C \text{ or } \delta^{15}N(‰) = \left[R_{sample}/R_{standard} - 1\right] \times 1000 \quad (5.1)$$

Where: $R = {}^{13}C/{}^{12}C$ or ${}^{15}N/{}^{14}N$

5.2.4 Data Analyses

As all stable isotopes values follow a non-normal distribution according to the Shapiro-Wilk and Bartlett tests ($p<0.05$), Kruskal-Wallis and Student-Newman-Keuls tests (Glantz 2002) were used to compare the differences in mean $\delta^{13}C$ and $\delta^{15}N$ between all samples collected from the lagoon ecosystem.

As $\delta^{15}N$ values provide an indication of the trophic position of a consumer, the following formula was used to estimate trophic level (TL) (Hobson & Welch 1992; Post et al. 2000):

$$TL = 2 + \left[\frac{\delta^{15}N_{trophicgroup} - \delta^{15}N_{primaryconsumer}}{3.4}\right] \quad (5.2)$$

where 3.4‰ is the assumed 15N trophic enrichment factor according to Minagawa and Wada (1984). Since $\delta^{15}N$ values ranged between 7.76 and 9.44‰ for zoo-plankton (the low primary consumer), trophic levels were defined as ranges rather than individual values of $\delta^{15}N$ (Carlier et al. 2007).

A simple linear regression model between $\delta^{13}C$ values (independent variable) and $\delta^{15}N$ values (dependent variable) was performed in order to establish trophic relationships in the different organism groups. All statistical analyses were performed using the Statistica Computer Program (StatSoft 1996).

5.3 Results

5.3.1 Characterization of Organic Matter Sources and Organisms

Ranges of $\delta^{13}C$ and $\delta^{15}N$ values of samples collected in EUL system are presented in Table 5.1 during two seasons. The C and N isotopic composition in the different organic matter sources were similar. The C:N ratios varied from 6.4 to 8.6 and 10.5 to 16.1 in SPOM and SSOM, respectively. Similar ranges of $\delta^{13}C$ values were determined in the dry and rainy seasons for SSOM and SPOM. However, there were seasonal differences ($p<0.05$) in ranges of $\delta^{15}N$ values for SSOM and SPOM. Considering the two seasons, the ranges of $\delta^{13}C$ values of SSOM and SPOM were significantly ($p<0.05$) different from most of organisms, except mangrove, oyster, mussel and snail. In contrast, the ranges of $\delta^{15}N$ for SSOM and for SPOM in both seasons were significantly ($p<0.05$) lower than all organisms. The ranges of $\delta^{13}C$ and $\delta^{15}N$ for phytoplankton values were similar ($p>0.05$) between the dry season and the rainy season. The range of $\delta^{13}C$ for mixed zooplankton was lower in the dry season than in the rainy season, while the range of $\delta^{15}N$ was higher in the dry season than in the rainy season. Copepods showed a narrow isotopic range in the rainy season. The $\delta^{15}N$ values of zooplankton were significantly ($p<0.05$) higher than those found for phytoplankton.

Five macroalgae species were sampled were 5 in the dry season (*Ulva lactuca, Ulva intestinalis, Ulva lobata, Gracilaria crispata* and *Caulerpa serticularioide*) and four species were sampled in the rainy season (*Gracilaria vermicullophylla, Gracilaria turgida, Chaetomorpha linum* and *Caulerpa serticularioides*). Some macroalgae species showed wide variations in isotopes composition. High variations of $\delta^{13}C$ and $\delta^{15}N$ were determined for *Ulva* sp.in the dry season. Gracilaria sp. also observed wide ranges for $\delta^{13}C$ and $\delta^{15}N$ in the rainy season. However, slightly variations in $\delta^{13}C$ and $\delta^{15}N$ were determined in *C. linum*. Only *C. serticularioides* was collected in both seasons, and no significant seasonal differences were observed in the ranges of $\delta^{13}C$ and $\delta^{15}N$. In contrast, the isotopes values for species of mangrove were homogeneous in both seasons. In general, significant differences ($p<0.05$) in $\delta^{13}C$ mean values between macroalgae (-20.55 ± 4.86‰) and mangroves (-27.36 ± 1.19‰) were detected.

A total of 218 samples of aquatic organisms were collected. In phytoplankton samples, diatoms (from 2 to 100%), dinoflagellates (from 1 to 19%) and chlorophyta (from 0 to 97%) predominated. Similar results were previously reported in adjacent areas of the Estero de Urías lagoon by Alonso-Rodríguez et al. (2000). In the mixed zooplankton samples, predominant groups were ctenofora (from 2 to 72%), porcellanides (from 3 to 70%) and copepods (from 17 to 45%). Two samples of copepods were isolated from zooplankton during the rainy season and were analyzed separately.

Among filter-feedering organisms, barnacle presented similar $\delta^{13}C$ and $\delta^{15}N$ composition in the dry season and the rainy season. However, the $\delta^{13}C$ composition for oyster and mussel varied from the dry season to the rainy season. The $\delta^{15}N$ values for oyster and mussel were similar between seasons. With respect to omnivorous organisms, the shrimp *L. vannamei* showed similar seasonal $\delta^{13}C$ ranges and their range in the dry season was lower than found for the shrimp

F. californiensis. The polychaete *S. benedicti* and the snail *L. aberrans* also presented similar $\delta^{13}C$ composition between seasons. The crab *C. arcuatus* also presented similar $\delta^{13}C$ ranges between seasons and these values were similar to those in both shrimp species. The $\delta^{15}N$ ranges of omnivores were different between species (the highest for the shrimp *L. vannamei* and the crab *C. arcuatus*) and the species showed similar values between seasons.

The juvenile flathead mullet showed a narrow $\delta^{13}C$ range in both seasons. However, their $\delta^{15}N$ ranges were higher in the dry season than the rainy season. The adult flathead mullet showed significant differences in the $\delta^{13}C$ ranges, more enriched in the dry season. Their $\delta^{15}N$ ranges showed slightly seasonal variations. The carnivore fish species showed enriched and homogeneous $\delta^{13}C$ composition and no significant differences ($p<0.05$) were determined in their ranges between the dry season and the rainy season. Similarly, carnivorous fish species also presented narrow $\delta^{15}N$ composition in both seasons. The cormorant, the only tertiary consumer collected, showed slightly seasonal variations in the $\delta^{13}C$ and $\delta^{15}N$. Among all organisms, the cormorant showed the highest ($p<0.05$) $\delta^{15}N$ mean value (17.97±0.96‰) in both seasons.

5.3.2 Stable Isotopes Composition and Trophic Levels of the Food Web

There are clear trends in isotopes composition in the SPOM and organisms collected in the EUL (Fig. 5.2). There was a defined pattern toward enrichment in $\delta^{13}C$ and $\delta^{15}N$ from SPOM to plankton, marine plants and filter-feedings organisms. The same increments were observed in successive organisms, with the major isotope composition in the cormorant bird. A significant ($p<0.01$) linear correlation between $\delta^{13}C$ and $\delta^{15}N$ values was determined, with the predictive equation of $\delta^{15}N=(0.75 \times \delta^{13}C)+26.47$ ($R2=0.53$; $n=294$).

The estimated number of trophic levels in the EUL were five (Fig. 5.2). The lowest $\delta^{15}N$ mean values ($p<0.05$) were detected in SSOM (5.72±3.01‰), SPOM (5.61±2.07‰) and phytoplankton (6.12±0.64‰). According to this, detritus, seston and phytoplankton seem to be at the base of the food web in the EUL (i.e., phytoplankton organisms represented the lowest trophic level of studied food web). Mangroves and macroalgae presented high N composition, with $\delta^{15}N$ values ranging between 9.02 to 11.31‰ and 8.02 to 13.86‰, respectively. The second trophic level contained the zooplankton species (9.84±1.23‰ for mixed zoo-plankton and 9.33±0.18‰ for copepods), oyster (8.14±1.55‰), mussel (8.81±0.84‰), barnacle (8.30±0.17‰) and snail (8.60±1.13‰).

The third trophic level was occupied by juvenile mullet, polychaete and shrimp and their $\delta^{15}N$ values means (11.25±0.11, 13.25±0.42 and 13.52±0.74‰, respectively) were significantly ($p<0.05$) higher than those in plankton and filter-feeding mollusks. Among crustaceans, the crabs had higher ($p<0.05$) $\delta^{15}N$ values means (15.71±0.55‰) than shrimp. Fish species were located in trophic level 4. In spite of differences in dietary sources, no significant differences ($p>0.05$) in $\delta^{15}N$ values means (16.22±0.16‰ for adult mullet, 16.36±0.61‰ for mojarra, 17.05±0.54‰ for snapper and 15.46±0.67‰ for grunt) were found. The cormorant was determined to have the highest ($p<0.05$) $\delta^{15}N$ mean value (17.97±1.07‰) was determined in the cormorant and thus was considered to occupy the highest trophic level and to be the top predator.

5.4 Discussion

5.4.1 Stable Isotopes Composition

The diffenences in $\delta^{13}C$ composition of SSOM and SPOM are indicative of diverse organic matter sources. Lower ^{13}C enrichment is possibly related to pluvial-terrestrial sources to the lagoon and higher enrichment ^{13}C is likely from marine sources (Cabana and Rasmussen 1996; Pinnegar and Polunin 2000; Garcia et al. 2002). Our results suggest that mangroves and macroalgae species are the main contributors to the pool of SSOM and SPOM in EUL. How-ever, these $\delta^{13}C$ values were not reflected in the phytoplankton neither zooplankton organisms. One explanation is that in this study, plankton samples were collected only from the water column, but there may be other plankton organisms in the benthos involved in the carbon fuel loop.

The primary producers (e.g. phytoplankton, macroalgae and mangroves), SSOM and SPOM are considered to be at the base of the food web (Pauly et al. 2000; Zetina-Rejón et al. 2003). Thus, we carried out a precise estimation of the temporal and spatial variation of $\delta^{13}C$ and $\delta^{15}N$ ratios for theses food sources. Results indicated a wide variation in $\delta^{13}C$ and $\delta^{15}N$ composition for the primary producers, SSOM and SPOM. Overall the $\delta^{13}C$ values and C:N ratios showed that organic matter in both food sources varies between terrestrial (by pluvial organic matter run-offs) and marine origin (Riera and Richard 1997; Carlier et al. 2007; Riera 2007). However, these signatures may be related also to seasonal variations in to $\delta^{13}C$ of mangrove and macroalgae species, organic matter that may ultimately be present in SSOM.

The role of SSOM and SPOM on benthic food web was discussed by Carlier et al. (2007). SPOM consists of a mixture of: (1) ^{13}C-depleted and more refractory material of terrestrial origin and (2) ^{13}C-enriched and more labile material of marine origin. Differences in isotopic values between SSOM and SPOM could be primarily due to different integration times with SPOM having short integration times and SSOM having long integration times. Since isotopic measurements of consumers are also associated also with rela-

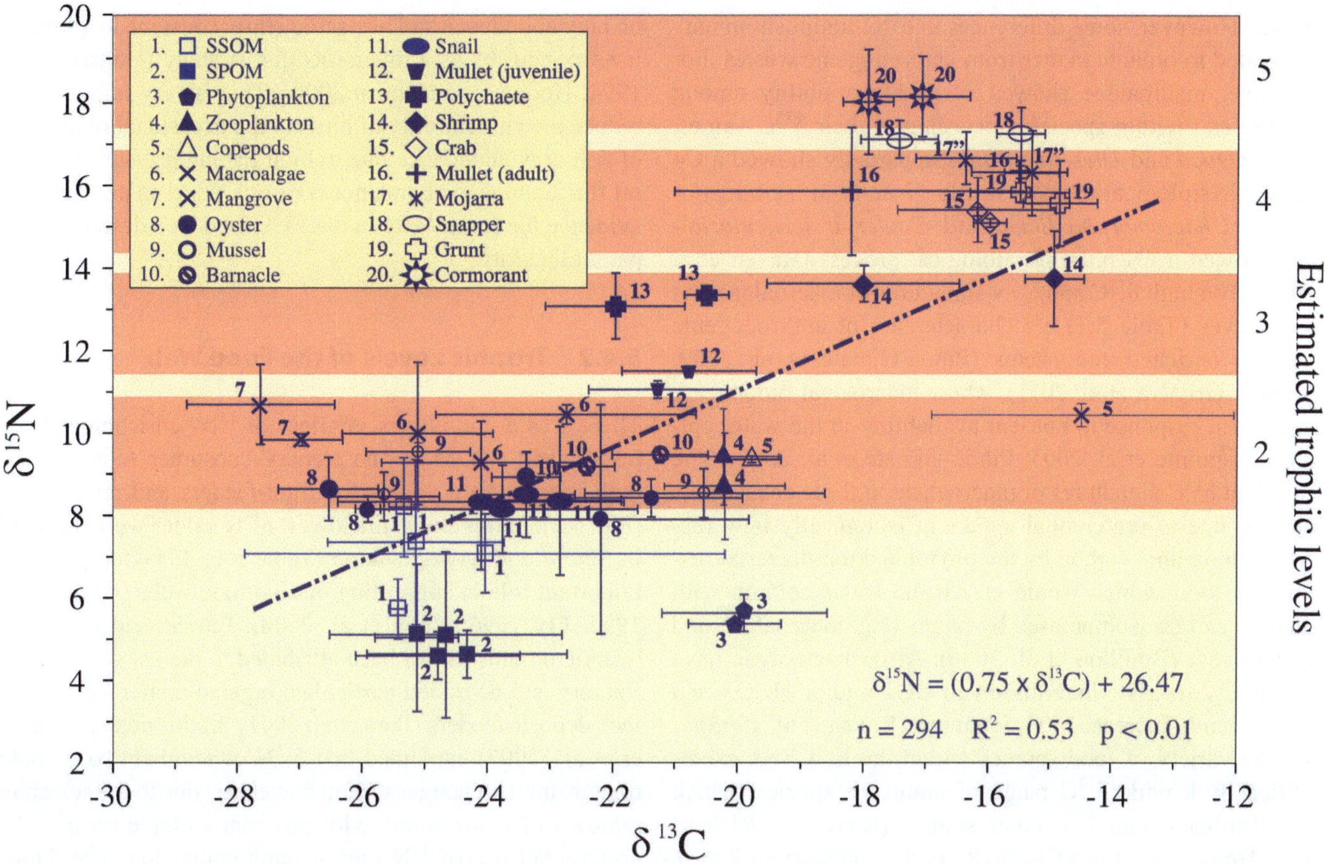

Fig. 5.2 Scatterplot of mean $\delta^{13}C$ vs $\delta^{15}N$ (‰) values of suspended particulate organic matter (SPOM) and organisms collected in the Estero de Urias lagoon system. *Vertical* and *horizontal* bars are one standard deviation. Shadowed red horizontal bands correspond to estimated trophic levels

tively long integration times (Tieszen et al. 1983; Hesslein et al. 1993), primary consumers would thus feed on a mixture of SPOM and SSOM. Here, re-suspension of organic matter is a key process for exchanges between the SPOM and SSOM pools. Consequently, SPOM and SSOM could act as single pools of organic matter or as a pool composed by both, depending on seasonal changes in coastal ecosystems. In this study, SPOM $\delta^{13}C$ values were similar to those in phytoplankton and zooplankton. This may be partly because interface-feeders depend on a detritus-based food web involving bacterial and meiofaunal intermediates, which raise the $\delta^{13}C$ of dissolved organic matter (Hobson and Welch 1992) and, transfer it to zooplankton via the plankton loop (Valiela 1995). Primary productivity of phytoplankton in coastal systems is known to be controlled by water column transparency (due to shallow depth, active tidal mixing and seasonal freshwater input) and anthropogenic activities mainly related to untreated domestic effluents (Hansson et al. 1997; Alonso-Rodríguez et al. 2000). Nutrients availability can cause phytoplankton blooms; this biomass may represent the highest percentages in SPOM (Choy et al. 2008) and frequently exhibit seasonal variations (Kling et al. 1992; Fry 1996; Choy et al. 2008). Mangrove areas are known to contribute greatly to detritus (Flores-Verdugo et al 1996; McLusky and Elliott 2004). In this study, mean $\delta^{13}C$ values of SSOM and mangroves were similar, indicative of its importance in detritus. The mean $\delta^{13}C$ values of SSOM and SPOM were also similar to oyster and mussel. Filter-feeders clearly fed directly on SSOM and SPOM, which can be deduced in two ways (Carlier et al. 2007). First, filter-feeders had similar $\delta^{13}C$ composition to SSOM and SPOM. Second, re-suspension and subsequent sedimentation results in frequent and intense exchanges between the SSOM and SPOM pools, which are especially common in marine coastal ecosystems and support benthos-water column coupling (Valiela 1995). Our $\delta^{13}C$ data (Table 5.1) suggest a direct transfer of organic carbon from SPOM, SSOM and phytoplankton to filter-feeding organisms.

Some primary and secondary consumers showed similar carbon isotopic com-position during both seasons, which implies that a relationship in diets exists between these organisms and that there are not substantial seasonal changes in car-bon sources. In contrast, the differences in $\delta^{13}C$ composition found in primary consumers (some filter feeders and crustaceans) may be related to changes in dietary sources associated with seasonal changes of organic matter. However, some differences in $\delta^{13}C$ composition may be related to organic matter from anthropogenic wastes. For example, macroalgae showed elevated variability among species and within species. According to their $\delta^{13}C$ values, *Ulva lactuca* and *Ulva intestinalis* apparently showed a C4 plant physiological process while *Gracilaria vermicullophylla*, *Chaetomorpha linum* and *Caulerpa serticularioides* showed a C3 plant physiological process (Raven et al 1995). The high $\delta^{13}C$ and $\delta^{15}N$ signatures of macroalgae and mangroves (Table 5.1) are characteristic of anthropogenic nutrient enriched ecosystems (Piñón-Gimate et al. 2009; Serrano-Grijalva et al. 2011), where macroalgal blooms are a frequent response to nutrient availability in the water column (Lapointe et al. 2005; Piñón-Gimate et al. 2009). The enriched $\delta^{13}C$ signatures of macroalgae in these ecosystems may be due to preferential uptake of isotopically light dissolved inorganic carbon by the phytoplankton-bacterial respiration loop, which would enrich the water column with heavier carbon isotope used by plants (e.g. macroalgae and mangroves) (Bouillon et al. 2000). Also, macroalgae have the ability to assimilate dissolved CO2 and/or HCO3 and subsequently increase $\delta^{13}C$ signatures (Raven et al. 1995).

The variety of food sources within the EUL was exemplified by a wide $\delta^{13}C$ range of omnivore species, which was similarly found in other studies (Riera and Richard 199; Hobson et al. 2002; Carlier et al. 2007; Riera 2007). Omnivores displayed a $\delta^{13}C$ signature very similar to that of SSOM suggesting that sediment organic matter is their main food source. However, many omnivorous organisms display opportunistic feeding strategies, consuming the most readily available organic material. Crustaceans were slightly enriched in $\delta^{13}C$ comparison to carnivorous fish (pelagic species), a phenomenon that appears to be widespread and has been documented in both marine and freshwater systems (France 1995; Hobson et al. 2002; Carlier et al. 2007; Riera 2007).

In spite of distinct feeding habits, fishes showed similar stable isotopes composition between species, and were located in the same trophic level. Amezcua-Linares (1996) reported a diversified diet of the same species of carnivorous fishes sampled for the current study. However, our $\delta^{13}C$ isotopic data suggest dietary similarities, possibly related to an abundance of prey. Many fish species demonstrate a degree of plasticity in prey item choice and consequently change their diet seasonally and/or in response to prey availability, which is frequent in some estuarine ecosystems (Cabral 2000; Laffaille et al. 2001; Elliott and Hemingway 2002; Pasquaud et al. 2008). However, fishes were most enriched in ^{13}C, suggesting a possible marine- origin diet, which is reasonable based on the mobility of fishes and being that they are not exclusive estuarine residents (Hansson et al. 1997). Cormorants occupied the top- trophic level, based on their feeding habits. There is evidence that bird diets depend on prey abundance and birds frequently consume low trophic level organisms as a diversification strategy (Hobson et al. 1994; Hodum and Hobson 2000). This is very important to reduce the vulnerability of birds to the occasional reductions of fish/prey abundance due to high fishing pressure or natural fluctuations in abundance. Overall, our data also provide evidence for complexity in the EUL food web despite its apparent linearity.

5.4.2 Trophic Levels of the Food Web

There was a continuous gradient of ^{15}N- enrichment from SSOM and SPOM to the tertiary consumer (cormorants) with intermediate values for filter-feeders and crustaceans (Fig. 5.2). In this study, the lowest $\delta^{15}N$ values were detected in SPOM and phytoplankton (Table 5.1). Plankton plays an important role in supporting marine food webs (Kling et al. 1992; Fry 1996; Choy et al. 2008). Enrichment of ^{15}N-in benthic organisms has been attributed to the use of more refractory and degraded particulate organic matter by subsurface deposit-feeders (Iken et al. 2001). In this context, Carlier et al. (2007) explained that $\delta^{15}N$ of sediment particulate organic matter changes during diagenesis due to the selective removal of components with different isotopic ratios. The preferential use of ^{14}N during remineralization contributes to enrich organic substrates in ^{15}N. Denitrification results in an increase in $\delta^{15}N$ of the remaining nitrates under anoxic conditions and may contribute also to significant ^{15}N- enrichment of particulate organic matter deep in the sediment, and to ^{15}N- enrichment of subsurface deposit-feeders. Subsequently, trophic interactions enrich ^{15}N- in top predators through the food web. However, most species with similar trophic levels have differences in ^{15}N- enrichment due to differential ^{15}N turnover rates (DeNiro and Epstein 1981; Peterson and Fry 1987; Mizutani et al. 1992), which may have occurred in this study among crustacean species.

Our results of $\delta^{15}N$ values in the EUL food web indicated five trophic levels (Fig. 5.2). This is consistent with assessments of previous studies that investigated marine food webs with seabirds as top predators in temperate coastal environments (Hobson and Welch 1992; Hobson et al. 1994; Hodum and Hobson 2000). In estuarine ecosystems, some studies found $\delta^{15}N$ fractionation ranges from 1.9 to 4.3 ‰ between phytoplankton and zooplankton (Goering et al. 1990; Montoya et al. 1990; Kling et al. 1992; Hansson et al. 1997), which was similar to our study (1.93–3.71 ‰). The $\delta^{15}N$ values of phytoplankton in our study (range 5.17–7.10 ‰; mean 6.1 ‰) were similar to previously reported values (Goering et al. 1990; Montoya et al. 1990; Kling et al. 1992; Hansson et al. 1997). The $\delta^{15}N$ values of SPOM in our study were similar also to values found in phytoplankton, which is congruent with other studies that have reported $\delta^{15}N$ composition of particulate organic matter as the $\delta^{15}N$ base value for phytoplankton (Montoya et al. 1990; Kling et al. 1992; Hansson et al. 1997).

Based on $\delta^{15}N$ values, zooplankton, mussels, oysters, barnacles and snails occupied the second trophic level (Fig. 5.2). There was great variability in $\delta^{15}N$ among species, due to the complexity of relationships in the food web by specialized consumers, which has been noted by others (McClelland and Valiela 1998; Choy et al. 2008). Mangroves and macroalgae species also apparently occupied trophic level two, with mean $\delta^{15}N$ values higher than expected for primary producers. Other authors reported alteration in the $\delta^{15}N$ composition of mangrove and macroalgae in ecosystems impacted by shrimp aquaculture activities (Jones et al. 2001; Costanzo et al. 2004; Kon et al. 2009; Serrano-Grijalva et al. 2011). Man-groves typically had higher mean $\delta^{15}N$ signatures than macroalgae but our data showed similar and even higher $\delta^{15}N$ values in macroalgae than mangroves. This is the result of high nitrification by bacteria, which oxidize organic matter; bacteria prefer the lighter ^{14}N for metabolism, which lead to enrichment of ^{15}N in the free nitrogen pool (Costanzo et al. 2004). Alternatively, ^{15}N enrichment may result from volatilization of oxidized nitrogen species (Heaton 1986).

Trophic level three were occupied by juvenile mullet, shrimp and polychaetes. This is in accordance with their benthic- derived food sources (Carlier et al. 2008). Crabs were located in trophic level four and their ^{15}N- enrichment could be related to ^{15}N turnover rates and/or due to their specific feeding habits. These results agree with Paul (1981), who reported that crab exhibit carnivorous feeding habits with a diet consisting of bivalve mollusks, crabs and fishes.

Lastly, mojarra, snapper and grunt fishes collected in the EUL occupied the trophic level four. These carnivorous and piscivorous feeders consume a diverse array of prey items including amphipods, decapods, isopods, cephalopods, bivalves, polychaetes, and other fishes (Amezcua-Linares 1996; Choy et al. 2008; Pasquaud et al. 2008). However, our $\delta^{15}N$ data suggest that snapper and grunt had a diverse diet in the current study as well. Knowledge of the dietary patterns of fish is extremely important for understanding the biological interactions that occur within an ecosystem (Choy et al. 2008; Pasquaud et al. 2008). The fishes studied in the EUL food web were strongly dependent on macrobenthos- and pelagic- derived nutrition. This suggests a considerable sharing of resources among organisms in the EUL, which may limit interspecific competition for access to food resources.

The highest trophic level in the in the EUL food web, level five, was occupied by the cormorant. However, there are a wide range of $\delta^{15}N$ values for this seabird, that overlap with $\delta^{15}N$ values of carnivorous fishes. The use of isotopic composition in bird ecology has limitations because they usually are migratory and exhibit a wide variety of possible preys (Hobson 1993). Consequently, they may not have the $\delta^{15}N$ composition that corresponds to their trophic level. For example, a common finding has been that juvenile bird exhibit higher $\delta^{15}N$ values than adult birds (Hobson 1993; Hodum and Hobson 2000); these variations may be linked to unusual fractionation factors due to differences in growth rates between age- classes (Forero and Hobson 2003).

The enrichment of $\delta^{15}N$ was significant in the sediment, macroalgae and man-groves of the EUL, likely as a result of shrimp-farming and sewage effluent previously documented in the ecosystem (Alonso-Rodríguez et al 2000; Páez-Osuna et al. 2003). Mangroves and macroalgae species showed high $\delta^{15}N$ values and thus these communities could be considered a biological indicator of the effects of nutrient enrichment by anthropogenic discharges on the ecosystem food web. The ultimate effect of nutrient enrichment is the structural changes that occur to the coastal lagoon ecosystems, such as a reduction in species diversity and changes to the top-down and bottom-up regulating forces in food webs (Polis et al. 1997). In this sense, this study contributes essential information about short- and long-term structural changes to coastal ecosystems of the Gulf of California.

At the global scale, coastal ecosystems fish communities are rapidly losing populations, species, or entire functional groups. However, there is evidence that diversity of organisms leads to increased ecosystem stability and recovery potential of ecosystems, and that anthropogenic activities decrease diversity (Worm et al. 2006). Is widely recognized that agriculture, aquaculture and industrial development are responsible for the majority of aquatic habitat destruction around world, with aquaculture development playing an important role in recent years (Páez-Osuna et al. 2003; Piñón-Gimate et al. 2009; Berlanga-Robles et al. 2011; Serrano-Grijalva et al. 2011). In this study we do not evaluate biodiversity nor the loss of species in the EUL. However, we found alteration in C and N composition of some species that were likely derived from anthropogenic activities within the ecosystem. In Mexico there are many regulations on coastal management that have be applied more strictly (Quijano-Poumián and Rodríguez-Aragón 2004; Berlanga-Robles et al. 2011). Future management decisions must buffer the impacts of anthropogenic activity on species diversity and foster the resistance and recovery of ecosystem services to ensure sustainability. There is a need to restore marine biodiversity through sustainable fisheries management, pollution control, maintenance of essential habitats, and the creation of marine reserves. To have a true sustainability, some areas must be restored in agreement with Mexican laws. Future human developments (e.g., aquaculture), must require apriori ecologic and economic evaluations of coastal to avoid additional impacts and to provide connectivity among wetlands and maintain the water and sediment flow within the intertidal zone. It is necessary to understand that human activities impact more than their intended targets. Studies like ours help to elucidate that fisheries resources are affected not only by the fishery itself, but also by other anthropogenic activities that can disturb the food web and ultimately, the fishery

5.5 Conclusions

The $\delta^{13}C$ and $\delta^{15}N$ composition of SSOM and SPOM reflected both terrestrial and marine organic- matter sources in the EUL ecosystem. Sediment suspended organic matter and SPOM results indicate that dissolved and particulate organic matter as, as recycled through the detrital pool, are the main food sources utilized by primary producers and primary consumers. Additionally, primary consumers preferentially consumed distinct portions of the heterogeneous SSOM and SPOM pools. In spite of the occurrence of a main pathway of organic matter transfer, our data also suggest the existence of a direct transfer of SPOM to phytoplankton and then zooplankton, and a direct transfer of SSOM to filter-feeding organisms.

Stable isotopic ratios varied widely in the EUL food web due to many complex relationships. Nevertheless, there was a significant noticeable gradient of ^{15}N-enrichment throughout the EUL food web. High $\delta^{13}C$ and $\delta^{15}N$ signatures of macroalgae were determined, which provides evidence that anthropogenic nutrient enrichment the EUL ecosystem. Based on ^{15}N, we estimated that the EUL food web is composed of five trophic levels. Except for mangroves and macroalgae, organisms were generally organized in the food web consistent with their recognized feeding habits. Our results additionally suggest that isotopic variability among species was related more to functional feeding groups than to habitats or taxonomic groups. Further, fishes were strongly dependent on macrobenthos- and pelagic- derived nutrition. The top predator in the EUL food web was the cormorant, which was consistent with its feeding habits as a tertiary consumer. In all, our results provide a valuable contribution to the knowledge of the trophic structure of a subtropical coastal lagoon ecosystem impacted by anthropogenic nutrients in Mexico.

References

Alonso-Rodríguez R, Páez-Osuna F, Cortés-Altamirano R (2000) Trophic conditions and stoichiometric nutrient balance in subtropical waters influenced by municipal sewage effluents in Mazatlán Bay (SE Gulf of California). Mar Pollut Bull 40:331–339

Amezcua-Linares F (1996) Peces demersales de la plataforma continental del Pacífico central de México. ICMYL- UNAM, Mexico

Barletta M, Barletta-Bergan A, Saint-Paul U (1998) Description of the fishery structure in the mangrove dominated region of Bragança (State of Pará-north Brazil). Ekotropica 4:41–53

Berlanga-Robles CA, Ruiz-Luna A, Hernández-Guzmán R (2011) Impact of shrimp farming on mangrove forest and other coastal wetlands: the case of Mexico. In: Sladonja B (ed) Aquaculture and the environment. InTech. doi: 10.5772/2463, pp 17–22

Bouillon S, Mohan ChP, Sreenivas N et al (2000) Sources of suspended matter and selective feeding by zooplankton in an estuarine mangrove ecosystem, as traced by stable isotopes. Mar Ecol Prog Ser 208:79–92

Cabana G, Rasmussen JB (1996) Comparison of aquatic food chains using nitrogen isotopes. Proc Natl Acad Sci 93:10844–10847

Cabral HN (2000) Comparative feeding ecology of sympatric *Solea solea* and *S. senegalensis*, within the nursery areas of the Tagus estuary, Portugal. J Fish Biol 57:1550–1562

Carlier A, Riera P, Amouroux JM et al (2007) Benthic trophic network in the Bay of Banyuls-sur-Mer (northwest Mediterranean, France): an assessment based on stable carbon and nitrogen isotopes analysis. Est Coast Shelf Sci 72:1–15

Carlier A, Riera P, Amouroux JM et al (2008) Food web structure of two Mediterranean lagoons under varying degree of eutrophication. J Sea Res 60:264–275

Choy EJ, An S, Kang CK (2008) Pathways of organic matter through food webs of diverse habitats in the regulated Nakdong River estuary (Korea). Est Coast Shelf Sci 78:215–226

Costanzo SD, O'Donohue MJ, Dennison WC (2004) Assessing the influence and distribution of shrimp pond effluent in a tidal mangrove creek in north-east Australia. Mar Pollut Bull 48:514–525

DeNiro MJ, Epstein S (1981) Influence of diet on the distribution of nitrogen isotopes in animals. Geochim Cosmo Acta 42:495–506

Elliott M, Hemingway K (2002) Fishes in Estuaries. Blackwells, London

Flores-Verdugo F, Briseño-Dueñas R, González-Farias F et al (1996) Balance de carbono en un ecosistema lagunar estuarino de boca efímera de la costa noroccidental de México. Temas Selectos de Oceanografía Biológica en México, Universidad Autónoma de Baja California, Mexico

Forero MG, Hobson KA (2003) Using stable isotopes of nitrogen and carbon to study seabird ecology: applications in the Mediterranean seabird community. Sci en Mar 67:23–32

France RL (1995) Differentiation between littoral and pelagic foodwebs in lakes using stable carbon isotopes. Limn Ocean 40:1310–1313

Fry B (1996) 13C/12C fractionation by marine diatoms. Mar Ecol Prog Ser 134:283–294

Garcia E, Duarte CM, Middelburg JJ (2002) Carbon and nutrient deposition in a Mediterranean seagrass (*Posidonia oceanica*) meadow. Limn Ocean 47:23–32

Glantz SA (2002) Primer of Biostatistics. McGraw-Hill, USA

Goering J, Alexander V, Haubenstock N (1990) Seasonal variability of stable carbon and nitrogen isotope ratios of organisms in a North Pacific Bay. Est Coast Shelf Sci 30:239–260

Green-Ruiz C, Páez-Osuna F (2006) Heavy metals and clay minerals in surface sediments of Mazatlán bay, Mexican subtropical Pacific coast. Int J Env Poll 26(1–3):201–219

Hansson S, Hobbie JE, Elmgren R et al (1997) The stable nitrogen isotope ratio as a marker of food-web interactions and fish migration. Ecology 78:2249–2257

Heaton THE (1986) Isotopic studies of nitrogen pollution in the hydrosphere and atmosphere: a review. Chem Geol 59:87–102

Hesslein RH, Hallard KA, Ramlal P (1993) Replacement of sulfur, carbon, and nitrogen in tissue of growing broad whitefish (Coregonus nasus) in response to a change in diet traced by $\delta^{34}S$, $\delta^{13}C$, and $\delta^{15}N$. Can J Fish Aqua Sci 50:2071–2076

Hobson KA, Welch HE (1992) Determination of trophic relationship within a high arctic marine food web using $\delta^{13}C$ and $\delta^{15}N$ analysis. Mar Ecol Prog Ser 84:9–18

Hobson KA (1993) Trophic relationship among high artic seabirds: insights from tissue-dependent stable-isotope models. Mar Ecol Prog Ser 95:7–18

Hobson KA, Piatt J.F, Pitocchelli J (1994) Using stable isotopes to determine sea-bird trophic relationships. J Animal Ecol 63:786–798

Hobson KA, Fisk A, Karnovsky N et al (2002) A stable isotope ($\delta^{13}C$, $\delta^{15}N$) model for the North water food web: implications for evaluating trophodynamics and the flow of energy and contaminants. Deep-Sea Res II 49:5131–5150

Hodum PJ, Hobson KA (2000) Trophic relationships among Antartic fulmarine pretels: insights into dietary overlap and chick provisioning strategies inferred from stable- isotope ($\delta^{15}N$ and $\delta^{13}C$) analyses. Mar Ecol Progs Ser 198:273–281

Iken K, Brey T, Wand U et al (2001) Food web structure of the benthic community at the Porcupine Abyssal Plain (NE Atlantic): a stable isotope analysis. Prog Ocean 50:383–405

Jones AB, O'Donohue MJ, Udy J et al (2001) Assessing ecological impacts of shrimp and sewage effluent indicators with standard water quality analyses. Est Coast Shelf Sci 52:91–109

Kling GW, Fry B, O'Brien WJ (1992) Stable isotopes and planktonic trophic structure in arctic lakes. Ecology 73:561–566

Kon K, Kawakubo N, Aoki J et al (2009). Effect of shrimp farming organic waste on food availability for deposit feeder crabs in a mangrove estuary, based on stable isotope analysis. Fish Sci 75:715–722

Laffaille P, Lefeuvre JC, Schricke MT, et al (2001) Feeding ecology of group sea bass, *Dicentrarchus labrax*, in salt marshes of Mont Saint Michel Bay (France). Estuaries 24:116–125

de la Lanza Espino G, Salaya-Ávila JJ, Varsi E (1994) Manejo y aprovechamiento acuícola de lagunas costeras en América Latina y el Caribe. FAO Technical Report No.10, Rome. Available via DIALOG. http://www.fao.org/docrep/field/003/AB485S/AB485S05.htm. Accessed 8 Dec 2012

Lapointe B, Barile PJ, Littler MM, et al (2005) Macroalgal blooms in southeast Florida coral reefs II. Cross-shelf discrimination of nitrogen sources indicates widespread assimilation of sewage nitrogen. Harm Algae 4:1106–1122

Maranger R, Caraco N, Duhamel J et al (2008) Nitrogen transfer from sea to land via commercial fisheries. Nature Geosci 1:111–113

McClelland JW, Valiela I (1998) Linking nitrogen in estuarine producers to land-derived sources. Limn Ocean 43:577–585

McLusky DS, Elliott M (2004) The estuarine ecosystem: ecology, threats and management. Oxford University Press, Oxford

Minigawa M, Wada E (1984) Stepwise enrichment of ^{15}N along food chains: further evidence and the relation between ^{15}N and animal age. Geoch Cosmo Acta 48:1135–1140

Mizutani H, Fukuda M, Kabaya Y (1992) ^{13}C and ^{15}N enrichment factors of feathers of 11 species of adult birds. Ecology 73:1391–1395

Montoya JP, Horrigan SG, McCarthy JJ (1990) Natural abundance of $\delta^{15}N$ in par-ticulate nitrogen and zooplankton in the Chesapeake Bay. Mar Ecol Prog Ser 65:35–61

Ochoa-Izaguirre MJ, Carballo JL, Páez-Osuna F (2002) Qualitative changes in macroalgal assemblages under two contrasting climatic conditions in a subtropical estuary. Bot Marina 45:130–138

Páez-Osuna F, Guerrero-Galván SR, Ruiz-Fernández AC et al (1997) Fluxes and mass balances of nutrients in a semi-intensive shrimp farm in northwestern Mexico. Mar Poll Bull. 34:290–297

Páez-Osuna F, Gracia A, Flores-Verdugo F, et al (2003) Shrimp aquaculture development and the environment in the Gulf of California ecoregion. Mar Poll Bull 46:806–815

Pasquaud S, Elie P, Jeantet C, et al (2008) A preliminary investigation of the fish food web in the Gironde estuary, France, using dietary and stable isotope analyses. Est Coast Shelf Sci 78:267–279

Paul RKG (1981) Natural diet, feeding and predatory activity of the crabs *Callinectes arcuatus* and *C. toxotes* (Decapoda, Brachyura, Portunidae). Mar Ecol Prog Ser 6:91–99

Pauly D (1988) Fisheries research and the demersal fisheries of Southeast Asia, In Gulland JA (ed) Fish Population Dynamics (2nd edition). Wiley Interscience, Chichester pp. 329–348

Pauly D, Yañez-Arancibia A (1994) Fisheries in coastal lagoons, In Kjerfve B (ed) Coastal Lagoon Processes. Elsevier Science Publishers, Amsterdam, p. 377–399

Pauly D, Christensen V, Walters C (2000) Ecopath, Ecosim and Ecospace as tools for evaluating ecosystem impact of fisheries. ICES J Mar Sci 57:697–706

Persson A, Hansson LA, Brönmark Ch, et al (2001) Effects of enrichment on simple aquatic food webs. Am Nat 157:654–669

Peterson BJ, Fry B (1987) Stable isotopes in ecosystem studies. Annu Rev Ecol Evol Syst 18:293–320

Pinnegar JK, Polunin NVC (2000) Contributions of stable-isotope data to elucidating food webs of Mediterranean rocky littoral fishes. Oecologia 122:399–409

Piñón-Gimate A, Soto-Jiménez MF, Ochoa-Izaguirre MJ et al (2009) Macroalgae blooms and $\delta^{15}N$ in subtropical coastal lagoons from the Southeastern Gulf of California: Discrimination among agricultural, shrimp farm and sewage effluents. Mar Pollut Bull 58:1144–1151

Polis GA, Anderson WB, Holt RD (1997) Toward an integration of landscape and food web ecology: the dynamics of spatially subsidized food webs. Ann Rev Ecol Syst 28:289–316

Post DM, Pace ML, Hairston Jr NG (2000) Ecosystem size determines food-chain length in lakes. Nature 405:1047–1049

Raven JA, Walker DI, Johnston AM et al (1995) Implications of ^{13}C natural abundance measurements for photosynthetic performance by marine macrophytes in their natural environment. Mar Ecol Prog Ser 123:193–205

Quijano-Poumián M, Rodríguez-Aragón B (2004) El marco legal de la zona costera. In: Rivera-Arriaga E, Villalobos GJ, Azuz-Adeath I et al (eds) El manejo costero en México. Universidad Autónoma de Campeche, SEMARNAT, CETYS-Universidad, Universidad de Quintana Roo, pp 69–84

Riera P, Richard P (1997) Temporal variation of $\delta^{13}C$ in particulate organic matter and oyster *Crassostrea gigas* in Marennes-Oléron Bay (France): effect of freshwater inflow. Mar Ecol Prog Ser 147:105–115

Riera P (2007) Trophic subsidies of *Crassostrea gigas*, *Mytilus edulis* and *Crepidula fornicata* in the Bay of Mont Saint Michel (France): a $\delta^{13}C$ and $\delta^{15}N$ investigation. Est Coast Shelf Sci 72:33–41

Serrano-Grijalva L, Sánchez-Carrillo S, Angeler DG et al (2011) Effects of shrimp-farm effluents on the food web structure in subtropical coastal lagoons. J Exp Mar Biol Ecol 402:65–74

Soto-Jiménez MF, Páez-Osuna F (2001) Distribution and normalization of heavy metal concentrations in mangrove and lagoonal sediments from Mazatlan Harbor (SE Gulf of California). Est Coastal Shelf Sci 53:259–274

Soto-Jimenez MF, Paez-Osuna F, Bojorquez-Leyva H (2003) Nutrient cycling at the sediment–water interface and in sediments at Chiricahueto marsh: a subtropical ecosystem associated with agricultural land uses. Water Res 37:719–728

StatSoft (1996) STATISTICA for Windows computer program, version 6.0. StatSoft, Tulsa

Tieszen LL, Boutton TW, Tesdahl KG et al (1983) Fractionation and turnover of stable carbon isotopes in animal tissues: implications for δ13C analysis of diet. Oecologia 57:32–37

Valiela I (1995) Marine ecological processes. Springer Verlag, New York

Vega-Cendejas ME, Arreguín-Sánchez F (2001) Energy fluxes in a mangrove ecosystem from a coastal lagoon in Yucatán, Peninsula, México. Ecol Mod 137:119–133

Wilkie ML, Fortuna S (2004) Status and Trends in Mangrove area Extent World-wide. Forest Resources Development Service, Forestry Department. FAO, Rome. Available via DIALOG. http://www.fao.org/do. Accessed 8 Dec 2012

Worm B, Barbier EB, Beaumont N et al (2006) Impacts of biodiversity loss on ocean ecosystem services. Science 314:787–790

Zetina-Rejón MJ, Arreguín-Sánchez F, Chávez EA (2003) Trophic structure and flows of energy in the Huizache-Caimanero lagoon complex on the Pacific coast of Mexico. Est Coast Shelf Sci 57:803–815

Changes in the Hydrological Regime of Coastal Lagoons Affect Mangroves and Small Scale Fisheries: The Case of the Mangrove-Estuarine Complex of Marismas Nacionales (Pacific Coast of Mexico)

6

Francisco Flores-Verdugo, Felipe Amezcua, John M. Kovacs, David Serrano and Manuel Blanco-Correa

Abstract

The estuarine system of Marismas Nacionales (Pacific coast of Mexico) is a Biosphere Reserve that is considered to be the most extensive mangrove region in the American Pacific, and an important region for artisanal fisheries since prehispanic times. Significant hydrological changes occurred in this region after 1976 when an artificial channel was constructed to connect the sea to the main body of the estuary. The channel was originally build to be 40 m wide by 2 m deep, but erosion of the sand barrier (consequence of strong ebb currents) caused the formation of a channel that is presently more than 700 m wide and 20 m deep. The consequent hydrological shift caused the mortality of more than 15,000 ha of mangroves; more than 33 % of the mangroves were affected by 1999, and it is considered that the process of deterioration is still happening. On the other hand the opening of the artificial channel improved the shrimp and finfish fisheries in the region: the number of fishing communities increased, as did the shrimp and fish landings, and large populations of pershell (*Atrina maura*) settled in the region. Several publications state the importance of mangrove as a key habitat for sustainable fisheries, but environmental conditions in some mangrove areas can be extremely difficult for the survival of fishes, crustaceans and mollusks (i.e. due to oxygen depletion). Thus, it seems that the role of mangroves is more related to the support of fisheries in surrounding habitats such as tidal channels, intertidal sand banks, seasonal flood plains, coastal lagoons and adjacent marine areas than "in situ" mangroves.

F. Amezcua (✉) · F. Flores-Verdugo
Instituto de Ciencias del Mar y Limnología, Universidad Nacional Autónoma de México, Av. Joel Montes Camarena s/n, 82040 Mazatlan, Sinaloa, México
e-mail: famezcua@ola.icmyl.unam.mx

F. Flores-Verdugo
e-mail: verdugo@ola.icmyl.unam.mx

J. M. Kovacs
Department of Geography, Nippising University, 100 College Dr, North Bay, ON P1B 8L7, Canada
e-mail: johnmk@nipissingu.ca

D. Serrano
Facultad de Ciencias del Mar, Universidad Autónoma de Sinaloa, Paseo Claussen s/n, Centro, 82000 Mazatlán, SIN, México
e-mail: davidsshh99@yahoo.com

M. Blanco-Correa
Investigación y Posgrado, Universidad Autónoma de Nayarit, CEMIC1, Ciudad de la Cultura Amado Nervo, 63190 Tepic, NAY, México
e-mail: mblanco@nayar.uan.mx

F. Amezcua, B. Bellgraph (eds.), *Fisheries Management of Mexican and Central American Estuaries*, Estuaries of the World, DOI 10.1007/978-94-017-8917-2_6, © Springer Science+Business Media Dordrecht 2014

This chapter describes the mangrove deterioration process in the region and proposes the hypothesis that mangroves combined with freshwater inputs, tidal channels and coastal lagoons with tidal influence and seasonal floodplains are required for a mangrove forest to be considered an adequate fish habitat, and therefore an important fishing region. Riverine, fringe and overwash mangrove forests are more beneficial to fisheries than basin and dwarf mangroves that play different ecological functions other than supporting fisheries.

Keywords
Mangrove forest · Hydrology · Estuarine systems · Small scale fisheries · Marismas Nacionales

6.1 Introduction

Mangrove forests and their environmental functions and ecological services are well recognized by human communities living adjacent to these ecosystems, as well as to governmental agencies, nongovernmental organizations (NGO) and the scientific community (Saenger 2002). For example, from the 45,000 t of landed shrimp that Mexico produces annually, it is assumed that about 7–10 % come from coastal lagoons with mangrove forests of Sinaloa and Nayarit; these states are considered one of the most productive shrimp regions in the country (Chapa-Saldaña and Soto-Lopez 1969).

This work describes the environmental changes that occurred in the mangrove-estuarine complex of Marismas Nacionales, the most extensive mangrove region in the American Pacific, as a consequence of changes in the hydrological regime caused by the opening of an artificial channel from 1974 to 1976. The impact and repercussion to the local fisheries is also discussed.

6.2 Mangrove Zonation and Hydroperiod

Flooding conditions and soil salinity have a direct impact on the settlement of mangroves and other wetland vegetation (Lopez-Portillo and Ezcurra 1989; Rico-Gray and Palacios 1996; Lewis 2005; Flores-Verdugo et al. 2007). Each mangrove species settles according to the microtopographic level and flooding conditions (hydroperiod), which causes mangrove zonation to be vertically distributed. Accordingly, red (*Rhizophora mangle*) and white (*Laguncularia racemosa*) mangroves settle at the lowest elevations, black mangrove (*Avicennia germinans*) settles at intermediate elevations, and buttonwood mangrove (*Conocarpus erectus*) settles at the highest elevations. There is not a clear zonation between red and white mangrove; however, red and white mangroves mix together and define a clear zonation apart from black mangrove (Flores-Verdugo et al. 2007). Previous studies undertaken in similar ecosystems in the area have found that red and white mangroves develop in the range of −15.8–+55.2 cm of mean sea level (MSL) and black mangroves settle in the range of 60.7–>68.3 cm above MSL (Monroy-Torres 2013). Tidal forces can also induce residual currents, overtides, tidal asymmetry and amplification or attenuation of oscillation inside coastal lagoons and seasonal flood plains (termed “marismas”) that can all affect mangrove zonation (Serrano et al. 2013).

6.3 Site Description

The estuarine complex of Marismas Nacionales is located in the alluvial plain of the States of Nayarit and southern Sinaloa on the Central Pacific coast of Mexico. This region is also known as Baluarte-Teacapán-Agua Brava-Las Haciendas-San Blas mangrove-estuarine complex, which are the names of all the different coastal lagoons that compose this large ecosystem (Fig. 6.1). The Teacapán and Agua Brava lagoons are the main coastal lagoons in the region. The San Blas lagoon has a unique geomorphological feature that consists of approximately 280 sub-parallel sand ridges formed by successive accretion of narrow beach ridges during the Holocene transgression (3,600–4,750 years ago). For a more complete description of the geomorphic history see Curray et al. (1969).

This large complex comprises approximately 175,300 ha of mangrove wetlands, saltworts (*Salicornia* spp. and *Batis maritima*), vegetated or unvegetated extensive seasonal flood plains, including some saltpans, coastal lagoons and tidal channels (Blanco-Correa et al. 2012). An estimated 80,000 ha is composed solely of mangroves. This complex is also largely composed of white mangrove.

The climate oscillates from dry subtropical to tropical sub-humid with summer rains occurring during the hurricane period (García 1988). The annual rainfall is between 800 and 1,500 mm, with the driest months from November to June and the wet months from July to September. The average annual temperature is 27 °C. Of the five rivers that discharge to the system, only two flow year-round (i.e., Acaponeta and San Pedro rivers).

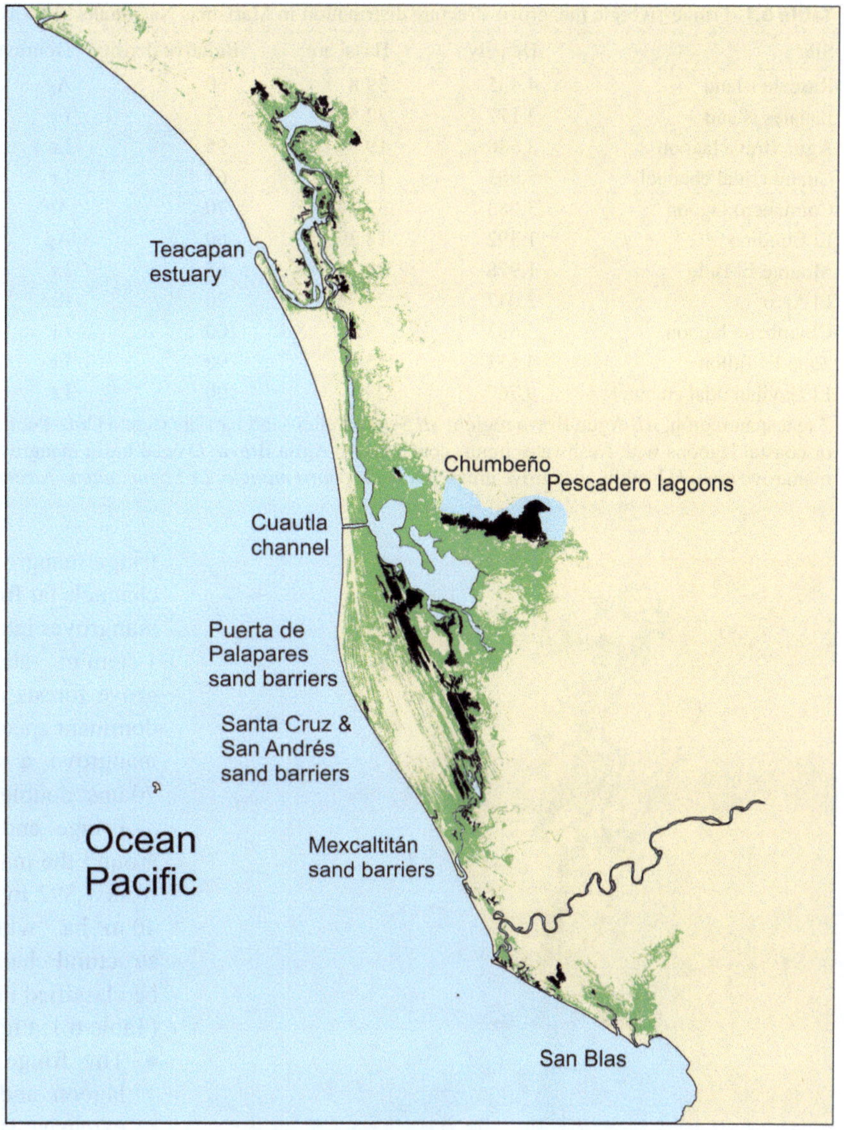

Fig. 6.1 Toponimics of Marismas Nacionales (Nayarit and south of Sinaloa). *Green* corresponds to mangroves and seasonal flood plains (marismas). *Black* corresponds to dead mangrove after the opening of the Cuautla channel

Marismas Nacionales is an important wintering region for migratory birds in the Pacific route (Alaska, Canada and USA) as well as for endangered species such as crocodiles (*Crocodylus acutus*) and jaguar (*Panthera onca*). Human populations have been established in the region since 3,500 BP (preclassic period) and have subsisted on seafood and wildlife (birds, wild boar and deer). Evidence of human settlements include the presence of numerous shell middens that were built with shells of *Tivela* spp, oysters (*Crassostrea* spp) and ark clamps (*Andara grandis*), as well as a 100 m long, 80 m wide, 25 m high shell pyramid located near Teacapán.

From 1974 to 1976 the federal government, through the ministry of Hydric Resources, opened an artificial channel at the Agua Brava coastal lagoon (named the Cuautla Channel) to connect the main water body to the ocean with the purpose of improving fisheries (mainly the shrimp fishery). Originally, the Cuautla Channel was planned to be 3 km long, 40 m wide and 2 m deep, but the tidal effect caused an uncontrolled opening of the channel, which is now more than 700 m wide and 20 m deep. This event has had deleterious effects on mangroves as 33 % of the mangroves were severely affected, and from these, about 18 % suffered mortality.

6.4 Mangrove Forest Distribution and Hurricanes

The region has five of the physiognomic types of mangrove forests as described by Lugo and Snedaker (1974): riverine, fringe, overwash, basin and dwarf mangrove. Dwarf mangrove forests predominate in the seasonal flood plains in 2 modalities: (1) a dense dwarf mangrove forest is located adjacent to a fringe mangrove forest with densities higher than 10,000 stems ha^{-1} and basal areas less than 10 m^2 ha^{-1}, and (2) a dwarf mangrove forest contains a narrow strip of

Table 6.1 Fringe-riverine mangrove structure distribution in Marismas Nacionales, Mexico

Site	Density	Basal area	Relative density	Dominant sp	Type	Reference
Rascale island	4,435	39.8	60	Ag	T	Flores-Verdugo et al. 1997
Panales island	2,177	22.53	33	Lr	T	Flores-Verdugo et al. 1997
Agua Brava lagoon	3,688	19.54	95	Lr	AB	Flores-Verdugo et al. 1997
Cuautla tidal channel	5,963	13.91	67	Lr	AB	Flores-Verdugo et al. 1997
Caimanero lagoon	3,083	37.14	70	Ag	SB	Flores-Verdugo et al. 1997
El Cundino	1,392	18.39	60	Ag	SB	Flores-Verdugo et al. 1997
Mogote El Bule	1,978	28.76	65	Lr	RTC	Flores-Verdugo et al. 1997
El Arco	2,047	22.74	70	Rm	RTC	Flores-Verdugo et al. 1997
Chumbeño lagoon	3,633	9.2	100	Lr	D	Kovacs et al. 2006
Tapo El Tallón	4,633	15.6	90	Lr	P	Kovacs et al. 2006
El Gavilan tidal channel	9,767	29.5	100	Lr	H, RTC	Kovacs et al. 2006

T teacapan región, *AB* Agua Brava región, *SB* Semiparallel sand barriers (Santa Cruz-Puerta de Palapares), *RTC* riverine forest of tidal channels or coastal lagoons with freshwater inputs connected to Agua Brava, *D* dead basin mangrove area (Pescadero-El Chumbeño), *P* poor conditions mangrove area, *H* healthy mangrove areas, *Rm Rhizophora mangle*, *Lr Laguncularia racemosa*, *Ag Avicennia germinans*

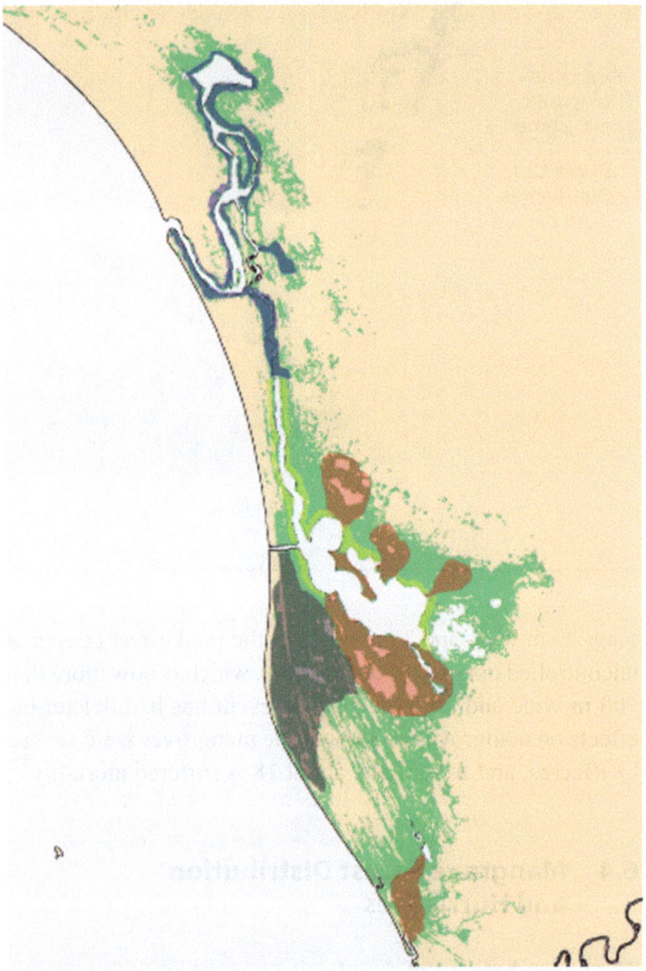

Fig. 6.2 Riverine-fringe mangrove distribution in Marismas Nacionales. *Dark blue* Teacapan region. *Soft gray* parallel sand barriers mangroves. *Pink* riverine mangroves in tidal channels and lagoons surrounding Agua Brava or rivers. *Light green* fringe mangrove of Agua Brava lagoon. *Dark green* basin, dwarf mangrove and seasonal floodplains. The widht of the riverine-fringe mangrove is overestimated for graphic purposes

fringe mangrove forest (3–5 m wide) in the tidal-intertidal channels far from the main water bodies and disperses dwarf mangroves into seasonal flood plains with densities less than 1 stem m^{-2} and basal areas less than 1 m^2 ha^{-1}. Dwarf mangrove forests range in height from less than 1–1.5 m. The dominant species in the dwarf mangrove forest is the black mangrove, a species tolerant to soil salinities as high as 70 ups; double the ups of marine water.

Fringe and riverine mangrove forests are distributed around the main estuarine systems with densities that vary from 1,392 to 6,830 stems ha^{-1} and basal areas from 10 to 40 m^2 ha^{-1} with an exceptional 77 m^2 ha^{-1} in rare cases. The structural characteristics of the fringe mangrove forest can be classified into 4 subsystems (Flores-Verdugo et al. 1997) (Table 6.1; Fig. 6.2):

- The fringe mangrove forest surrounding Agua Brava lagoon and other associated estuarine systems forms extensive stands of predominantly white mangrove with high relative densities of 50 to 90% that is occasionally mixed with red mangrove varying from 6 to 8 m in height. Extensive regions of dead mangroves are found also in fringe forests (Kovacs et al. 2001, 2005; Fig. 6.1).
- The region of Santa Cruz and Puerta de Palapares contains more than 127 semi-parallel sand barriers with wetland-mangrove forest complexes between the sand barriers (locally known as "cañadas"). In this subsystem it is possible to observe the classical zonation of the mangrove forest with white mangrove mixed with red mangrove in the lower elevations, black mangrove in intermediate elevations, and buttonwood at the highest elevations. Black mangroves are the dominant species with relative densities of 60–70% and high basal areas (>12 m^2 ha^{-1}).
- Riverine mangrove forests that are a mix of red and white mangroves. The tallest trees (>20 m) and highest basal areas (>20 m^2 ha^{-1}) are found along tidal and delta channels and coastal lagoons that open into the Agua Brava estuarine system, where freshwater inputs exist.

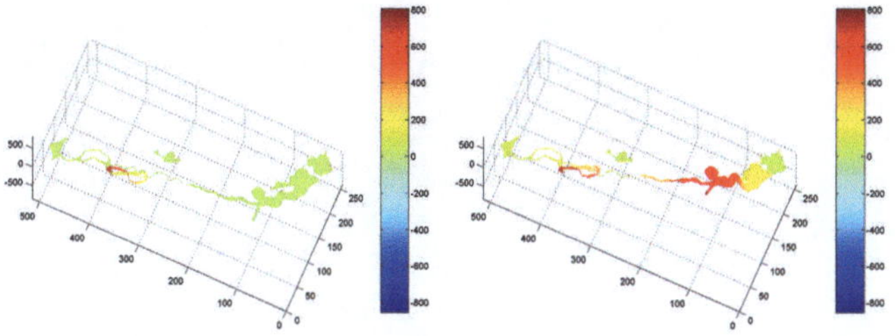

Fig. 6.3 Tidal influence simulation in Marismas Nacionales during high tides (in mm) before and after the opening of the Cuautla channel. 0 mm corresponds to mean sea level (msl)

- Mangroves surrounding the estuary of Teacapán, where the dominant species is black mangrove, with relative densities that vary from 75 to 92%. Mixed red and white mangroves are located adjacent to the main tidal channels. Farther inland, a gradient can be observed of extensive to disperse dwarf black mangroves in the seasonal flood plains containing with saltworts and unvegetated areas.

Regarding the relationship between mangroves and hurricane events, hurricane Rose hit the region in October 1994 with wind speeds of 167 km h^{-1}, causing extreme damage to the mangrove forests in the area. According to Kovacs et al. (2001b), by the year 2000, 44% of the mangroves were recovered and well vegetated, whilst 28% had the main stem broken or uprooted; red and black mangroves were the least affected. It seems that large diameter trees are more susceptible to hurricanes, because a similar observation occurred in Jamaica and Dominica after hurricanes Gilbert and Davis, respectively, affected the mangrove forests (Wunderle et al. 1992; Lugo et al. 1983).

6.5 Mangrove Response to Hydrological and Salinity Changes

Before 1974, the estuarine complex of Marismas Nacionales was connected to the sea through the Teacapán estuary, which is a 75-km long meandering channel located in the north part of the complex. The effect of tides on some parts of the inland wetlands was minimal, such as in the Agua Brava lagoon, which at the time was predominantly a freshwater body with low oxygen levels.

When the Cuautla Channel was opened in 1976, the hydrological pattern was drastically changed; water and soil salinities also changed in the regions of Pescadero and El Chumbeño (Fig. 6.1). Before the opening of the constructed channel, the tidal influence was limited to the northern region in the area of the Teacapán estuarine system (Fig. 6.3a). However, the presence of more than 10,000 ha of flood areas with direct tidal effects produced high velocity ebb flows that induced notorious erosion in the sand barriers. Several attempts to control the channel erosion with rocks and jetties were undertaken but all of them failed to significantly reduce the erosion. In 1985 the channel was 300 m wide and 18 m deep. After hurricane Rose the channel was 500 m wide and 20 m deep. Currently, the channel is more than 700 m wide and 20 m deep (Fuentes-Mariles and Jimenez-Espinoza 2002). Consequently, the tides now influence the northern region of Teacapán as well as the Agua Brava lagoon (Fig. 6.3b).

The increase of marine water intrusion through the Cuautla Channel increased the salinity in the coastal lagoons, from freshwater-oligohaline water (<4 ups) before 1974 to marine (35 ups) and hyperhaline (>100 ups). This increased salinity has had particularly adverse impacts on white mangroves, which were previously the dominant species, but are sensitive to salinity increases (Rollet 1974; Pool et al. 1977). The hydroperiod also changed significantly, from a near-stagnant freshwater swamp to a tidal flooding region with marine saltwater and a low tidal season (spring and beginning of summer). During very low tides, the soil is temporally exposed to the air and evaporation rate increases, causing water salinities greater than 100 ups. Similar hydrological changes due to hurricane impacts have been observed at other mangrove sites like the Yucatán in the Mexican Caribbean when Hurricane Isidore caused significant water salinity changes in 2002 (Batllori and Febles 2007).

Sediment distribution also changed due to construction of the Cuautla Channel. Silt and clay sediments that had accumulated for years have been consistently removed by tidal flows, leaving some places subject to erosion and subsidence. Salinity and subsidence were considered the main cause of the mortality of 15,000 ha of mangroves (18.7% of the total), which occurred primarily in the east region of the Agua Brava coastal lagoon (Fig. 6.1); a similar quantity of mangroves were adversely affected, but did not die (mainly in the estuarine systems of Pescadero and El Chumbeño; Kovacs et al. 2001, 2004). Mangroves in the lagoons between the semi-parallel sand barriers of Puerta de Palapares and Santa Cruz also deteriorated and may be related to the intrusion of sand from the ocean through the Cuautla Channel. However, this deterioration of mangroves was not detected in the region until after 1977 (Rollet 1974; Pool et al. 1977).

6.6 Mangroves, Hydrology and Fisheries: An Integrative Hypothesis

Several authors have highlighted the importance of mangroves as habitat and nursery areas for fish and shrimp, and therefore as important habitat for fisheries (Mumby et al. 2004; Afendi and Chong 2006; Chong 2006). Turner (1977, 1991) estimated an annual loss of approximately 800 kg of fish and shrimp per hectare of destroyed mangrove. Other authors have calculated significant losses (90%) in fisheries yields with mangrove losses in the Philippines in a

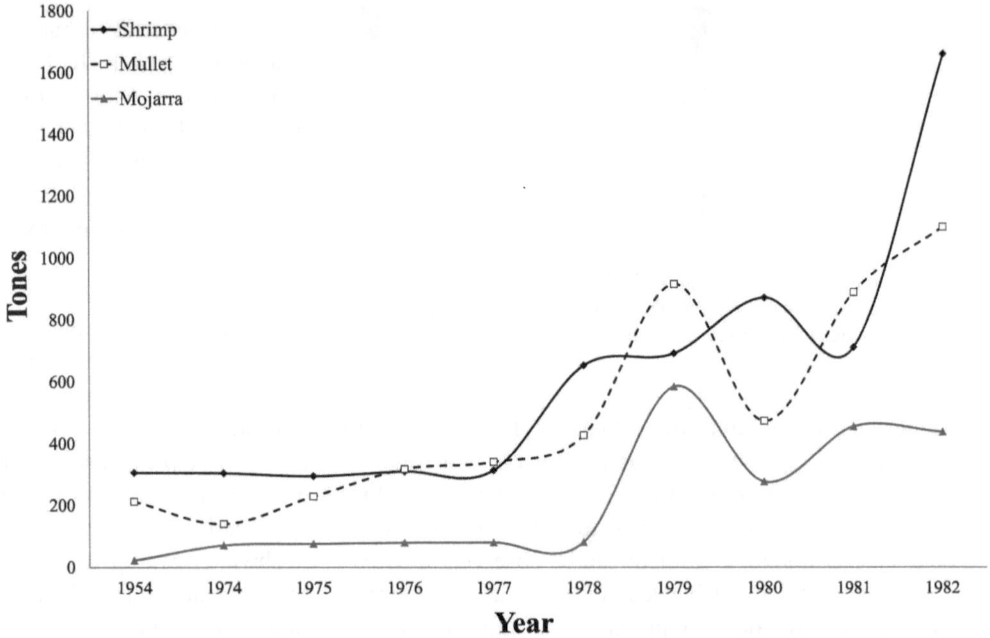

Fig. 6.4 Shrimp, mullet and mojarra landings in the years previous and after the opening of the Cuautla channel

mangrove-cleared area (Altamirano 2006). Aburto-Oropeza et al. (2008) indicated that, particularly for the Gulf of California, fisheries landings were positively related to the local abundance of mangroves and, in particular, to the productive area in the mangrove–water fringe that is used as nursery and/or feeding grounds by many commercially harvested species. However, other authors consider that the relationship between mangroves and fisheries is based on circumstantial evidence and that more experimental and quantitative studies are required to support this argument. Overall, the linkage between fisheries and mangroves remains poorly understood (Blaber 2006).

The importance of mangroves as a feeding ground for fish has been subject to debate in several regions. For example, research in Tanzania with stable isotope ratios of carbon δ^{13}Creveal that fringing mangroves (with lower flooding periods) are less important to the ecosystem food chain and the main carbon basal resources for fishes than submerged mangroves, which are in Teacapán and in the Agua Brava lagoon, where the flooding periods are higher (Lugendo et al. 2007).

In the nearby coastal lagoon of Huizache-Caimanero (a 17,500 ha system with very few mangroves), located approximately 50 km north of our study site, there was a strong correlation between rainfall and shrimp fishery productivity; years with heavy rainfall had higher yields of shrimp (and fishes) than dry years (Chapa-Saldaña and Soto-Lopez 1969). Based on these results, a program was developed to construct man-made channels from the adjacent rivers (Presidio and Baluarte) directly to the lagoon. This resulted in a threefold increase in yield in the season following construction, in comparison to normal years.

Relatedly, the construction of the Cuautla Channel caused several hydrological changes that induced extensive damage to mangrove areas. However, in contradiction to popular belief, the abundance of fishing resources increased. Landings of whiteleg shrimp (*Litopenaeus vannamei*), mullet (*Mugil* spp.) and mojarra (Gerreidae Family) were 294–313, 139–340 and 22–80 t year^{-1}, respectively, from 1954 to 1977. But from 1978 to 1982, once the channel was built and widened by the effect of the tides, landings increased to 652–1659, 427–1100 and 81–487 t year^{-1}, respectively, which represented increases of 5.5, 5 and 5.8 times, respectively. Fish landings peaked in 1982 (Dirección General de Pesca e Industrias Conexas 1954; Departamento de Pesca 1976, 1977; Departamento de Pesca 1980 a,b, 1981; Secretaria de Pesca 1982, 1984, 1985) (Fig. 6.4). It is necessary to consider that although the channel was opened in 1976, it was shallow and narrow and hydrological changes did not occur until the following year when tidal effects widened the channel.

Forced outwelling that occurred with the opening of the Cuautla Channel also gave origin to the colonization and settlement of bivalves. In 1985, a 5-km long bank of penshell, *Atrina maura*, a species unknown in the estuarine complex until the Cuautla Channel was built, appeared in the adjacent marine region and other smaller banks inside the lagoon. These smaller colonies were overharvested in less than a year because of their relatively small size (less than 3 × 1 km).

Based on these observations, it is possible to conclude that the influence of tides and freshwater inputs are of key importance for the relationship between mangroves and fisheries. During flood tides, water with high oxygen, low salinity, and low particulate organic matter (POM) enters the mangrove forests. The flooding also allows access of the area to aquatic species that use the mangroves as feeding grounds. During freshwater-only inputs, the water also

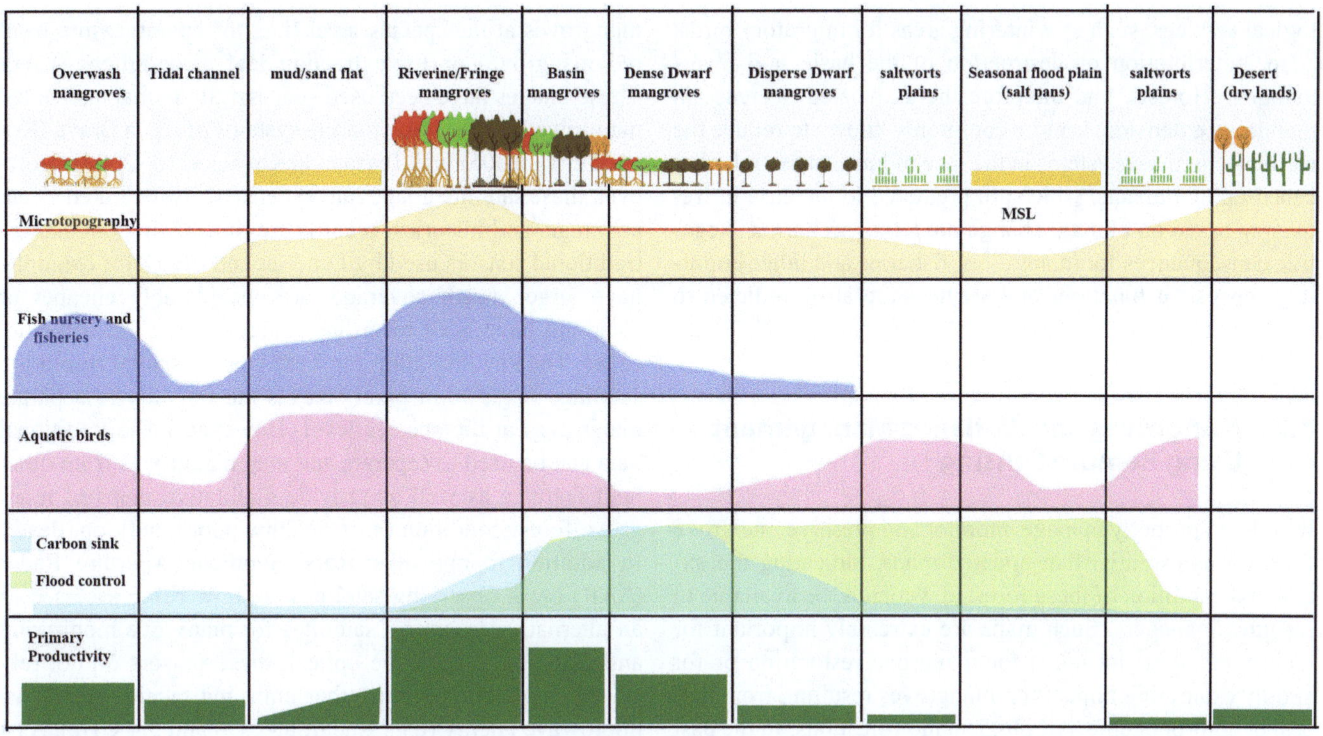

Fig. 6.5 Mangrove physiognomic types and other wetland distribution according to the microtopography and tidal influence and their relation with some of the ecological services they provide. Microtopography profile in a scale from 0 to 2 m. *MSN* mean sea level. Red mangrove (*Rhizophora mangle*) in *red* color, white mangrove (*Laguncularia racemosa*) in *green*, black mangrove (*Avicennia germinans*) in *dark green* and bottomwood mangrove (*Conocarpus erectus*) in clear *brown*. Mangrove mean heights are: overwash 3–4 m, riverine > 10 m, fringe 8 m, basin 4–6 m, dense dwarf mangrove < 1.5 m, disperse dwarf < 1 m. Saltworts: *Salicornia* spp, *Batis maritima*. For ecological services intervals are relative rates between 0 and 100 %

contains a relatively high quantity of nutrients that flood the mangrove areas. During ebb tides, the water oxygen content is lower and contains high POM and higher salinities that move from the mangroves to the neighboring water bodies (tidal channel, coastal lagoon, intertidal mud flats and other habitats). It is well known that during flood tides, juveniles and larvae stages of fish and invertebrates enter the mangroves and during ebb tides, adults and pre-adults move from the mangroves to adjacent water bodies. This creates a net flow of biomass from mangroves to adjacent habitats and improves local fisheries (Afendi and Chong 2006).

A hypothetical model is shown in Fig. 6.5. This model represents the mangrove: water body ratio or tidal influence that would optimize mangroves for fisheries. In this hypothetical model, which is based solely on the observations presented in this chapter, extra freshwater inputs represent higher fisheries yields. In general, a higher spatial heterogeneity (or connectivity) contributes to the creation of favorable conditions for fish, shrimp, and other fishing resources, as well as for mangroves and the surrounding habitats (e.g., lagoons, tidal channels, intertidal mudflats, freshwater inputs and seasonal floodplains).

The opening of the Cuautla Channel caused severe destruction and deterioration of a large percentage of the mangroves in the surrounding system; however, these changes increased fishery yields, and consequently increased the relationship between people and the environment. A similar situation was observed in the Mexican Caribbean when hurricanes Gilbert and Isidore in 1988 and 2002, respectively, changed the coastline by breaching sand barriers in the Yucatan peninsula and connecting the sea with coastal wetlands. This transformational event caused pronounced hydrological changes that ultimately resulted in the introduction of a large diversity of fishes. This situation also caused an improvement for an impoverished human population (Batllori-Sampedro and Febles-Patron 2009). Although a formal study on the socioeconomic aspect of the fishers in the area of Marismas Nacionales has not yet been undertaken, anecdotal evidence indicates that after the opening of the channel, the number of fishers and fishers groups (i.e., cooperatives) has increased.

In conclusion, the types of mangrove forests that have the greatest benefit to fisheries yield are the fringe and riverine mangrove forest types, while dwarf mangrove forests have no direct impacts to fisheries yields. However, basin and dwarf mangroves play an important role as carbon sinks, and together with unvegetated seasonal flood plains play an important role as flooding buffer areas and have other ecological services, such as wintering areas for migratory birds. Also, deterioration or destruction of the basin and dwarf mangrove forests, and therefore the associated decrease in mangrove extension, is more commonly known to reduce the resilience of these systems in the face of hurricanes and other catastrophic impacts, which are predicted to increase in frequency based on climate-change modeling scenarios. Negative consequences for mangroves, fisheries and other important ecosystem functions and services are also predicted to be affected (Kjerfbe and Macintosh 1997; Smith et al. 2012).

6.7 Mangroves and Wetlands Management Using Remote Sensing

In order to properly manage, monitor and preserve mangrove forests it is essential that updated maps, indicating the extent and condition of these forested wetlands, be available to resource managers. Such maps are extremely important for identifying ideal locations for mangrove restoration or for simply examining impacts on mangroves resulting from natural or anthropogenic hydrological modifications. In the past such monitoring activities were extremely costly and time consuming, often involving traditional techniques of aerial photographic interpretation and/or extensive field collection of biophysical (e.g. tree height, tree diameter). This latter approach is logistically difficult given the remoteness of mangrove forests and the harsh environmental conditions associated with these forests (e.g. tidal fluctuations, loose substrate). Consequently, there have been numerous attempts, mainly in Northwest Mexico, to use earth observational satellites to replace this aspect of mangrove forest management (Berlanga-Robles and Ruiz-Luna 2002; Kovacs et al. 2001, 2004, 2005, 2006, 2008, 2009).

Earth observational satellites can provide repetitive coverage of the most remote mangrove forests and because these are digital data, they can be ready for processing (i.e. manipulated) to provide digital classification maps of mangrove types (e.g. dwarf mangrove, tall mangrove). Traditional optical remote sensing platforms such as LandSat MSS, LandSat TM, SPOT and more recently the higher spatial resolution satellites (e.g., 1 m on-the-ground pixel size) such as IKONOS, QuickBird and GeoEye have been used. These sensors collect information from the visible area of the electromagnetic spectrum (~400–700 nm) as well as the non-visible near infrared (~760–900 nm). This later region of the electromagnetic spectrum is extremely useful for monitoring the health of vegetation including mangroves. In particular, very high reflectance in the infrared is indicative of relatively healthy plants and, conversely, low reflectance in the infrared occurs for unhealthy or senescent plants.

Very high spatial resolution images, such as the ones obtained from IKONOS or QuickBird, are required to map mangroves at the species level (i.e., to separate white from red mangrove) or for extracting leaf area indices (LAI). These images have been used successfully to map estimated mangrove LAI for the estuarine systems of Agua Brava (Kovacs et al. 2005) and Teacapan (Kovacs et al. 2009). However, these satellite images are expensive, have limited swath coverage, and historical records are limited. In contrast, the traditional sensors used by LandSat and SPOT for example, have larger swath coverage, are considerably cheaper or free, and have been recording images of the Earth for decades. The key limitation for these coarser spatial resolution satellites (e.g., 30-m pixel size) is the inability to separate mangroves at the species level. However, these traditional data can be used to separate and map mangroves from other land features as well as provide qualitative maps of mangrove forest condition (e.g., healthy, poor condition, dead). In addition to optical sensors, Synthetic Aperture Radar (SAR) Earth observing satellites are now being assessed as an alternative to optical satellites for mangrove monitoring and mapping. Unlike the optical, these sensors do not rely on the Sun's energy but rather emit and receive their own microwave energy (e.g., RadarSat-2 C-band 5.4 GHz). As a result they can process mangrove images at night and, unlike optical sensors, can obtain images in cloudy conditions. In addition, SAR can provide information on the geometry and dielectric properties of the targets it is measuring.

However, only a few studies, many of them conducted in Northwest Mexico (Kovacs et al. 2006, 2008), have so far examined SAR in relation to mangrove forest monitoring. The results suggest that SAR is particularly useful, either alone or in conjunction with optical satellite data, for monitoring the health of mangrove forests undergoing degradation. Specifically, the healthy trees are white (i.e., radar intensity) in the SAR image because the leaves deflect the emitted radar signal back to the satellite. In contrast, in the absence of leaves (i.e., corner reflectors) there is no signal (i.e., dark in image) from the dead stands as the particular emitted radar signal (L-band, HV polarimetry) simply deflects away off the ground, never to return to the satellite.

Several authors using low-resolution LANSAT images have reported a series of disparities with values that made it impossible to evaluate the degree of mangrove deforestation at the national level, partly due to the lack of a significant field validation until 2005–2006 when 770,000 ha of mangroves were estimated (CONABIO 2009).

Until recently the majority of mangrove maps of Marismas Nacionales were obtained from satellite imagery using a simple mangrove classification scheme that did not distinguish mangrove species or condition and may not be useful for conservation and management purpose (Kovacs et al. 2011). The majority of studies undertaken based on these images focused simply on mapping mangroves at very generic and aggregated levels and thus made it difficult to determine if the condition of a certain area of mangroves was improving over time. In fact, some of these studies have hindered efforts to conserve and manage these ecosystems. For example, whilst studies that were validated with field observations show that the system is undergoing extensive degradation (Kovacs el al. 2001, 2009), studies completed using low quality imagery report that some mangrove areas of Marismas Nacionales have recently expanded (Berlanga and Ruiz-Luna 2002). Arguably, these mangrove expansions have not occurred.

6.8 Mangrove Restoration

Mangrove restoration in Marismas Nacionales and other regions has failed, or at best has obtained meager results. This may be because restoration efforts have focused on direct planting, or nursery mangrove plantations, without adequately considering the hydrological characteristics of the region. According to our observations, the hydrological characteristics of mangrove systems are essential for healthy mangroves and the associated ecosystem. Mangrove restoration programs that have focused on hydrological restoration rather than reforestation have occurred in Colombia (Sanchez-Páez et al. 1998, 2004), Florida (Lewis 1982, 2005) and other places. In Mexico, similar restoration has occurred in the Yucatan peninsula (Agraz-Hernández et al. 2007; Batllori-Sampedro et al. 2008; de la Gala-Mendez and Nakawa 2009), northwestern Mexico (Agraz-Hernández 1999) and southern Mexico (Reyes-Chargoy and Tovilla-Hernández 2002).

The hydrological distribution of tides and other factors are not homogeneous in an area as complex and extensive as Marismas Nacionales; however, quantification of the variables that describe the traits of this ecosystem are needed for proper restoration. Blanco-Correa et al. (2012) classified Marismas Nacionales into several hierarchical hydrological units based on tidal region. The first step in the classification process is to understand the functional ecology of each tidal region in order to define strategies that enable sustainable management and conservation of the available resources. Next, a restoration program is formed that is specific to each tidal region and its interdependent ecosystem functions.

Recently, several hydrologic restoration projects have been developed in Marismas Nacionales by the National Commission for Protected Natural Areas (CONANP) in collaboration with fishing communities, the National Commission for Fisheries (CONAPESCA), the National Forestry Commission (CONAFOR); educational and research institutions such as the State University of Nayarit (UAN), the Graduate Studies Center of Montecillo-IPN, the National Autonomous University of Mexico (UNAM) and the University of Nippising (Canada); and national and foreign research institutions including non-governmental organizations (NGOs) such as Pronatura and the World Wildlife Fund (WWF). These restoration projects have included the creation of nurseries to rehabilitate the hydrology of Marismas Nacionales, constructing channels to reconnect natural tidal channels and constructing artificial-island terraces (Anonymous 2010).

The Cuautla Channel has been affecting Marismas Nacionales for more than 30 years. Further channel erosion is expected, but the rate of erosion will likely decrease as the channel becomes wider. In several areas of the system new generations of mangroves have adapted to the new hydrological conditions. If the channel is closed in order to protect the mangroves, as has been proposed by several environmental agencies and NGOs, it is possible that mangroves established after the opening of the Cuautla Channel will be affected. Consequently, the productive fisheries and socioeconomic dynamics may also be negatively impacted.

An alternate and potentially more sustainable approach to mitigating mangrove deterioration may be to concentrate efforts on the hydrological rehabilitation of channeled and terraced sites where swaths of dead mangroves occur. This has the potential to restore some of the ecosystem services and functions of mangroves, rather than simply rehabilitating the mangrove forests that existed previously (Anonymous 2010). Restoring hydrological function is not likely to return Marismas Nacionales to its original state of a monospecific forest of white mangrove. Although this represents a change in the previously held paradigm of mangrove restoration, a new paradigm has been defined that mangrove wetlands should include forests, waterways, mudflats and salt pans (Kjerfbe and Macintosh 1997); that is, a diversity of habitats. Ideally, a diverse habitat consisting of fringe mangrove forests of red, white and black mangrove in the dredge spoils, combined with open waters and mudflats that allow the inclusion of several other mangrove services, such as fisheries, may be the more ideal situation. This diversity of habitats was not well represented in the original monospecific forests of the basin. In all, we propose that hydrological rehabilitation should be focused on reducing soil and water salinity and increasing water movement to mitigate the subsidence processes.

Acknowledgments We thank D. Benitez-Pardo for his help in processing the information.

References

Aburto-Oropeza O, Ezcurra E, Danemann G et al (2008) Mangroves in the Gulf of California increase fishery yields. PNAS 105(30):10456–10459

Afendi N, Chong VC (2006) Shrimp ingress into mangrove forests of different age stands, Matang mangrove forest reserve, Malasya. Bull Mar Sci 80(3):915. (First International symposium on mangroves as fish habitat, Miami. April 19–21, 2006)

Agraz-Hernández CM (1999) Reforestación de manglares en ecosistemas lagunares-estuarinos de la costa nooccidental de México. Dissertation, Universidad Autónoma de Nuevo León

Agraz-Hernández CM, Osti-Saénz J, Jiménez-Zacarías JJ et al (2007) Restauración con manglar: criterios y técnicas hidrológicas de reforestación y forestación. Universidad Autónoma de Campeche, Centro EPOMEX, Comisión Federal de Electricidad and Comisión Nacional Forestal, Mexico

Altamirano JP (2006) Declining mangroves and fisheries in the Batan estuary, Panay island, Central Philippines. In: First international symposium on mangroves as fish habitat, Miami, 19–21 April 2006

Anonymous (2010) Manual comunitario de restauración de manglares: plan de acción regional para la restauración de los manglares de Nayarit. SEMARNAT, CONAFOR, British embassy, DEFRA, Mexico

Batllori E, Febles JL (2007) Changes in the hydrogical characteristics of Chabihau coastal wetlands, Yucatan, Mexico, associated with hurricane Isidora impact. Indian J Mar Sci 36(3):183–192

Batllori-Sampedro E, Febles-Patron JL (2009) Adaptive management response of rural fishery community due to changes in the hydrological regime of a tropical coastal lagoon. J Hum Ecol 26(1):9–18

Batllori-Sampedro E, Ruiz-Barranco HL, Novelo-López J et al (2008) Programa de restauración, conservación y manejo de los humedales costeros de Yucatán (2008–2011). Secretaria de Desarrollo Urbano y Ecología (SEDUMA), Gobierno del Estado de Yucatán, Mexico

Berlanga-Robles CA, Ruiz-Luna A (2002) Land use mapping and change detection in the coastal zone of northwest Mexico using remote sensing technics. J Coast Res 18:514–522

Blaber SJM (2006) Mangrove and fishes: isssue of diversity, dependence and dogma. Bull Mar Sci 80(3):457–472. (First international symposium on mangroves as fish habitat. Miami, Florida, April 19–21 April 2006)

Blanco-Correa M, Flores-Verdugo F, Ortiz-Perez MA et al (2012) Diagnóstico funcional de marismas nacionales: Informe final. Universidad Autónoma de Nayarit, Comisión Nacional Forestal, Tepic

Chapa-Saldaña H, Soto-López R (1969) Resultados preliminares del estudio ecológico y pesquero de las lagunas litorales del sur de Sinaloa. In: Ayala-Castañares A, Phleger FB (eds) Coastal lagoons international simposia proceedings. UNAM-UNESCO, Mexico

Chong VC (2006) Mangrove and fisheries linkages. In: First international symposium on mangroves as fish habitat. Miami, 19–21 April 2006

Curray JR, Emmel FJ, Crampton PJS (1969) Holocene history of a strand plain, lagoonal coast, Nayarit, Mexico. In: Ayala-Castañares A, Phleger FB (eds) Coastal lagoons international simposia proceedings. UNAM-UNESCO, Mexico

CONABIO (2009) Manglares de México: Extensión y distribución. Comisión Nacional para el Conocimiento y uso de la Biodiversidad. CONABIO, Mexico

de la Gala-Mendez JR, Nakawa M (2009) Manual práctico para la rehabilitación del ecosistema de manglares en Yucatán. CONANP, Ria Celestún-JICA, Mexico

Departamento de Pesca (1976) Estadísticas pesqueras 1976. Departamento de Pesca, Mexico

Departamento de Pesca (1977) Estadísticas pesqueras 1977. Departamento de Pesca, Mexico

Departamento de Pesca (1980a) Estadísticas pesqueras 1978. Departamento de Pesca, Mexico

Departamento de Pesca (1980b) Estadísticas pesqueras 1979. Departamento de Pesca, Mexico

Departamento de Pesca (1981) Estadísticas pesqueras 1980. Departamento de Pesca, Mexico

Dirección General de Pesca e Industrias Conexas (1954) Anuario estadístico de actividades pesqueras en aguas territoriales Mexicanas. Secretaria de Marina, Mexico

Flores-Verdugo FJ, Gonzalez-Farías F, Blanco-Correa M et al (1997) The Teacapan-Agua Brava-Marismas Nacionales mangrove ecosystem on the Pacific coast of Mexico. In: Kjerfve B, Drude L, Diop ES (eds) Mangrove ecosystem studies in Latin America and Africa. UNESCO/ISME/USDA, Paris, pp 35–46

Flores-Verdugo F, Moreno-Casasola JP, Agraz-Hernández CM et al (2007) La topografía y el hidroperiodo: dos factores que condicionan la restauración de los humedales costeros. Bol Soc Bot Méx 80:33–47

Fuentes-Mariles OA, Jiménez Espinosa M (2002) Observaciones sobre la erosión playera cerca del poblado palmar de Cuautla, Nayarit. SEGOB-CENAPRED, Mexico

García E (1988) Modificaciones al sistema de clasificación climática de Köppen para adaptarlo a las condiciones de la República Mexicana. Offset Larios, Mexico

Kjerfbe B, Macintosh DJ (1997) The impact of climate change on mangrove ecosystems. In: Kjerfbe B, de Lacerda LD, Salif Diop EH (eds) Mangrove ecosystems studies in Latin America and Africa. UNESCO/ISME/USDA, Paris, pp 1–7

Kovacs JM, Wang J, Blanco-Correa M (2001) Mapping mangrove disturbances using multi-date Landsat TM imagery. Environ Manage 27:763–776

Kovacs JM, Malkczewski J, Flores-Verdugo F (2004) Examining local knowledge of hurricane impact in a mangrove forest using an Analytical Hierarchy Process (AHP) Approach. J Coast Res 20(3):792–800

Kovacs JM, Wang J, Flores-Verdugo F (2005) Mapping mangrove leaf area index at the species level using IKONOS and LAI-2000 sensors for the Agua Brava Lagoon, Mexican Pacific. Estuar Coast Shelf Sci 17:30–37

Kovacs JM, Vanderberg CV, Flores-Verdugo F (2006) Assessing fine beam RADARSAT-1 backscatter from white mangrove (*Laguncularia racemosa* (Gaertner)) canopy. Wetl Ecol Manage 14:401–408

Kovacs JM, Malczewski J, Flores-Verdugo F (2008) Mapping the condition of mangroves of the Mexican Pacific using C-band ENVISAT ASAR and Landsat optical data. Ciencias Marinas 34(4):407–418

Kovacs JM, King JML, Flores-de-Santiago F et al (2009) Evaluation the condition of mangroves forest of the Mexican Pacific based on an estimated leaf area index mapping approach. Environ Monit Assess 157:137–149

Kovacs JM, Liu Y, Zhang C et al (2011) A field based statistical approach for validation a remotely sensed mangrove forest classification scheme. Wetl Ecol Manage 19:409–421

Lewis RR III (1982) Mangrove Forests. In: Lewis RR III (ed) Creation and restoration of coastal plant communities. CRC Press, Boca Raton, pp 153–172

Lewis RR III (2005) Ecological engineering for successful management and restoration of mangrove forest. Ecol Eng 24:403–418

Lopez-Portillo J, Ezcurra E (1989) Response of three mangrove to salinity in two geoforms. Funct Ecol 3:355–361

Lugendo BR, Nagelkerken I, Kruitwagen G et al (2007) Relative importance of mangroves as feeding habitat for fishes: a comparison between mangrove habitats with different settings. Bull Mar Sci 80(3):497–512

Lugo AE, Apllefield M, Pool DJ (1983) The impact of hurricane David on the forest of Dominica. Can J Forest Res 13:201–211

Lugo AE, Snedaker SC (1974) The ecology of mangroves. Annu Rev Ecol Syst 5:39–64

Monroy-Torres M (2013) Determinación de la distribución de tres especies de manglar con relación al hidroperiodo como apoyo en estrategias de reforestación. Dissertation, Universidad Nacional Autónoma de México

Mumby PJ, Edwards AJ, Arias-González JE et al (2004) Mangrove enhance the biomass of coral reef fishes communities in the Caribbean. Nature 427:533–536

Pool DJ, Snedaker SC, Lugo A (1977) Structure of mangrove forests in Florida, Puerto Rico, Mexico and Costa Rica. Biotropica 9:195–212

Reyes-Chargoy MA, Tovilla-Hernández C (2002) Restauración de áreas alteradas de manglar con *Rhizophora mangle* en la costa de Chiapas. Madera y Bosques 8:103–114 (Número Especial)

Rico-Gray V, Palacios M (1996) Salinidad y nivel del agua como factores en la distribución de vegetación en la ciénega del NW de Campeche, México. Acta Bot Mex 34:53–61

Saenger P (2002) Mangrove ecology, silviculture and conservation. Kluwer Academic , Dordrecht

Sanchez-Páez H, Ulloa-Delgado G, Alvarez-León R (1998) Conservación y uso sostenible de los manglares del Caribe Colombiano. Ministerio de Medio Ambiente de Colombia, Asociación Colombiana de Reforestadores and Organización Internacional de Maderas Tropicales, Colombia

Sanchez-Páez H, Ulloa-Delgado G, Arsenio-Tavera HA (2004) Manejo integral de manglares por comunidades locales del Caribe de Colombia. Ministerio de Ambiente, Vivienda y Desarrollo Territorial de Colombia. Dirección de Ecosistemas, Corporación Nacional de Investigación y Foemjto Forestal and Organización Internacional de Maderas Tropicales, Colombia

Secretaria de Pesca (1982) Estadísticas pesqueras 1981. Departamento de Pesca, Mexico

Secretaria de Pesca (1984) Estadísticas pesqueras 1982. Departamento de Pesca, Mexico

Secretaria de Pesca (1985) Estadísticas pesqueras 1983. Secretaria de Pesca, Mexico

Smith JB, Strzepek KM, Cardini J et al (2012) Coping with climate variability and climate change in La Ceiba, Honduras. Clim Change 108:457–470

Rollet B (1974) Introduction a l'étude des mangroves du Mexique. Bois et Foréts des Tropiques 156:3–26

Serrano D, Ramirez-Felix E, Valle-Levinson A (2013) Tidal hydrodynamics in a two-inlet coastal lagoion in the Gulf of California. Cont Shelf Res 63:1–12

Turner RE (1977) Intertidal vegetation and comercial yields of penaeid shrimp. Trans Am Fish Soc 106:411–416

Turner RE (1991) Factors affecting the relative abundance of shrimp in Ecuador. In: Olsen S, Arriaga L (eds) A sustainable shrimp mariculture industry for Ecuador. Technical Report Series TR-E-6, International Coastal Resources Management Project, Univ. of Rhode Island, N.Y., Gobierno de Ecuador, USAID, pp 121–139

Wunderle JM, Lodge DJ, Waider RB (1992) Short-term effects of hurricane Gilbert on terrestrial bird populations on Jamaica. Auk 109:148–166

Effect of Environmental Factors on Zooplankton Abundance and Distribution in River Discharge Influence Areas in the Southern Gulf of Mexico

7

Román Rodolfo Vera-Mendoza and David Alberto Salas-de-León

Abstract

From April 8th to 19th 2010 we studied the hydrodynamic patterns and zooplankton biomass in the Coatzacoalcos River discharge influence area to determine the zooplankton biomass, abundance, and distribution, and its dependence on physical variables. The physical variables were temperature, salinity, relative density (σT) and the speed and direction of currents. Zooplankton biomass was obtained from organism samples taken using zooplankton close-open systems and absolute intensity acoustic data from an Acoustic Doppler Current Profiler (ADCP). The Coatzacoalcos River forms an anticyclonic eddy to the east of the plume and a smaller cyclonic eddy is formed to the west. Currently, velocities (103 mm s^{-1}) measured using the ADCP indicate a highly dynamic area. The vertical positions of the thermocline, pycnocline and halocline indicate that they are modulated by a cold, wind-mixed layer. Zooplankton biomass, abundance and distribution were modulated by physical parameters including salinity, which affected the temporal and spatial distribution of zooplankton biomass. The largest zooplankton biomass concentrations were sampled near the coast and to the west of the Coatzacoalcos river plume, in the mixed base layer.

Keywords

Zooplankton · Gulf of Mexico · Coatzacoalcos · Fisheries productivity · ADCP

7.1 Introduction

Variability in fish productivity is related to the variation in physical and chemical processes that affect the productivity of plankton. For example, nutrient rich upwelling areas off the Peruvian coast cause very high primary production that helps create one of the richest fisheries in the world. Additionally, when the El Niño weather phenomenon occurred in northern Chile, there was not a dramatic decrease in primary production and zooplankton, but rather a change in zooplankton species composition that affected the efficiency of trophic interactions of the top predators (Thiel et al. 2007).

Regions of freshwater influence (ROFI) are of particular significance to fish productivity due to their intermediate hydrodynamic regime between coastal-estuarine and marine areas. Thermal stratification of ROFI's (Fig. 7.1) can be related to the force of the flow caused by differences in horizontal density gradients, vertical mixing caused by tides, and winds and advection due to tidal and coastal currents. Numerical models on these forcing processes have demonstrated characteristics of the shape of ROFI's and the importance that the pattern of currents and densities can have on the distribution and abundance of zooplankton (Dauvin et al. 1998). Due to

R. R. Vera-Mendoza (✉) · D. A. Salas-de-León
Instituto de Ciencias del Mar y Limnología, Universidad Nacional Autónoma de México, Circuito Exterior, Ciudad Universitaria, 04510 México, D.F., México
e-mail: roman_vera@hotmail.com

D. A. Salas-de-León
e-mail: david.alberto.salas.de.leon.@gmail.com

F. Amezcua, B. Bellgraph (eds.), *Fisheries Management of Mexican and Central American Estuaries*, Estuaries of the World, DOI 10.1007/978-94-017-8917-2_7, © Springer Science+Business Media Dordrecht 2014

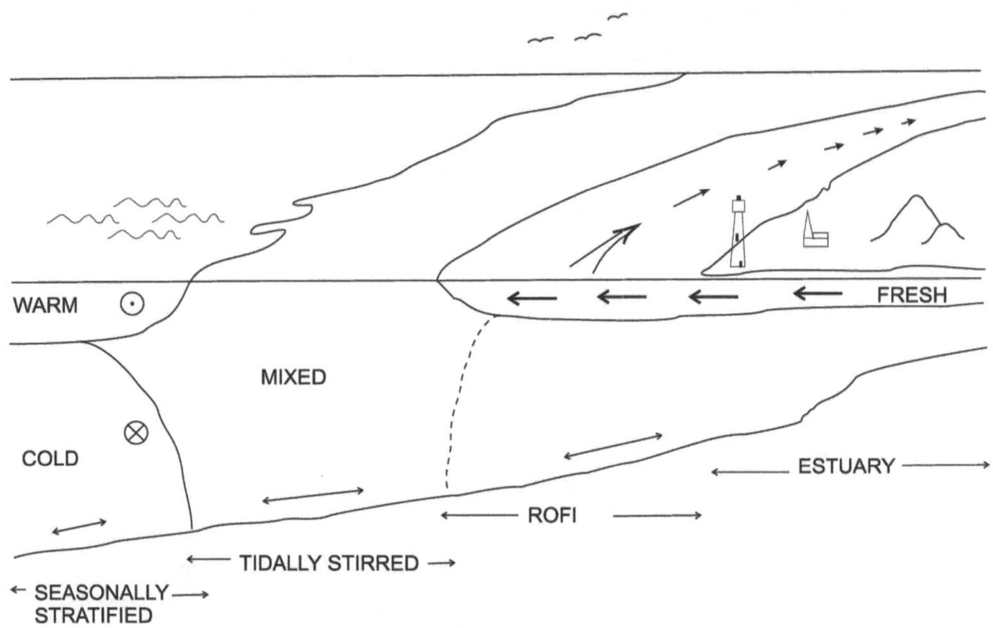

Fig. 7.1 Schematic of the characteristic regimes of shelf and estuary based on Simpson 1997

these physical characteristics, coastal aquatic ecosystems are considered one of the most productive natural systems in the world (Contreras 1993).

Currently there is much emphasis on the study of plankton dynamics and organization because any short or long term alteration of environmental factors that influences plankton dynamics will result in functional changes (Wickstead 1979). Several studies indicate that the behavior of zooplankton is related to physical and chemical variables; however, there is a paucity of information regarding these dynamics in the Coatzacoalcos River Estuary in the southern Gulf of Mexico. Flores-Coto et al. (1988) and Salas-de-León et al. (1998) have noted that in the southern Gulf of Mexico, the highest values of zooplankton biomass and abundance were recorded in coastal areas influenced by fluvial lagoon systems at depths between 20 and 60 m. In these areas, spatiotemporal variation of zooplankton biomass and abundance is affected by mesoscale currents (Salas-de-León et al. 1998; Lara-López 2003).

In the marine region of influence of the Coatzacoalcos River, Bozada and Paéz (1987) observed that the highest densities of icthyoplankton were found in neritic coastal areas where surface temperatures ranged from 26.2 to 28.9 °C and salinities varied from 27.1 to 37 ups. The period of greatest icthyoplankton concentration was in spring and early winter.

The objective of our study was to more completely describe zooplankton dynamics of the Coatzacoalcos River ROFI. We hypothesized that the abundance and biomass of zooplankton would be greatest in the sea adjacent to the Coatzacoalcos River mouth. Our ultimate goal was to describe how changes in environmental factors affected zooplankton and document functional relationships that could be used to predict changes in zooplankton biomass and abundance.

7.2 Materials and Methods

The study area was located in the immediate coastal region at the mouth of the Coatzacoalcos River between Lat 18° and 19° N and Long 94° and 95° W. The Coatzacoalcos River is one of the largest rivers in Mexico with an average discharge of 493 $m^3\ s^{-1}$ and a maximum discharge of 6,737 $m^3\ s^{-1}$ (Riverón-Enzástiga 2008). The maritime area of the Coatzacoalcos River plume is part of the Bay of Campeche whose circulation is predominantly cyclonic (Vázquez-de la Cerda 1979; Monreal-Gómez and Salas-de-León 1990) and is caused by the formation of a gyre in the east part of the bay during the months of August and September. The gyre continues until December and moves in a westerly direction.

Surface temperatures of the study area range between 24° and 29 °C, with no marked seasonal differences (Villalobos and Zamora 1975; Padilla-Pilotze et al. 1986). There are also contributions from inland waters that are usually colder and establish vertical thermal gradients in the coastal zone (Villalobos and Zamora 1975; Czitrom-Bauss et al. 1986; Monreal-Gomez et al. 1992).

During the fall and winter a broad surface-mixed layer forms due to the effect of the North winds, which cause a decrease in temperature of the top 150 m of the water column (Monreal-Gómez and Salas-de-León 1990; Gio-Argáez 2000). During summer, the depth of the mixed layer is thinned by high surface temperatures, which creates a strong vertical gradient and strong thermoclines (Licea and Luna 1999). Salinity varies more than temperature. Low salinity waters reach the Campeche Bank in spring and summer from the northeast to the Yucatan Current (Bogdanov 1969).

Biological and physical data were collected at various sampling locations located throughout the area of influence

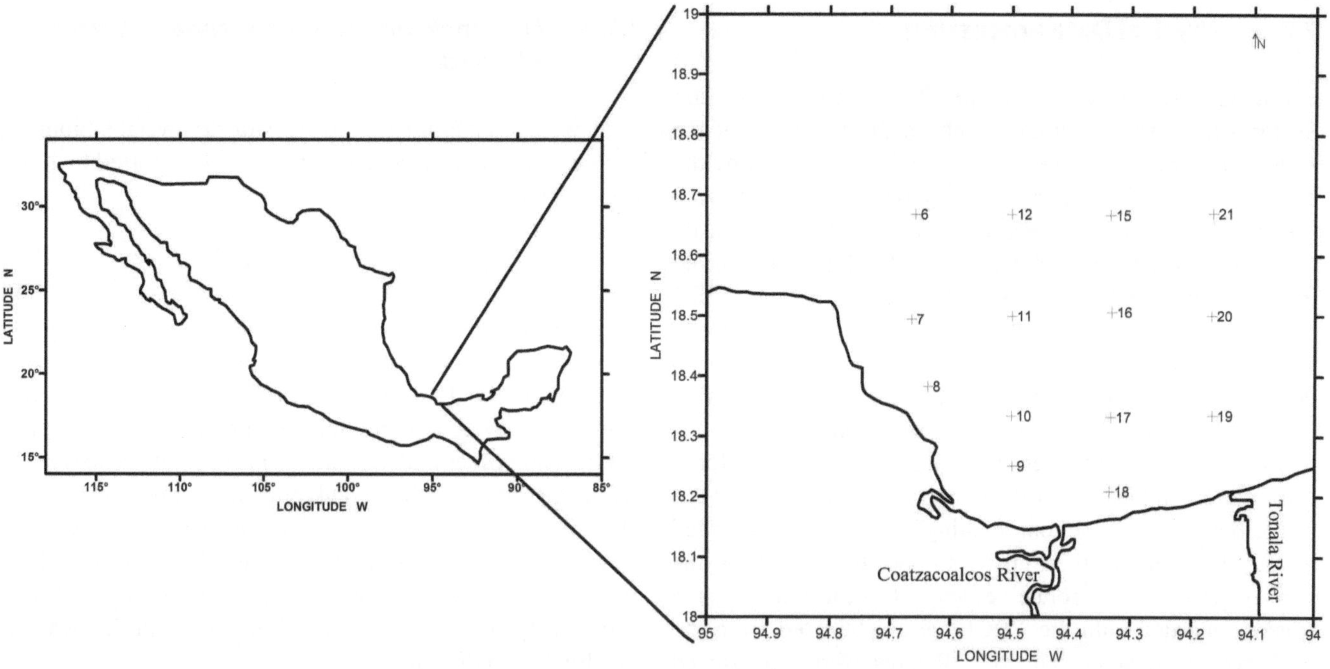

Fig. 7.2 Location of sampling stations at the Coatzacoalcos river mouth and the adjacent area

of the Coatzacoalcos River plume (Fig. 7.2). Sampling was conducted during an oceanographic survey from April 8th to 19th, 2000 in the southern Gulf of Mexico from the research vessel "Justo Sierra", which is operated by the National Autonomous University of Mexico. This data was collectedly conjointly as part of the project "Ocean Processes and Mechanisms of Biological Production in the Southern Gulf of Mexico" (i.e., "PROMEBIO III").

7.2.1 Biological Samples

Zooplankton samples were obtained using a net of 500 μm mesh and with a mouth diameter of 75 cm. The net was trawled at depths of 0–10, 10–20, 20–30, 40–60, 60–80 and 80–100 m, for a period of 15 min at each depth.

7.2.2 Physical Data

Temperature (°C), pressure (db) and conductivity (microSiemens cm^{-1}) were obtained with a conductivity, temperature and depth (CTD) sonde (Oceanic General Neil Brown Mark IV model); the pressure sensor was modified according to recommendations of the World Ocean Circulation Experiment (WOCE) 1990–2002. The CTD was scheduled to record temperature, pressure and conductivity every 0.25 s. The CTD was lowered into the water column at a rate of 1 m s^{-1}; as such, data were obtained at every 25-cm increment within the water column. With the resulting data, we calculated the relative density or sigma-T (σT) in kg m^{-3} and salinity in UPS using the polynomials proposed by UNESCO (Fofonoff and Millard 1983).

7.2.3 Acoustic Data

Acoustic data were collected using an Acoustic Doppler Current Profiler (ADCP; RD Instruments, RD-VM0075 model Mark II) that was designed for fixed installation in ships. The frequency used was 75 kHz at a rate of 0.7 pulses per second; with a sound wave 16-m long.

7.2.4 Biological Data Processing

Zooplankton samples were weighed for biomass estimation (wet weight) using an analytical balance and a suction pump to remove excess water following the method recommended by Zavala-Garcia and Flores-Coto (1989). The estimation of biomass using dry weight is more accurate; however, this method involves the destruction of samples for other analyses. Thus, to find the relationship between wet weight, volume displaced, dry weight and the amount of carbon, we used formulas by Wiebe (1988) that can be used to convert among the above variables. Wet weight was transformed into dry weight using Wiebe's equation (1988):

$$Log10(D_W) = 2.107 + 1.053\,Log10(W_W) \qquad (7.1)$$

Where DW is dry weight and WW is wet weight.

7.2.5 Physical Data Processing

For acoustic data collection, the DAS (Data Acquisition System) program was used to control the functions of the ADCP and profile parameters. Binary files produced by the ADCP were converted to ASCII with the LOGDAS program (version 1.03), which created an array of Echo or Acoustic Gain Control (AGC). Each variable in the formula required transformation to Absolute Acoustic Intensity (AAI, in dB) as well as position data, depth and other control parameters. This array was then processed in a Turbo Basic program to determine AAI (see below) and generate arrays containing latitude, longitude, time, date, depth and biomass.

In areas with depths less than 600 m, the ADCP stored information on the east-west and north-south apparent speed of the bottom (i.e., "bottom tracking"), which corresponded to the true speed of the ship with opposite signs. With the aim of getting the absolute velocity of ocean currents, the speeds recorded with the ADCP were subtracted from the background apparent velocities. Results were constructed with flat distribution of currents in vector form.

Given that ADCP readings were recorded at every 16–20 m depth, the absolute velocities of ocean currents were estimated by the ADCP to depths of 20, 36, 52, 68 and 84 m in order to coincide as much as possible to the depth levels of the zooplankton samples (i.e., 10–20, 20–30, 40–60, 60–80 and 80–100 m).

7.2.5.1 Estimation of Absolute Acoustic Intensity (AAI) from ADCP Data

To transform the echo amplitude of the acoustic signal into AAI, the following formula was applied (RD Instruments 1996):

$$AAI = 10 Log_{10}\left\{\left[\frac{\left[4.47\times10^{-20} k_2 k_s (T_X + 273)\left(10^{\frac{K_C(E-E_r)}{10}} - 1\right)\right]}{cPK_1 10\frac{-2\alpha R}{10}}\right]\right\} \tag{7.2}$$

Where AAI is Absolute Acoustic Intensity (db), K_2 is the system noise factor, K_s is the system constant, T_x is the real-time temperature transducer (°C), K_c is the conversion factor for echo intensity (dB), E is the echo intensity (count), E_r is the electronic thermal noise reference (count), R is the interval addressing the cell depth (m), c is the speed of sound in each cell (m s^{-1}), P is the length of the transmitted pulse, K_1 is the real power of sound in water (W), and α is the sound absorption coefficient of water (dB m^{-1}).

7.2.6 Zooplankton Biomass Estimation from AAI Data

Data were transformed from AAI to dry weight biomass. Biomass was calculated using the formula reported by Lara-López (2003):

$$Log_{10}(DW) = 6.1115 + 0.05518 \times AAI \tag{7.3}$$

Where DW is the biomass dry weight. To observe the correspondence of zooplankton biomass data with respect to the different physical variables at different depths, horizontal scatter plots were drawn with contours delineating variable gradients. In order to determine significant differences in the spatial distribution of zooplankton biomass and the influence of the physical variables, a similarity matrix analysis was undertaken with information on the date and time of sampling, zooplankton biomass, water temperature, salinity and relative density. The similarity coefficient used in the analysis was Euclidean distance.

7.3 Results

7.3.1 Vertical Distribution of Zooplankton Biomass and Physical Variables

Thermoclines were found between depths of 35 and 45 m at stations 6, 11, 12, 15, 16, 20 and 21. The vertical relative density showed a pattern inversely proportional to temperature and because a halocline was not observed, it follows that the relative density was governed by temperature. At stations 7, 8, 9, 10, 17, 18 and 19, thermoclines were not observed. Greater zooplankton biomass was recorded at shallower depths. For example, at station 18, the overall maximum zooplankton biomass of 20.40 mg m^{-3} was recorded at a depth of 5 m whereas the minimum overall biomass of 0.14 mg m^{-3} was recorded at station 12 at 70-m depth (Table 7.1).

7.3.2 Horizontal Distribution of Physical Variables and Zooplankton Biomass

From 0 to 10 m depth in the Coatzacoalcos River plume, the highest concentrations of zooplankton biomass were recorded at stations 9 and 18 with values of 10.33 mg m^{-3} and 20.40 mg m^{-3}, respectively (Fig. 7.3a). A similar distribution was observed for temperature and relative density and coincided with the trend in the vertical distribution;

Table 7.1 Depth (m) of clines (thermocline, halocline and pycnocline), and lowest and highest values of zooplankton biomass (mg m^{-3}) by depth

Station	Termocline depth (m)	Halocline depth (m)	Pycnocline depth (m)	Highest biomass (mg m^{-3})	Depth of highest biomass (m)	Lowest biomass (mg m^{-3})	Depth of lowest biomass (m)
6	35		35	2.29	15	0.94	50
7				1.87	15	0.75	25
8				6.75	5	4.83	15
9				10.33	5	4.21	15
10				3.88	15	3.50	5
11	48		48	4.67	5	2.14	25
12	35		35	6.18	5	0.14	70
15	30		30	1.70	25	0.32	70
16	35		35	4.72	15	2.13	50
17				2.85	25	2.77	15
18				20.40	5	1.99	15
19				6.0	5	4.08	15
20	35		35	3.93	5	2.54	15
21	28		28	2.37	15	1.09	90

temperature influenced the relative density more than salinity. Temperatures were highest near the shore with values of 26.25, 25.75 and 26.55 °C (Fig. 7.3b). Salinity was also high near the coast, with values of 36.12 and 35.76 ups (Fig. 7.3c). Relative density distribution showed a pattern very similar to temperature, with maximum values of 23.80, 23.92 and 23.4 kg m^{-3} (Fig. 7.3d).

The horizontal distribution of zooplankton biomass was very homogeneous at depths of 10–20 m (Fig. 7.4a) with values ranging from 1.21 mg m^{-3} at station 15 to 4.82 mg m^{-3} at station 8. The lowest temperature (25.5 °C) was recorded near the coast and temperature increased towards the open sea (Fig. 7.4b). Salinity was low off the Tonala River mouth (35.75 ups) and a higher concentration was found in the Coatzacoalcos River plume (36.16 ups, Fig. 7.4c). The highest relative density occurred off the mouth of the Coatzacoalcos River (24.3 kg m^{-3}) and was lower in other areas (23.5 kg m^{-3}) (Fig. 7.4d). Again, this indicates that relative density depended mostly on temperature.

At a depth of 20–30 m, zooplankton biomass was very similar among the sampling stations with values ranging from 0.75 mg m^{-3} at station 7 to 4.07 mg m^{-3} at station 17 (Fig. 7.5a). This increase in biomass could be due to the location of station 17, which is near an area where depth increases abruptly. Consequently, this depth variation may contain favorable conditions for the concentration of zooplankton due to upwelling that occurs in this zone. To the west of the Coatzacoalcos River offshore from the Tonala River, the highest temperature was found (26.3 °C) whereas cooler water (25.3 °C) was found offshore from the Coatzacoalcos River mouth; the temperature difference indicates higher discharge of the Coatzacoalcos River (Fig. 7.5b). Salinity was highest in the eastern portion of the study area (36.15 ups) and decreased toward the west (35.8 ups) (Fig. 7.5c). The relative density distribution was the highest far from shore in the western portion of the study area (24.05 kg m^{-3}) and lowest near the coast in the eastern portion of the study area (23.86 kg m^{-3}) (Fig. 7.5d).

There were few sampling stations at depths of 40–60, 60–80 and 80–100 m due to the bathymetry of the area, so figures with contours could not be made (Fig. 7.6a, b, c). The distribution of zooplankton biomass at these three sampling depths varied from 0.13 mg m^{-3} at station 12 at a depth of 60–80 m to 2.13 mg m^{-3} at station 16 at a depth of 40–60 m. Temperature ranged from 21.28 °C at a depth of 80–100 m to 24.07 °C at a depth of 40–60 m. Salinity ranged from 36.20 ups at a depth of 40–60 m to 36.28 ups at a depth of 60–100 m. The relative density was lowest (24.59 kg m^{-3}) at a depth of 40–60 m and highest (25.40 kg m^{-3}) at a depth of 80–100 m.

7.3.2.1 Zooplankton Biomass Data According to AAI

In general, the highest biomass according to AAI occurred to the east of the mouth of the Coatzacoalcos River, except at 84 m depth to the west of the mouth where the continental shelf is very narrow. The highest value of zooplankton biomass was 14.1 g m^{-3} at 20 m (Fig. 7.7a), 30.1 g m^{-3} at 36 m (Fig. 7.7b), 24.1 g m^{-3} at 52 m (Fig. 7.7c), 36.1 g m^{-3} at 68 m and 36.1 g m^{-3} at 84 m (Fig. 7.7d, e). Note that blank areas in Fig. 7.7 indicate that zooplankton biomass was immeasurable at that sampling location.

7.3.2.2 Currents

Between latitude 18.6° and 19° N, a predominant circulation pattern rotating from west to east was measured moving at a speed of approximately 1,200–2,800 mm s^{-1} (Fig. 7.8). Off the Coatzacoalcos River mouth (latitude 18.2–18.5 °N and longitude 94–94.5 °W), anticyclonic rings moved at speeds of 200–600 mm s^{-1}. In general this pattern of current flow was observed at all depths (20, 36, 52, 68 and 84 m) from ADCP data.

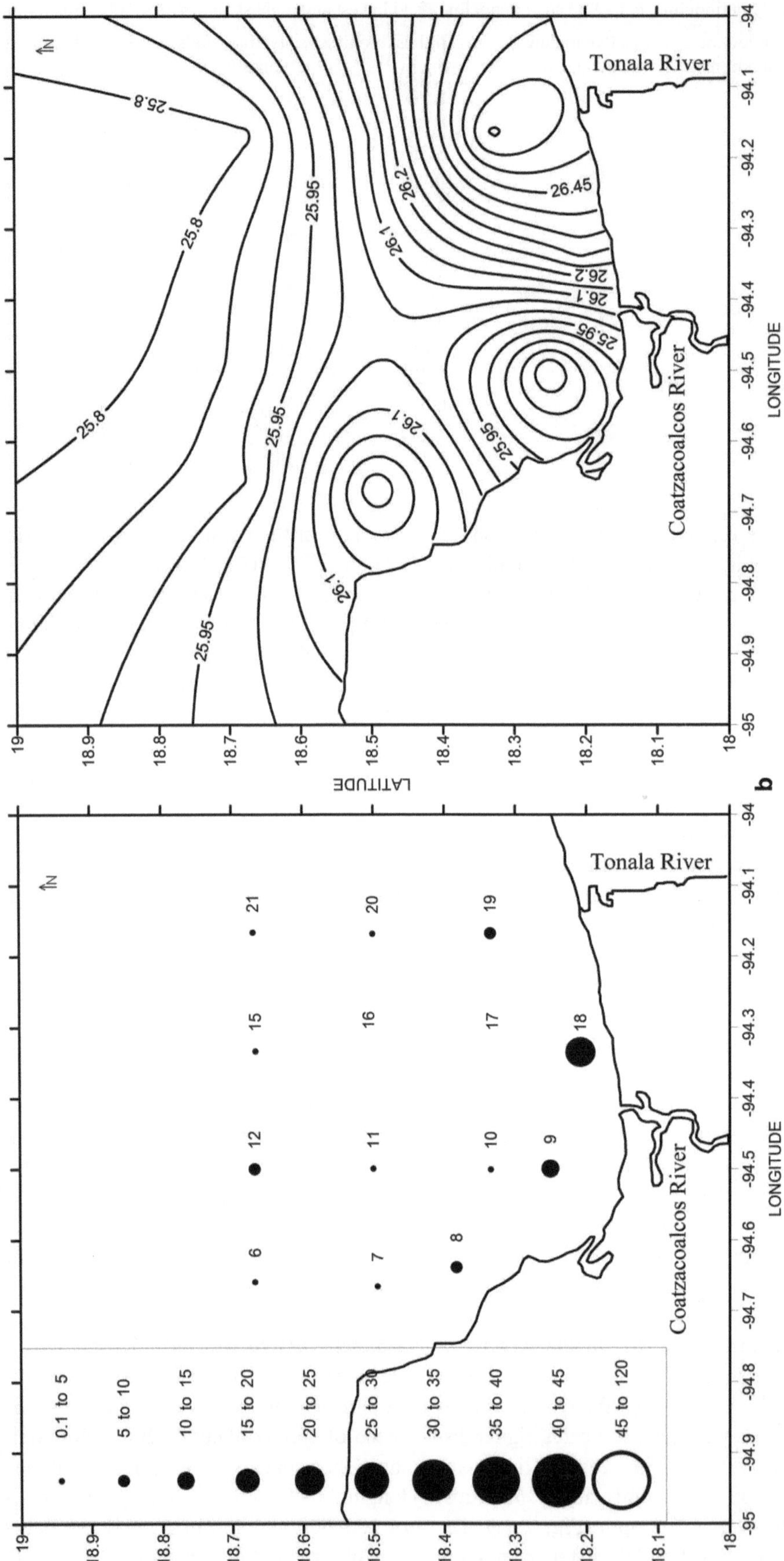

Fig. 7.3 Horizontal distribution: **a** zooplankton biomass (mg m^{-3}), **b** temperature (°C), **c** salinity (ups) and **d** density (kg m^{-3}), at depths of 0–10 m

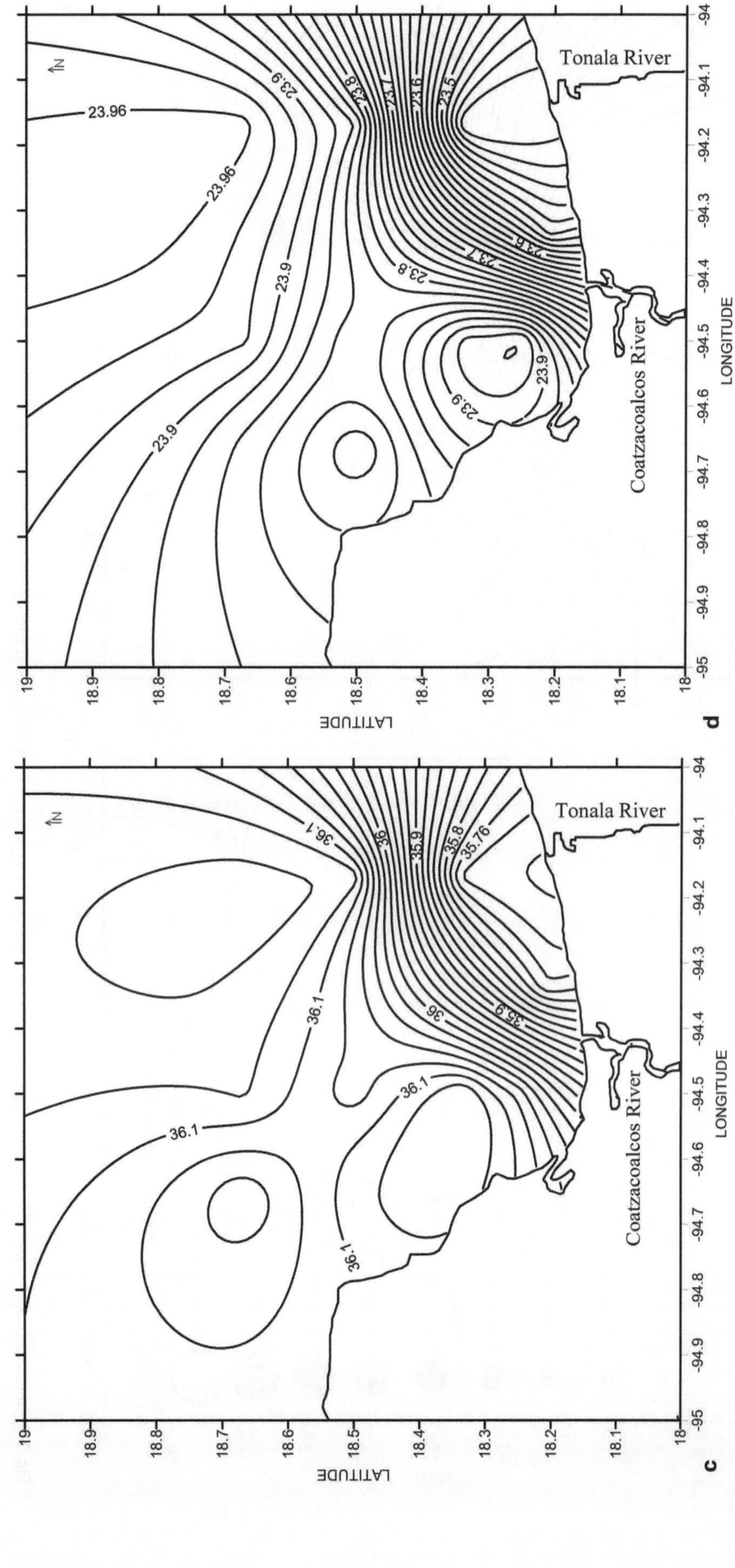

Fig. 7.3 (continued)

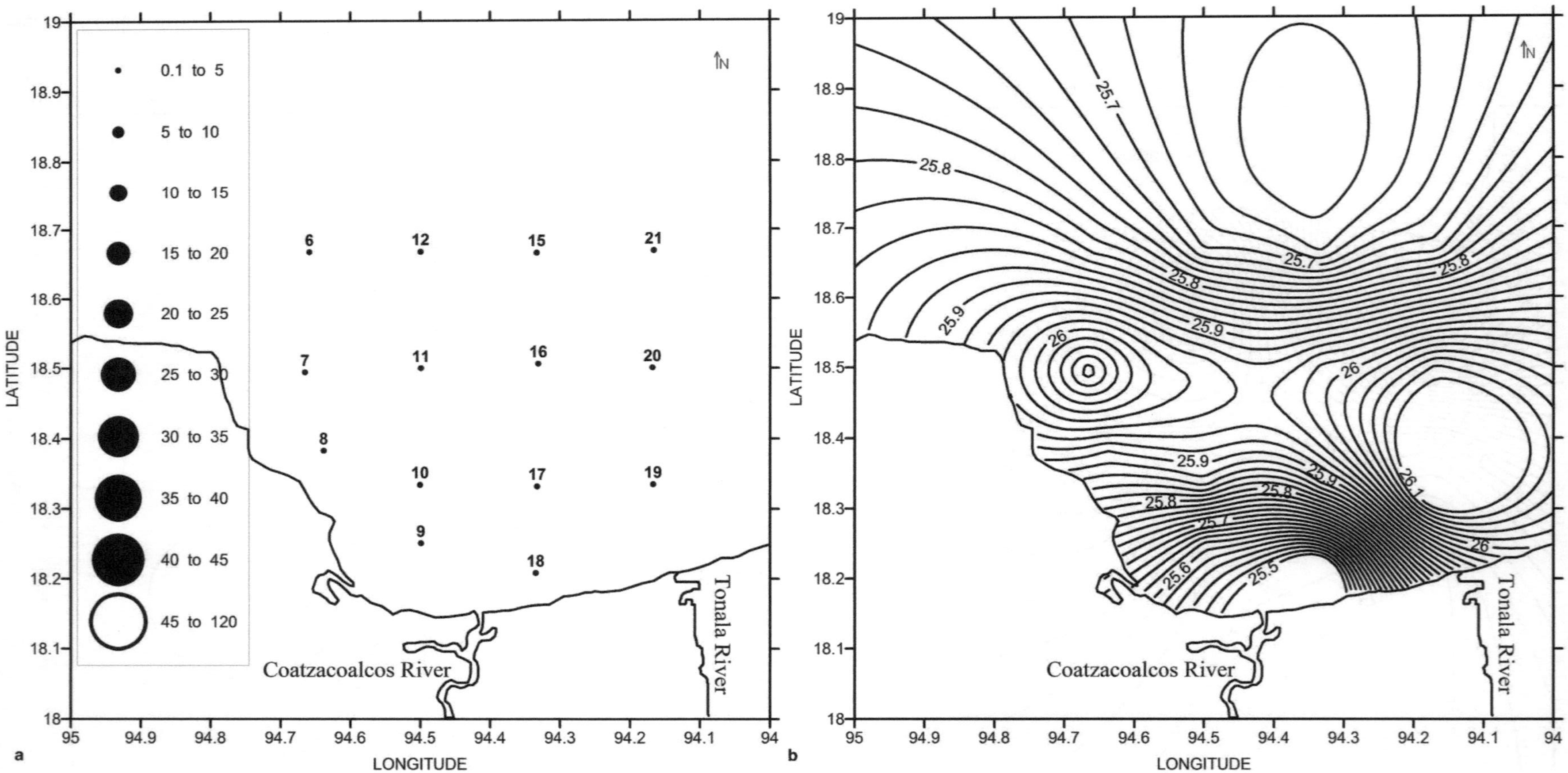

Fig. 7.4 Horizontal distribution: **a** Zooplankton biomass (mg m^{-3}), **b** temperature (°C), **c** salinity (ups), and **d** density (kg m^{-3}), at depths of 10–20 m

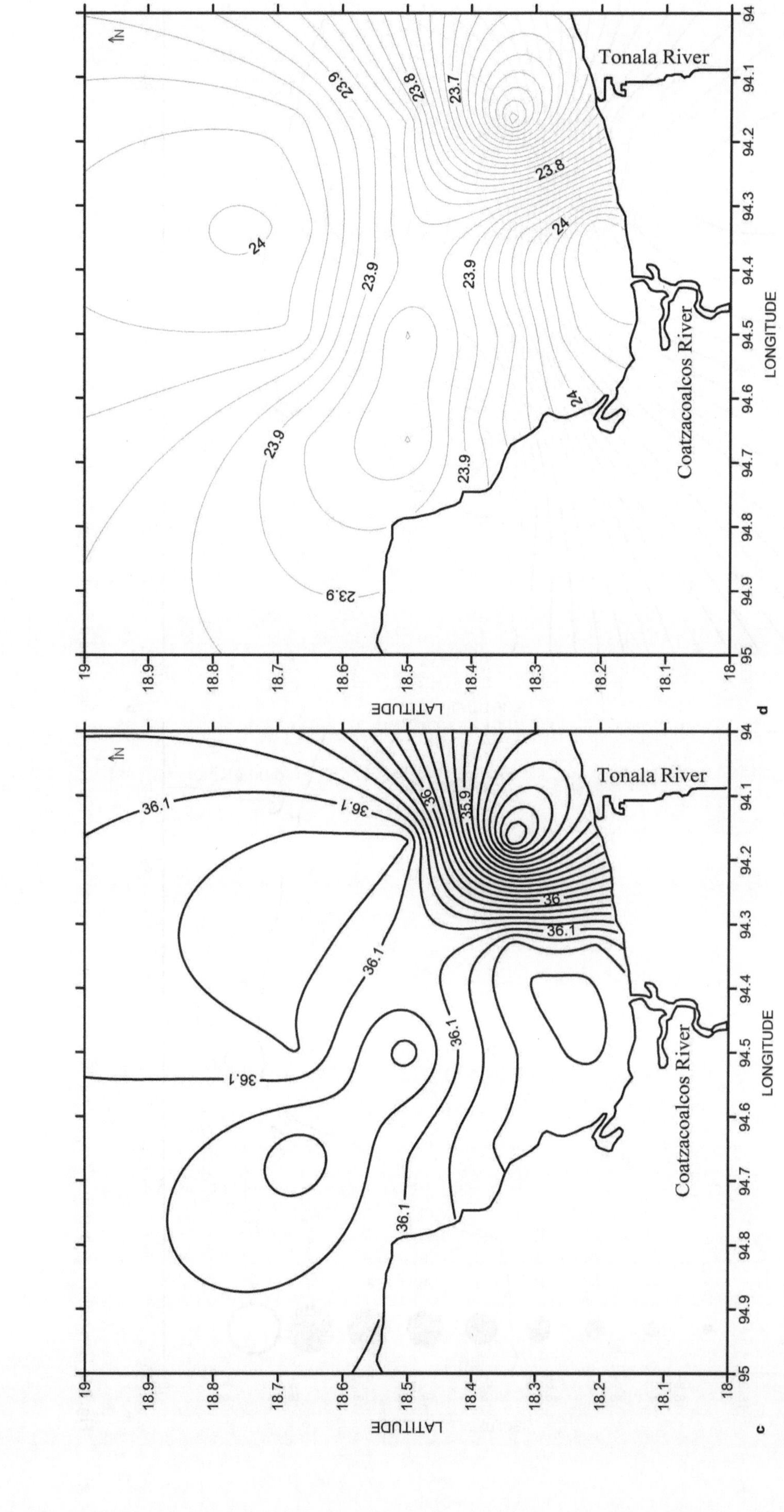

Fig. 7.4 (continued)

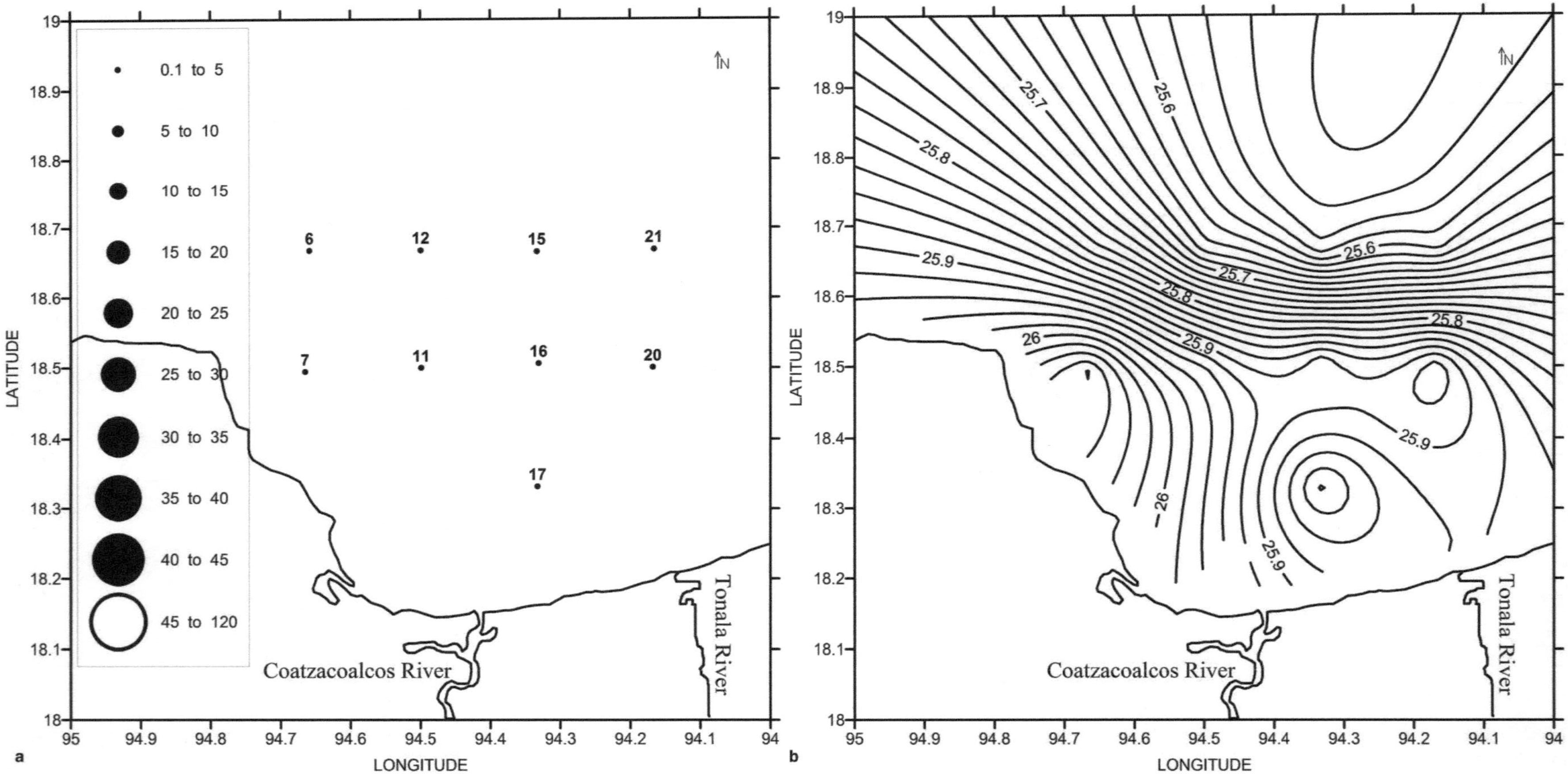

Fig. 7.5 Horizontal distribution: **a** Zooplankton biomass (mg m^{-3}), **b** temperature (°C), **c** salinity (ups), and **d** density (dg m^{-3}), at depths of 20–30 m

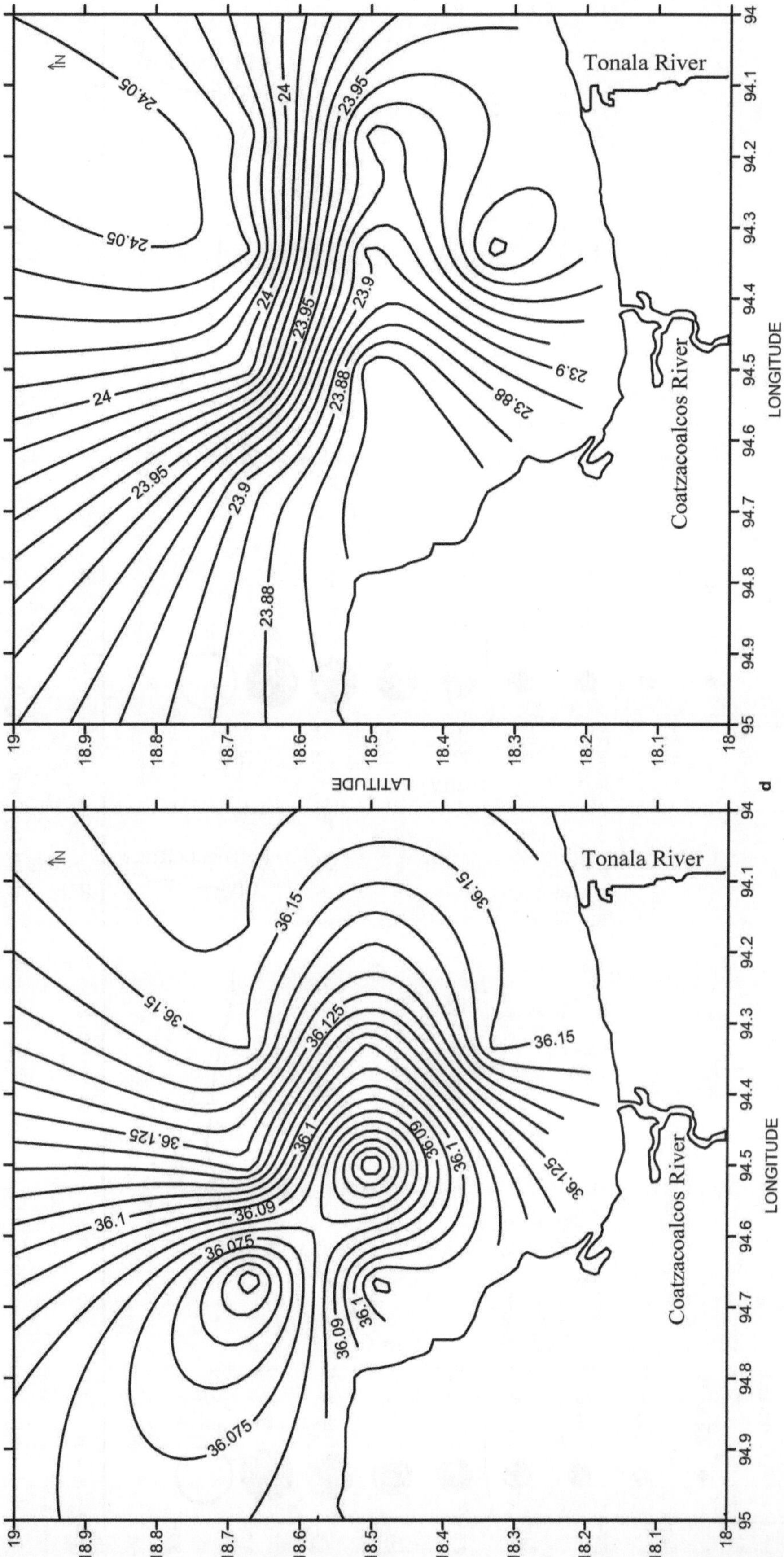

Fig. 7.5 (continued)

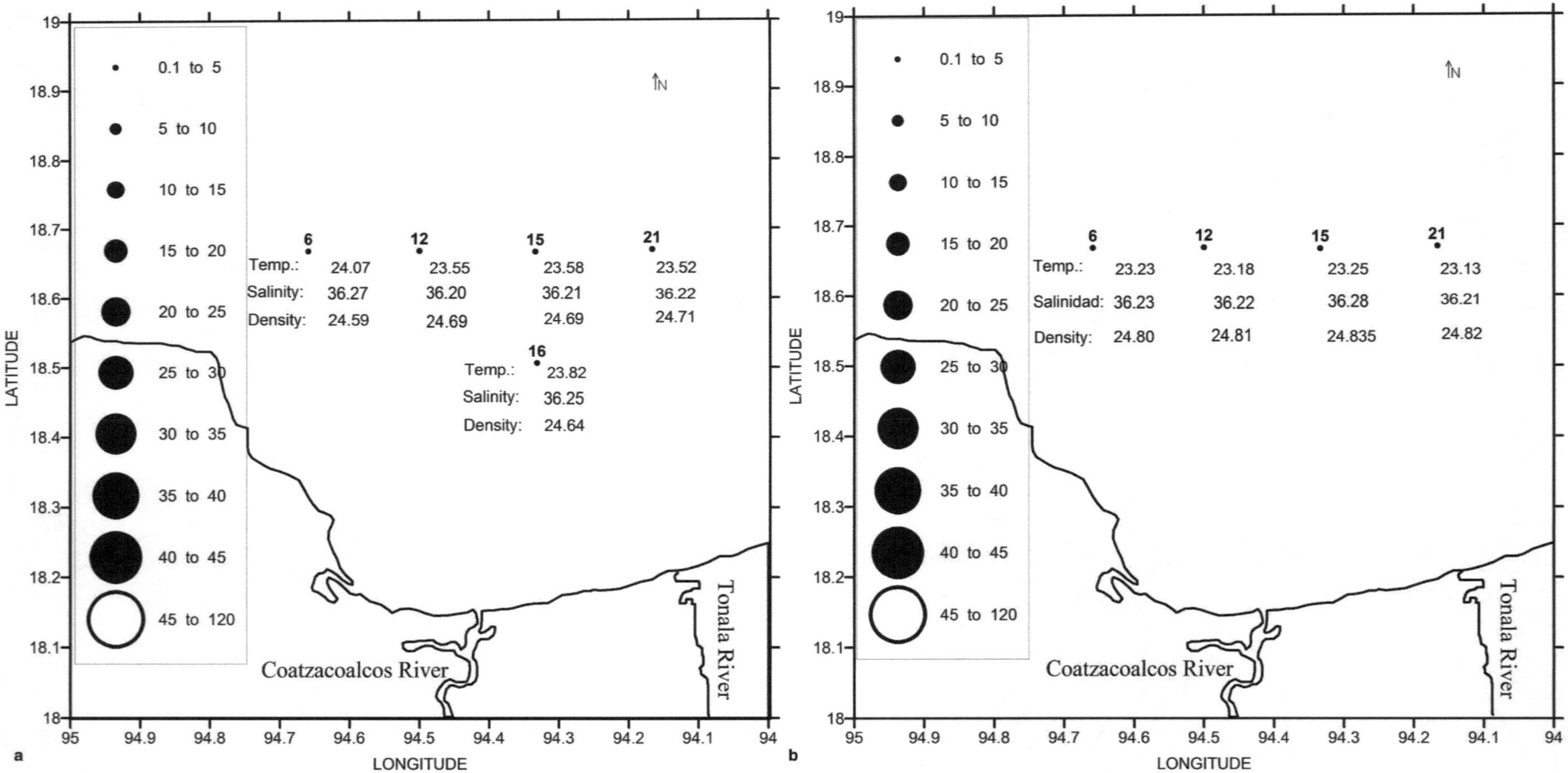

Fig. 7.6 Horizontal distribution: zooplankton biomass (mg m^{-3}), temperature (°C), salinity (ups) and density (kg m^{-3}) at depths of **a** 40–60 m, **b** 60–80 m, and **c** 80–100 m

Fig. 7.6 (continued)

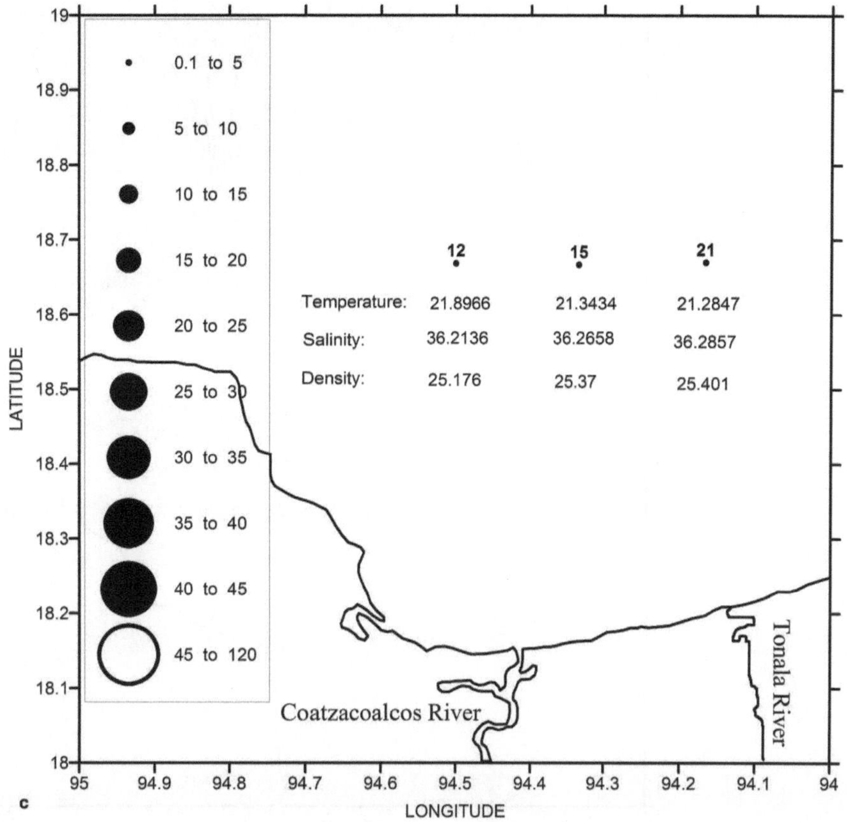

Table 7.2 Degree of similarity among variables at sampling stations. Sampling time is the time of day when the sample was taken

	Biomass (mg m^{-3})	Temperature (°C)	Salinity	Density (kg m^{-3})
Biomass (mg m^{-3})				
Temperature (°C)	17.298			
Salinity	22.302	22.107		
Density (kg m^{-3})	22.407	25.188	6.21	
Sampling time (h)	18.30	19.04	20.30	20.20

7.3.2.3 Multivariate Analyses

Density and salinity most influenced the abundance of zooplankton biomass among sampling stations (Table 7.2). However, because density is a function of (i.e., highly correlated with) salinity, temperature and pressure; salinity was regarded as the most influential variable that differentiated sites. Variables that had less influence were temperature and time of sampling, with values of 17.29 and 18.30, respectively. Together, all four variables accounted for 80.51 % of the total variance of zooplankton biomass; thus, 19.48 % was attributable to other causes or forcing mechanisms.

7.4 Discussion and Conclusions

A small plume of the Coatzacoalcos River was detected at 20–30 m depth, which was likely because of the temperature distribution; the plume extended toward the east of the mouths of the two rivers. This may be due to the absence of a large fresh water supply, and consequently the plume was detected because of temperature rather than salinity (Monreal-Gómez et al. 1992). Additionally, the surface layer is well mixed at the end of the winter storm season and leaves only a trace of the plume in the subsurface layers (Monreal-Gómez and Salas-de-León 1997). Finally, the shift of the plume to the east of the mouths of the rivers is a result of the Coriolis effects (Hickey and Banas 2003; Salas-de-León et al. 2008).

The currents measured with the ADCP showed a predominant direction of flow from west to east and the formation of a significant gyre in the northwestern portion of the study area, which was likely induced by the existence of a shallow area of only 22 m depth between depths greater than 100 m. Additionally, there was a well-defined anticyclonic gyre to the east of the mouth of the Coatzacoalcos River. Anticyclonic gyres are common in ROFIs in the northern hemisphere (Hickey and Banas 2003; Salas-de-León et al.

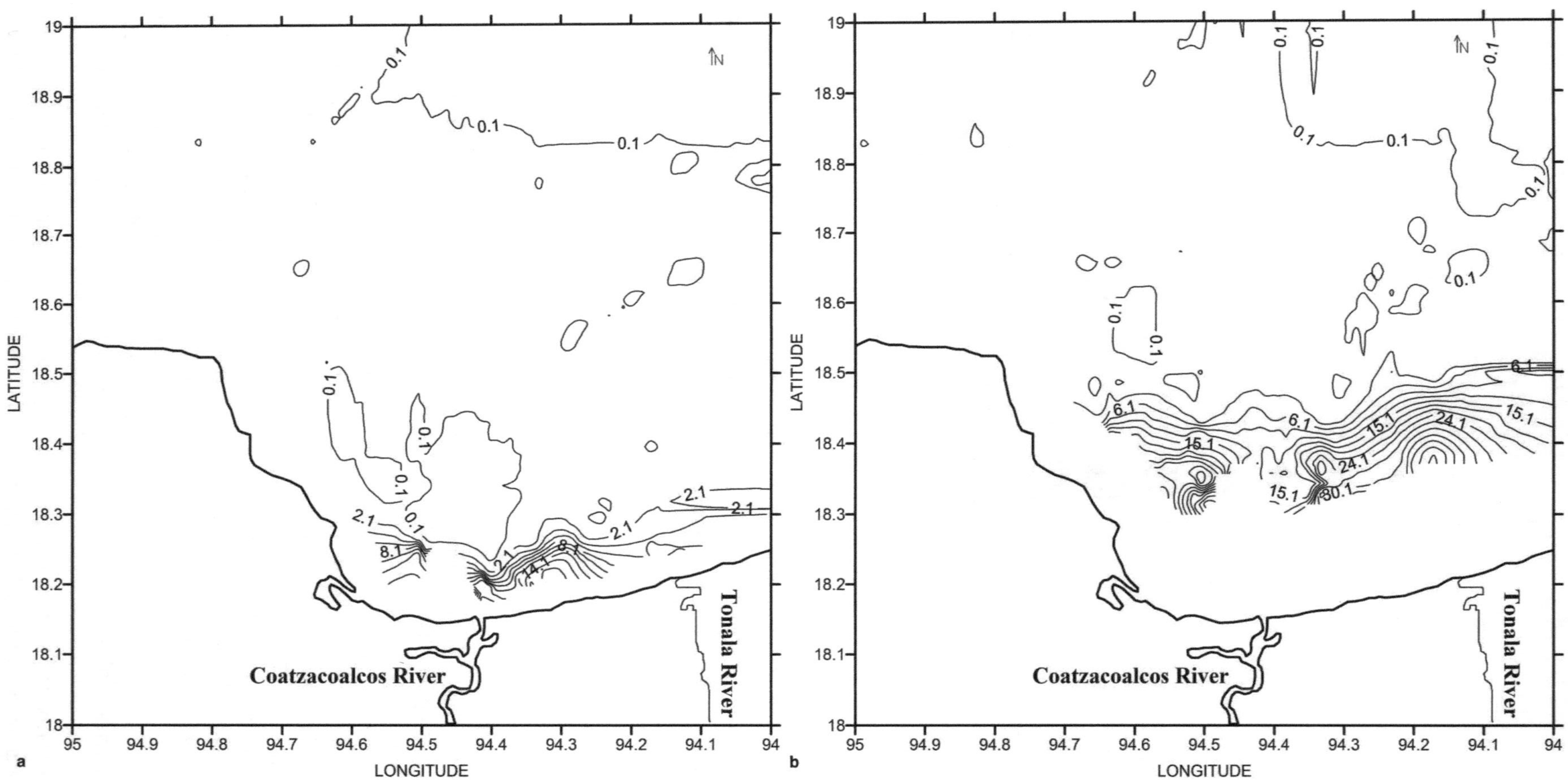

Fig. 7.7 Horizontal distribution of zooplankton biomass (g m^{-3}) obtained with an Acoustic Doppler Current Profiler (ADCP), at depths of: **a** 20 m, **b** 36 m, **c** 52 m, **d** 68 m, and **e** 84 m. Blank areas without biomass contours contained immeasurable amounts of zooplanktons

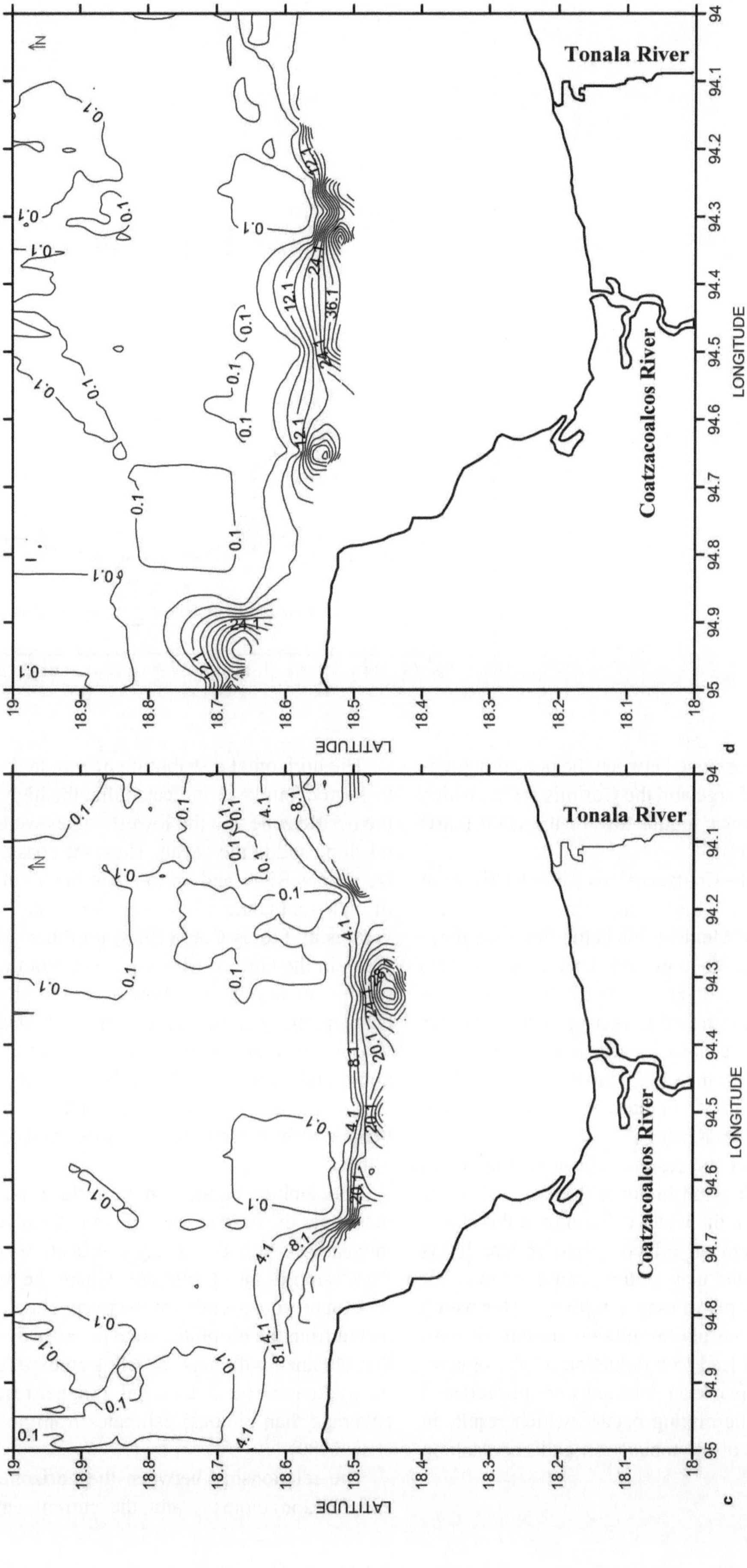

Fig. 7.7 (continued)

Fig. 7.7 (continued)

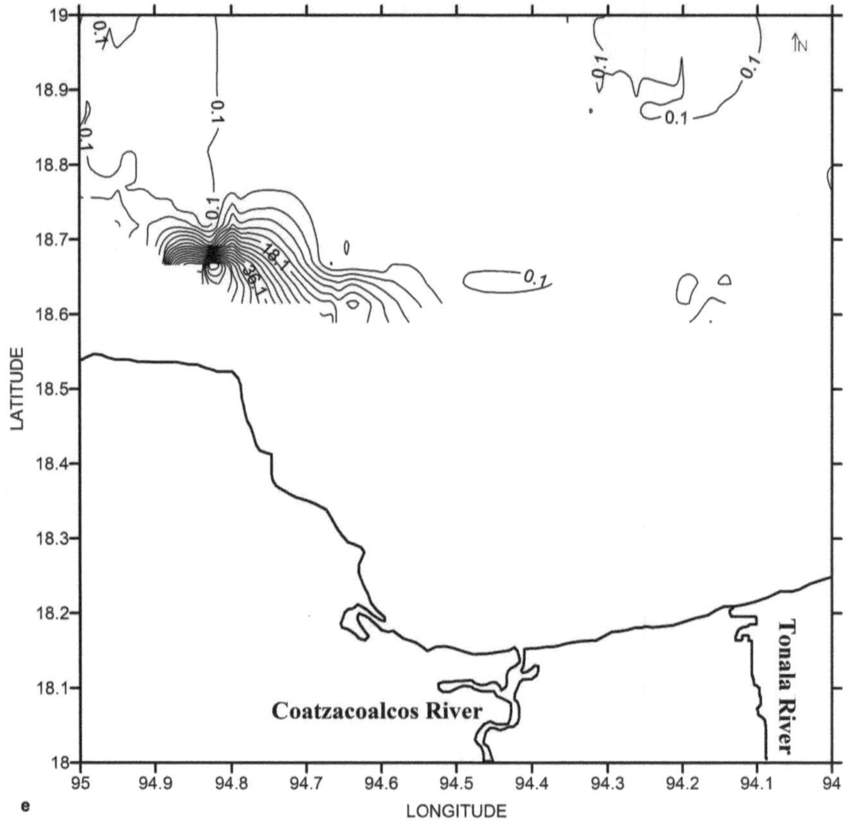

2008) and are due to the balance between the pressure gradient induced by river discharge and the Coriolis effect, which diverts water from the river mouth toward the right (east) (Salas-de-León et al. 2008).

Currents recorded in the Coatzacoalcos River ROFI in the current study were relatively high compared to other studies in this area of the Gulf of Mexico; this is the first time these high velocities have been documented. Díaz-Flores (2004) and Salas-de-León et al. (2004) reported 250 mm s^{-1} currents in a region to the north and farther offshore from our study site, where currents usually are less intense. However, Salas-de-León et al. (2008) measured current velocities that varied from 500 to 600 mm s^{-1} in the region near the mouth of the Grijalva–Usumacinta River.

At the offshore stations, our results indicate that the vertical distribution of zooplankton biomass was highest in the mixed layer and higher in the water column than the clines. Consequently, clines form an effective barrier that limits zooplankton movement and may, in turn, cause an epicontinental decrease in the supply of usable nutrients. This would have a direct effect on the photosynthetic activity of phytoplankton, which could lead to a reduction in phytoplankton populations and zooplankton. Nutrients are not retained under clines because little mixing occurs, which results in an increased population of phytoplankton and zooplankton (Wickstead 1979).

The horizontal distribution of zooplankton biomass at the 0–10 m depth layer indicated that the highest values were in the neritic zone and the lowest values were on the continental slope and in the ocean. Thus, the coastal area of the Coatzacoalcos River and up to about the 50 m isobath are areas of high zooplankton biomass. This is consistent with observations by Flores-Coto (1988) for other southern coastal regions in the Gulf of Mexico. Away from the coast there are stabile conditions that allow access to phytoplankton nutrients and the establishment of optimal conditions for the development of zooplankton (Hayward et al. 1986). Based on horizontal sampling at depths below 10 m, it was determined that zooplankton biomass was similar between sampling stations, which indicates similar horizontal oceanographic conditions.

Zooplankton biomass in the surface was less than the biomass near the base of the mixing layer according to acoustic intensities. These results are consistent with results obtained from samples taken with nets where the maximum biomass was found consistently on the pycnocline. The biomass estimated from the absolute sound pressure data was higher than that obtained with nets, which is an expected result because the hydroacoustic method has a higher resolution and wider coverage than biomass estimates from net sampling (Flagg et al. 1989).

The relationship between the horizontal distribution of zooplankton biomass and the current circulation patterns

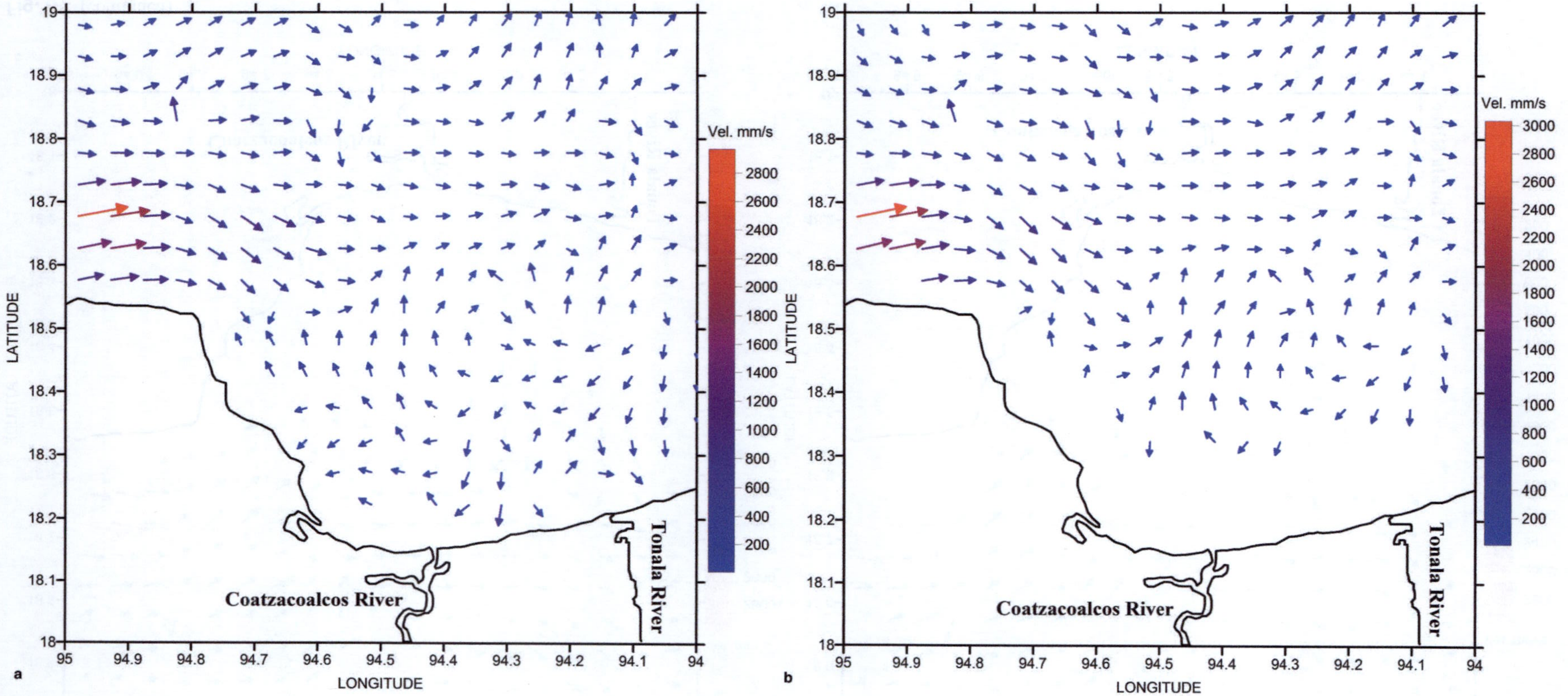

Fig. 7.8 Current circulation patterns (mm s^{-1}) obtained with an Acoustic Doppler Current Profiler (ADCP), at depths of: **a** 20 m, **b** 36 m, **c** 52 m, **d** 68 m, and **e** 84 m

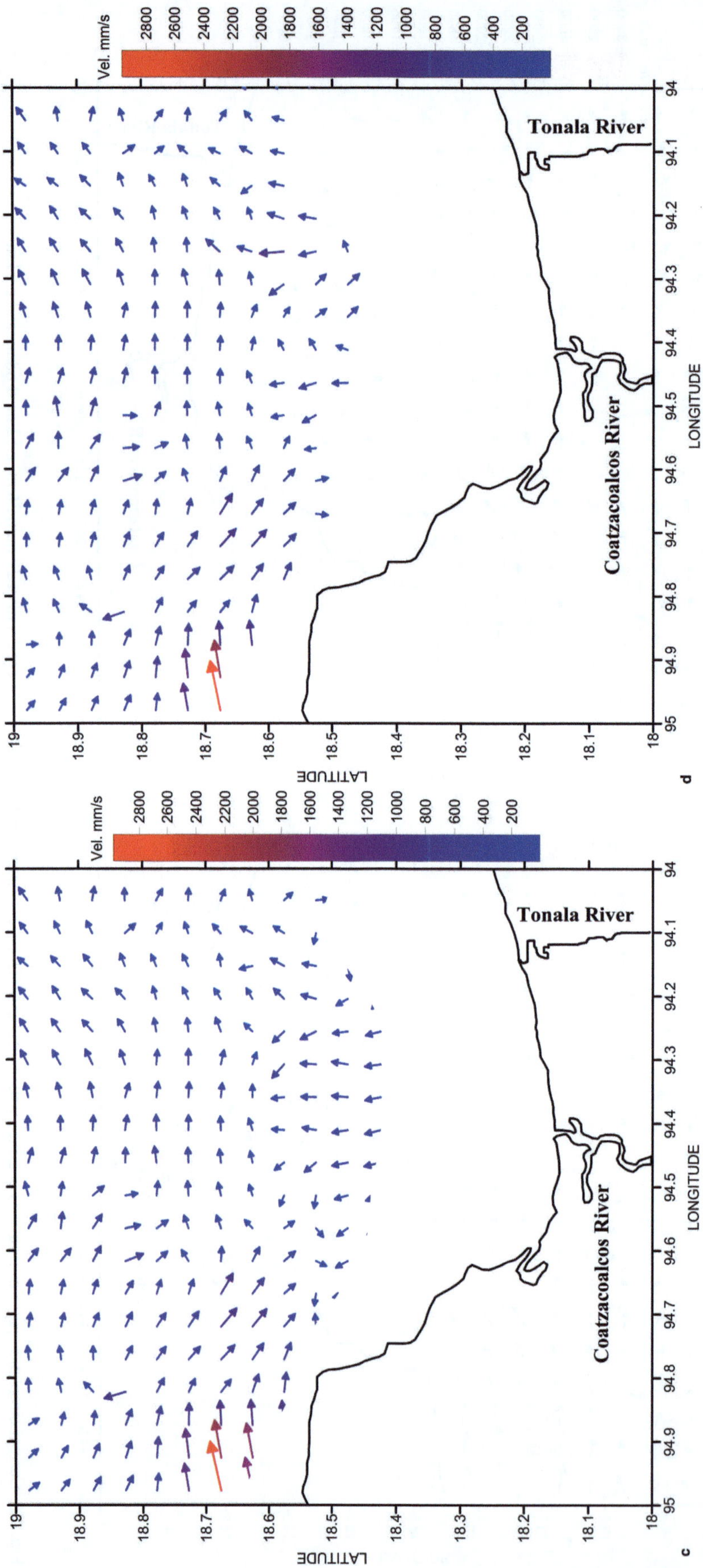

Fig. 7.8 (continued)

Fig. 7.8 (continued)

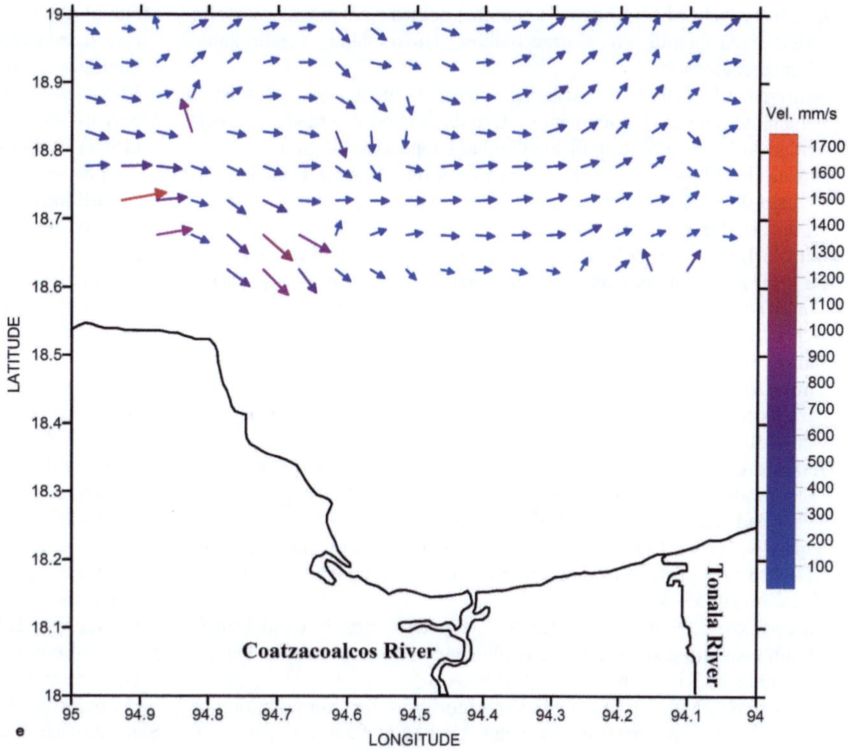

at the 20, 36, 52 and 68 m depths indicated higher biomass offshore of the Coatzacoalcos River plume and to the west. This may be due to the predominant current that flows from west to east and transports large quantities of nutrients that are used by the phytoplankton—thus creating better living conditions for zooplankton. The eastern side of the Coatzacoalcos River plume showed the lowest concentrations of zooplankton, which may be due to the presence of an anticyclonic eddy. The eddy creates a thermocline with less-dense water that accumulates in the center of the eddy and denser water, with a low phytoplankton concentration, at the periphery (Monreal-Gómez and Salas-de-León 1997).

Zooplankton abundance and distribution estimated from sampling with nets was only compared to the biomass estimated by acoustic intensity at the 10–20-m depth and the trends in distribution were similar. In all cases the biomass obtained with the acoustic sampling was higher and the highest concentration of biomass was located to the west of the Coatzacoalcos River plume. It is noteworthy that the balance between the pressure gradient and Coriolis effects forms a cyclonic gyre to the right of northern hemisphere plumes (Hickey and Banas 2003; Salas-de-León et al. 2008) and that this gyre is compensated by the formation of a cyclonic gyre to the west of the plume (Salas-de-León et al. 2008). These conditions can presumably form optimal conditions for the development of zooplankton in the western area of the Coatzacoalcos River ROFI.

With the information obtained in this study and given the relationship between zooplankton biomass and abundance of fish, we can estimate a distribution pattern of fishes similar to zooplankton (Flores-Coto et al. 1988). This relationship has been demonstrated by several authors in different geographical areas (Sherman et al. 1983; Krebs 1985; Leggett 1984; Rodríguez and Rubín 1991; Cowan and Shaw 1991; Kane 1993). However, one must consider that the composition, abundance and distribution of fish will not be perfectly correlated to zooplankton biomass due to differences in spawning areas and times, and physical and chemical processes that disperse or concentrate various fish species (Salas-de-León et al. 1998; Sanvicente–Añorve et al. 1998, 2006). Bozada (1987) confirmed that the biomass and abundance of zooplankton and fish are governed by the system's hydrodynamic conditions. Additionally, a greater abundance of zooplankton occurred during the dry season in the middle to lower Coatzacoalcos River and it was expected that this area would also have the highest abundance of fish (Bozada 1987).

References

Bogdanov DV (1969) Some oceanographic features of the Gulf of Mexico and Caribbean Sea. In: Bogdanov AS (ed) Soviet-Cuban-fishery research, 1st edn. Program for Scientific Translations, Tel Aviv, Israel, pp 13–15

Bozada L, Paéz M (1987) Fauna acuática del río Coatzacoalcos. Serie Medio Ambiente en Coatzacoalcos. Universidad Veracruzana, Coatzacoalcos

Contreras EF (1993) Ecosistemas costeros mexicanos. Comisión Nacional para el Conocimiento y Uso de la Biodiversidad y Universidad Autónoma Metropolitana Unidad Iztapalapa, Mexico

Cowan JH, Shaw RF (1991) Ichthyoplankton off west Louisiana in winter 1981–1982 and its relationships with zooplankton biomass. Contrib Mar Sci 32:103–121

Czitrom-Bauss S, Ruíz F, Alatorre MA et al (1986) Preliminary study of a front in the bay of Campeche, Mexico. In: Nihoul JCJ (ed) Marine interfaces ecohydrodynamics. Elsevier oceanographic series. Elsevier, Liege, pp 301–311

Dauvin JC, Thiébaut E, Wang Z (1998) Short-term changes in the mesozooplanktonic community in the Seine ROFI (Region of freshwater Influence) (easter English Channel). J Plankton Res 20(6):1145–1167

Díaz-Flores M (2004) Estudio de las corrientes en la Bahía de Campeche utilizando un perfilador acústico Doppler (ADCP). Dissertation, Universidad Nacional Autónoma de México

Flagg CN, Wirick CD, Smith SL (1989) On the use of the acoustic doppler current profiler to measure zooplankton abundance. Deep Sea Res 36(3):455–474

Flores-Coto C, Sanvicente-Añorve L, Pineda-López R et al (1988) Composición, distribución y abundancia ictioplanctónica del sur del Golfo de México. Univ Cien 5(9):65–84

Fofonoff NP, Millard RC Jr (1983) Algorithms for computations of fundamental properties of sea water. UNESCO Technical papers in marine Science 44, Paris

Gio-Argáez FR (2000) Distribución y sistemática de los ostrácodos de la Bahía de Campeche. Dissertation, Universidad Nacional Autónoma de México

Hayward D, Haas LW, Boon JD III et al (1986) Empirical models of stratification variation in the York River Estuary, Virginia, USA. In: Bowman J, Yentsch M, Peterson WT (eds) Tidal mixing and plankton dynamics. Springer-Verlag, Berlin, pp 346–367

Hickey BM, Banas NS (2003) Oceanography of the U.S. Pacific Northwest coastal ocean and estuaries with application to coastal ecology. Estuaries 26:1010–1031

Kane J (1993) Variability of zooplankton biomass and dominant species abundance on Georges Bank, 1977–1986. Fish Bull 91:464–474

Krebs CJ (1985) Ecología, estudio de la distribución y la abundancia. Oxford University Press, México

Lara-López AL (2003) Variación espacio-temporal de la biomasa zooplanctónica en las regiones de influencia del sistema Grijalva-Usumacinta durante 1999. Dissertation, Universidad Nacional Autónoma de México

Leggett WC (1984) Fish migrations in coastal and estuarine environments: a call for new approaches to the study of an old problem. In: McCleave JD, Arnold GP, Dodson JJ et al (eds) Mechanisms of migration in fishes. Plenum, New York, pp 159–178

Licea S, Luna R (1999) Spatio-temporal variation of phytoplankton on the continental margin in the SW Gulf of Mexico. Rev Soc Mex His Nat 49:89–99

Monreal-Gómez MA, Salas-de-León DA (1990) Simulación de la circulación de la Bahía de Campeche. Geofis Int 29(2):101–111

Monreal-Gómez MA, Salas-de-León DA (1997) Circulación y estructura termohalina del Golfo de México. In: Lavín MF (ed) Contribuciones a la oceanografía física en México. Unión Geofísica Mexicana, Mexico, pp 183–199

Monreal-Gómez MA, Salas-de-León DA, Padilla-Pilotze AR et al (1992) Hydrographic and estimation of density currents in the southern part of the Bay of Campeche, México. Cienc Mar 18(4):115–133

Padilla-Pilotze AR, Alatorre MA, Ruíz F et al (1986) Observaciones recientes de la estructura termohalina en el sur del Golfo de México. *J. Urrutia–Fucugauchi (Ed.) Unión Geofísica Mexicana. Memorias de la Reunión Anual,* p. 434–440

RD Instruments (1996) Acoustic Doppler current profilers—principles of operation: a practical primer. RD Instruments, San Diego

Riverón-Enzástiga ML (2008) Análisis de la variabilidad de las descargas de agua de los principales ríos del sur de la cuenca hidrológica del Golfo de México. Dissertation, Universidad Nacional Autónoma de México

Rodríguez JM, Rubín JP (1991) El ictioplancton y la biomasa del zooplancton en aguas del sur de Galicia, en abril de 1987. Bol Inst Esp Ocean 7:115–127

Salas-de-León DA, Monreal-Gómez AM, Sanvicente-Añorve L et al (1998) Influence de la circulation à long terme sur la distribution des organismes zooplanctoniques dans la Baie de Campeche, Mexique. Ocean Acta 21(1):87–93

Salas-de-León DA, Díaz-Flores MA, Monreal-Gómez MA (2004) Circulation and vorticity in the Southern Gulf of México. *Meteorological and Geophysical Fluid Dynamics, 1,* 229

Salas-de-León DA, Monreal-Gómez MA, Díaz-Flores MA et al (2008) Role of near-bottom currents in the distribution of sediments within the Southern Bay of Campeche, Gulf of Mexico. J Coast Res 24(6):1487–1494

Sanvicente-Añorve L, Flores-Coto C, Sánchez-Velasco L (1998) Spatial and seasonal patterns of larval fish assemblages in the southern Gulf of Mexico. B Mar Sci 62(1):17–30

Sanvicente-Añorve L, Soto LA, Espinosa-Fuentes ML et al (2006) Relationship patterns between ichthyoplankton and zooplankton: a conceptual model. Hydrobiologia 559:11–22

Sherman KJ, Green R, Goulet JR et al (1983) Coherence in zooplankton of a large northwest Atlantic ecosystem. Fish Bull 81:855–862

Simpson JH (1997) Physical processes in the ROFI regime. J Mar Sys 12(1):3–15

Thiel M, Erasmo CM, Enzo A et al (2007) The Humboldt current system of Northern and Central Chile. Ocean Mar Biol 45:195–344

Vázquez-de la Cerda AM (1979) Circulación del agua en el sureste del Golfo de México. Secretaría de Marina, Dirección General de Oceanografía, Mexico

Villalobos A, Zamora ME (1975) Importancia biológica de la Bahía de Campeche. In: Memorias del I Simposio Latinoamericano de Oceanografía Biológica, Secretaria de Marina, Mexico, 25–29 November 1974

Wiebe PH (1988) Functional regression equations for zooplankton displacement volume, wet weight, dry weight and carbon: a correction. Fish Bull 86(4):333–335

Wickstead JH (1979) Zooplancton marino. Cuadernos de Biología. Omega, España

Zavala-García F, Flores-Coto C (1989) Medición de biomasa zooplanctónica. An Inst Cien Mar Lim UNAM 16(2):273–278

Managing Artisanal Fisheries in Estuarine Systems Through the Use of Fishing Zones in the South Eastern Gulf of California

8

Mauricio Ramírez-Rodríguez, Felipe Amezcua and Angela Aguiar-Moreno

Abstract

Artisanal fisheries in the estuarine systems of Mexico are poorly regulated due to fishery complexity and the lack of information on the biology of exploited species, their habitat and fishing methods. Due to these unknowns and to utilize existing catch data (i.e., landing date, catch and landing site, catch per species, and price), this chapter proposes the use of zones as management units to understand fishing processes by region, and to identify possible spatiotemporal changes of the marine communities along the coast of Sinaloa (SE Gulf of California). A total of 97 species, primarily teleost fishes ($n=80$), composed the artisanal catch within 6 defined zones; the shrimp fishery was also important in all the zones. The importance of other fisheries differed according to zone, but in general, swimming crabs were most economically important in the North and demersal fishes were most important in the South. Three categories of target species were classified: high economic value and high abundance seasonal species; low abundance and high value resident species; and high abundance and low value resident species. Zonation will allow the identification of catch trends, which can be used for management; however, there still remains essential information that is needed for better management.

Keywords

Small scale fisheries · Fisheries management · Gulf of California · Fishing zones · Exploited species

8.1 Introduction

Artisanal fisheries are small-scale fisheries for subsistence or local markets that use traditional fishing techniques from small boats or skiffs. They occur around the world and are vital to livelihoods and food security for many families. Artisanal fisheries are complex and dynamic because of the diversity of environments where fishing occurs, fishing gears, socioeconomic variables and the number of fish species involved. This complexity makes proper management plans and sustainable development difficult to implement (Arreguín-Sánchez 2006; Espino-Barr and Cruz-Romero 2006; Ramírez-Rodríguez 2009).

In small-scale fisheries of Mexico where fishing occurs in both the open sea and estuarine systems, fishermen operate

M. Ramírez-Rodríguez (✉)
Centro Interdisciplinario de Ciencias Marinas, Instituto Politécnico Nacional, Av. IPN s/n, Col. Playa Palo de Santa Rita, 23096 La Paz, BCS, México
e-mail: mramirr@ipn.mx

F. Amezcua
Instituto de Ciencias del Mar y Limnología, Universidad Nacional Autónoma de México, Av. Joel Montes Camarena s/n C.P., 82040 Mazatlan, Sinaloa, México
e-mail: famezcua@ola.icmyl.unam.mx

A. Aguiar-Moreno
Posgrado en Ciencias del Mar y Limnología, Universidad Nacional Autónoma de México, Circuito Exterior s/n Ciudad Universitaria, 04510 México, DF, México
e-mail: angela@ola.icmyl.unam.mx

F. Amezcua, B. Bellgraph (eds.), *Fisheries Management of Mexican and Central American Estuaries*, Estuaries of the World, DOI 10.1007/978-94-017-8917-2_8, © Springer Science+Business Media Dordrecht 2014

skiffs called "pangas" with outboard motors and a load capacity of less than 10 t. Fishing gear is fitted in the boats according to the target species and include cast nets, gill nets, trawl nets, long lines, traps, spear guns, and hook-and-line. A typical crew consists of 2–3 fishermen who undertake fishing trips that last up to one day.

Management of artisanal fisheries is in accordance with the Law for Sustainable Fisheries and Aquaculture of Mexico (Ley de Pesca y Acuacultura Sustentables), issued by the Ministry of Agriculture, Animal Husbandry, Rural Development, Fisheries and Food of Mexico (i.e., Secretaría de Agricultura, Ganadería, Desarrollo Rural, Pesca y Alimentación; SAGARPA) (SAGARPA 2007). These fisheries are typically controlled through fishing licences and concessions that specify the number of boats and types of fishing gears authorized for a given target species. Specific regulations are associated also to each county, state or other geographic area. A few fisheries (i.e. shrimp, rays and sharks, swimming crabs, mullets) are regulated through fishing "norms" (Normas Oficiales de Pesca) such as the NOM-002-PESC-1993 which regulates the shrimp fishery, NOM-016-PESC-1994 which regulates the mullet fishery, NOM-029-PESC-2006 which regulates the sharks and rays fisheries, and the NOM-039-PESC-2003 which regulates the swimming crab fishery. Each norm establish the way the fisheries must be operated including the types of boats, fishing gears, fishing areas, fishing seasons and minimum catch sizes. The National Fishing Chart (Carta Nacional Pesquera, SAGARPA 2012) is a useful document that defines the status of some fishing resources (i.e. deteriorated, maximum sustainable level or unexploited populations that could be developed) and includes a series of guidelines to adequately manage these fisheries within the geographic limits of each state.

The complexity of artisanal fisheries, including the population dynamics of the exploited populations and their relation to environmental and anthropogenic factors, make it difficult to establish proper indicators of fishing pressure and the effect on the resource. Further, estimating the efficiency of the fishery in terms of economic value, currency generation and employment is also difficult due to fishery complexity. Lastly, even basic information about fishing effort and catch per species is scarce or nonexistent, which impedes the appropriate analyses of yields and landing trends that are necessary for adequate administration of artisanal fleets (Ramírez-Rodríguez 2011). These uncertainties lead to a precarious situation and indicate the need for a unique method to ensure proper fisheries management of artisanal fisheries.

This study proposes a potential solution to address the difficulty of managing artisanal fisheries through the implementation of fishing zones that are based on indicators of fleet operation. Ideally, a better understanding of the fishery can be achieved and an adequate management plan can be implemented (Ramírez-Rodríguez and Ojeda-Ruíz 2012). Besides this, knowledge of the fishing processes by region allows proper interpretation of possible spatiotemporal changes of the marine communities (Pitcher and Hart 1982). We considered the fishing fleet as the primary operational unit, defined as the group of boats that share the same resource with a similar level of technology and economic structure (Accadia and Franquesa 2006; Ramírez-Rodríguez and Ojeda-Ruíz 2012).

The study area was located along the continental shelf of the SE Gulf of California in the state of Sinaloa. In Sinaloa, artisanal fishing occurs along 656 km of coastline, which is important to both fishing and tourism. Eleven rivers flow into bays and lagoons along the coast that adds an additional 221,000 ha of coastal habitat (de la Cruz-Agüero 2002; Rubio-Rocha and Beltrán-Magallanes 2003). The continental shelf of the study area is 17,751 km^2 and is influenced by the California, Gulf of California and Pacific North Equatorial currents (Cano-Pérez 2000).

Due to the oceanographic characteristics of the littoral zone, as well as the wide variety of habitats, biodiversity in the coastal region of Sinaloa is high (Hendrickx and Brusca 2002; Contreras-Espinoza and Castañeda-López 2003; Amezcua–Linares 2009). Marine fauna in this area are generally tropical and are representative of the Mexican biogeographic province; however, the faunal community is also directly influenced by the California current, which brings fish species from the Cortez biogeographic province. This transition zone between the tropical Mexican province and the temperate Cortez province is called the Sinaloan gap (Briggs 1974, 1995; Castro-Aguirre et al. 1999). The Sinaloan gap has substrates that are predominantly composed of sand and silt and mangrove forests are located within several estuarine and coastal lagoon systems (van der Heiden and Findley 1988); these areas are considered important fishing grounds.

The artisanal fishing fleet in the state of Sinaloa is the largest along the Pacific coast of Mexico with 9,179 registered skiffs (SAGARPA 2011). The total annual catch from 1980–2011, increased from 86,738 to 337,864 t (SAGARPA 2011). The average landing biomass from the period 2002–2011 was 278,842 t.

The artisanal fleet participates in the catch of some of the most important fisheries in Mexico such as squid (*Dosidicus gigas*), shark (*Alopias* spp., *Carcharhinus* spp., *Mustelus* spp., *Sphyrna* spp., among others) and shrimp (Family Penaeidae). These fisheries represented approximately 20% of the total catch in Mexico from 1980–2011; approximately 71,000 t (Fig. 8.1). The shrimp fishery represents 88% of the Sinaloa artisanal catch by weight (approximately 62,500 t). The shrimp fishery is considered the most important in Mexico because of the revenue it produces and 14% of the shrimp catch comes directly from the artisanal fishermen

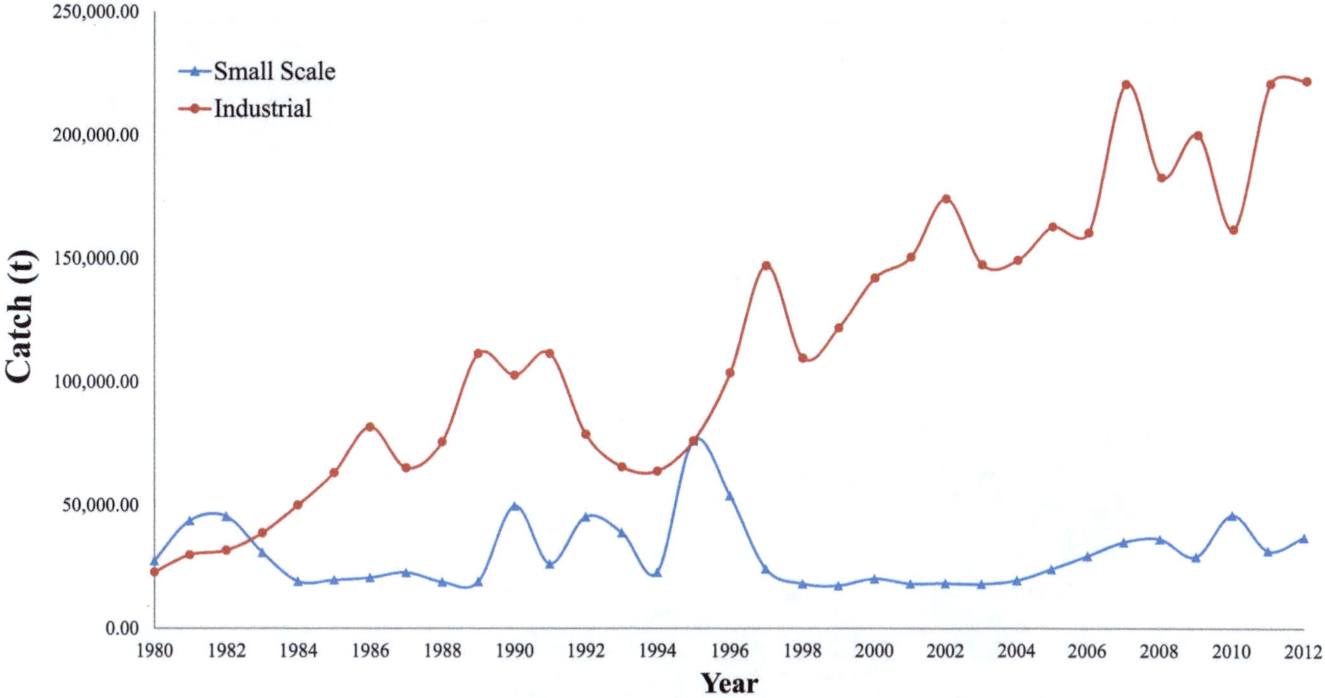

Fig. 8.1 Registered catch by fishery type in tons (t) in the state of Sinaloa

in estuarine systems. Other important species captured by the artisanal fleet in Sinaloa are clams (Family Veneridae), weakfish (Family Sciaenidae), swimming crab (Family Portunidae), mullet (Family Mugilidae), oysters (Family Ostreidae) and the Pacific sierra (Scomberomorus sierra) that represent 20 % of the total catch of the artisanal fishery. The remainder of the catch is composed of a large number of other species, mainly fin fish. In the State of Sinaloa the landings of the artisanal fishery represent 25 % of the total catch (approximately 53,900 t).

The annual catch records of fishermen indicate which groups of species are landed; however, these groups are usually different from year to year because fishermen use common names to identify species. Additionally, the common names of a large number of species are not registered and they are logged as "other species". Another problem is that annual fishing catch records as well as the official fishing statistics do not differentiate between fisheries operations along the coast versus fisheries in estuarine systems. These differences in record keeping are due to spatial and temporal variability of the targeted resources; differences in the types of fishing permits and fishing regulations; the interaction among the different artisanal fisheries and industrial fisheries, aquaculture, and recreational fishing. To help improve and standardize record-keeping, our study defines zones along the coast of Sinaloa that could be used as a basis for proper management plan of the artisanal fisheries in this region.

8.2 Methods

Fishery statistics from landing log books of the artisanal fishery in Sinaloa were available from 1997–2008. These were obtained from fisheries offices of SAGARPA. These offices are located in the towns of Los Mochis, Topolobampo, Guasave, La Reforma, Navolato, Culiacán, Mazatlán, Rosario and Escuinapa; all are coastal towns near the main fishing grounds and estuarine systems (Fig. 8.2). Catch data include landing date, fishery office, catch site, landing site, catch per species in terms of biomass (kg), and price in Mexican Pesos per kg. A code was assigned to each fishing ground according to Ramírez-Rodríguez et al. (2007).

To establish different zones along the coast of Sinaloa, first the fishing grounds were identified and ordered from north to south. After that, zones were defined by grouping fishing grounds related to the distance to geographic landmarks (i.e. coastal lagoons and bays) and the presence of at least one landing site or port used by fishers in the area.

Captured species were registered using their local or common names, which were then labeled by taxonomic Family and Genus using published lists of common and scientific names of species in the region (Miller and Lea 1972; Castro-Aguirre 1978; Thomson et al. 1979; Fisher et al. 1995; Escobar-Fernández and Siri 1997; Espino-Barr et al. 2003, 2004; Robertson and Allen 2006; Amezcua-Linares 2009; SAGARPA 2012). Economic and biomass value were then calculated for each species.

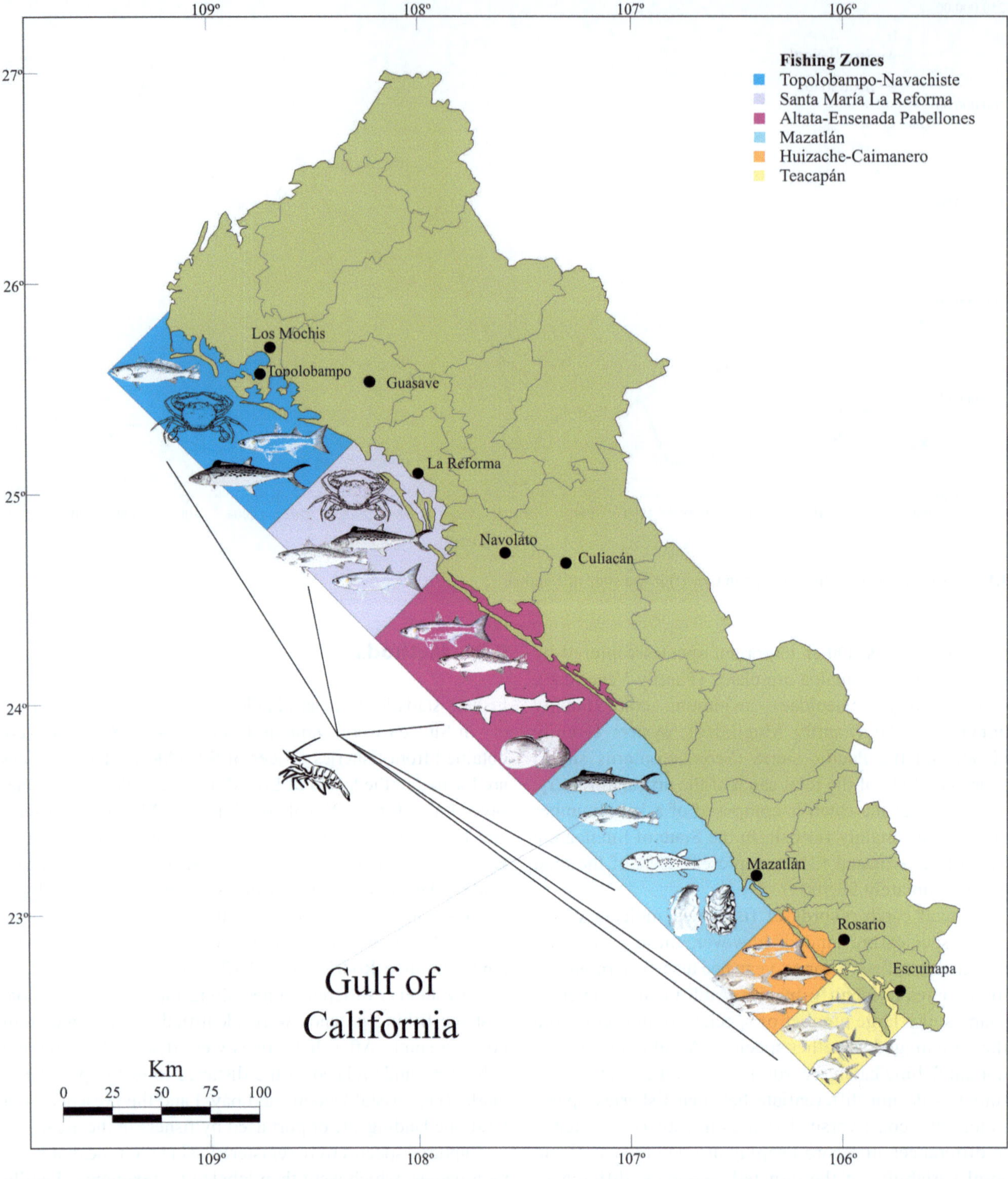

Fig. 8.2 Proposed fishing zones and the current main landing sites (where catch is reported) of the artisanal fishing fleets in Sinaloa, Mexico

To characterize the catch of every zone in every season, three indicators of artisanal fleet performance were calculated: the frequency of a given species in the landings data, the catch per species in terms of biomass (kg), and the economic value of the catch in the local currency (Mexican Pesos). The catch per unit effort was not used because effort data were not available. The Index of Relative Importance (IRI) was estimated monthly for each fishery in each zone using the formula:

$$\text{IRI} = \%_C + \%_V + \%_F \tag{8.1}$$

Where: C = catch in biomass (kg); V = value of the catch and F = frequency in the landings (Ramírez-Rodríguez and Ojeda-Ruíz 2012). Fisheries defined in the present study were classified not only by species but also by the set of skiffs that exploited the same target species in a specific area. Results were plotted by month and zone. The monthly results were then used to establish seasonal differences in fishery landings.

8.3 Results

A total of 128 fishing grounds were reported in the landing log books from the artisanal fisheries of Sinaloa. The main landing sites were Los Mochis, Topolobampo, Guasave, La Reforma, Navolato, Mazatlan and Escuinapa which showed a frequency of use for landing purposes above 6 % at each one of them. Six zones were defined based on the 7 landing sites (Fig. 8.2). Each zone had specific identifiable characteristics such as the presence of rivers, coastal lagoons and estuarine systems with mangrove forests, flood plains, marshes, sand dunes, beaches and channels. All of these habitats are affected by agricultural, urban and industrial drains.

The six defined zones were:

1. Topolobampo-Navachiste (25°45′N—109°25′W to 25°17′ N—108°30′): This zone includes 45 landing sites and the most frequently used are Topolobampo, Bahía de Santa María, Huitussi and Cerro Cabezón; these sites include the estuarine systems of Colorado, Santa María, Topolobampo, Ohuira, San Ignacio and Navachiste, which is part of the El Fuerte River delta. Lankford (1977) classified the estuarine area of San Ignacio and Navachiste as a II-A (I-C) type coastal lagoon (i.e. it has an intra-deltaic depression with a flooded valley with a barrier separating the system from the sea). The mangrove species present in this zone are red mangrove (*Rhizophora mangle*) and black mangrove (*Avicennia germinans*).
2. Santa María La Reforma coastal lagoon (25°17′ N —108°30′to 24°41′N—108°01′W): Thirteen landing sites are included in this zone with the primary sites being La Bocanita, La Reforma and Los Algodones. Santa Maria La Reforma is classified as a type III-A system, which is an inner-shelf coastal lagoon (Lankford 1977). This zone is located in the middle part of Sinaloa state and is populated by black mangrove, button mangrove (*Conocarpus erectus*), white mangrove (*Laguncularia racemosa*), and red mangrove. This lagoon connects to the Pacific Ocean through two channels that are each 5-km wide and 12–17-m deep.
3. Altata-Ensenada Pabellones (24°41′N—108°01′W′ to 23°55′ N—106°58′ W): A total of 21 landing sites were identified in this zone; Navolato, Las Arenitas, El Conchal and Cospita were the most frequently used by fishers. This zone includes the coastal lagoon of Altata, which according to Lankford (1977) is a III-A (I-D) type coastal lagoon characterized by flooded depressions and protected from the sea by sand barriers created by waves and currents during the last 5,000 years.
4. Mazatlan (23°55′ N—106°58′ W to 23°07′ N—106°19′ W): This zone includes 20 landing sites with Mazatlan, Playa Sur and Isla de la Piedra being the most frequently used. The main estuarine system is the Urias, a type III-B(III-a) coastal lagoon with a permanent connection to the sea (Lankford 1977). This system is adjacent to the city of Mazatlan, which is a source of urban sewage to the estuary. Industrial fishing ports and fleets for shrimp, tuna, shark, anchovies, etc. are located also along this shoreline of the estuary. Consequently, this system is one of the most impacted estuarine areas in the region; however, one section of the system south of Mazatlan is lower in contamination and the mangroves are likely not affected by human activities (Méndez 2002; Páez-Osuna et al. 1997). Urias estuary is inhabited by black mangrove, white mangrove and red mangrove.
5. Huizache-Caimanero (23°07′ N—106°19′ W to 22°49′ N—106°00′ W): In this zone, 15 landing sites were identified including the primary sites of Rosario, El Huizache and Los Cerritos. This is an estuarine system located in the southern part of the coastal plateau of Sinaloa between the Presidio and Baluarte rivers. Ecologically this system is a marine seasonal floodplain with two basins; Huizache is located in the north and Caimanero is located in the south. The basins are separated by a narrow channel about 250 m wide. According to Lankford (1977) this is a type III-A coastal lagoon with a total area of approximately 260 km^2, which can be reduced by 75 % during the dry season to a depth no greater than 1.5 m (Aquino-Guzman et al. 1983). The mangrove forests are only located along channels in the south and north areas connected to the Presidio and Baluarte rivers. These channels are inhabited by red mangrove, white mangrove and black mangrove. However, mangrove forests are absent in both of the lagoons (Chapa-Saldaña and Soto-López 1969; Amezcua-Linares 1977; Warburton 1978).
6. Teacapán (22°49′ N—106°00′ W to 22°32′ N—105°45′ W): This zone includes 14 landing sites with Escuinapa, Teacapán and La Brecha being the most important. This zone is located in the estuarine system of Teacapán, which is a type III-A (III-C) coastal lagoon with a permanent connection to the sea (Lankford 1977). This system hosts the largest mangrove forest along the Pacific coast in Mexico (i.e., accounts for 15–20 % of the total area of mangrove forests in Mexico) and is inhabited by white mangrove, red mangrove, black mangrove and button mangrove. This system connects to the sea through a 1-km wide channel.

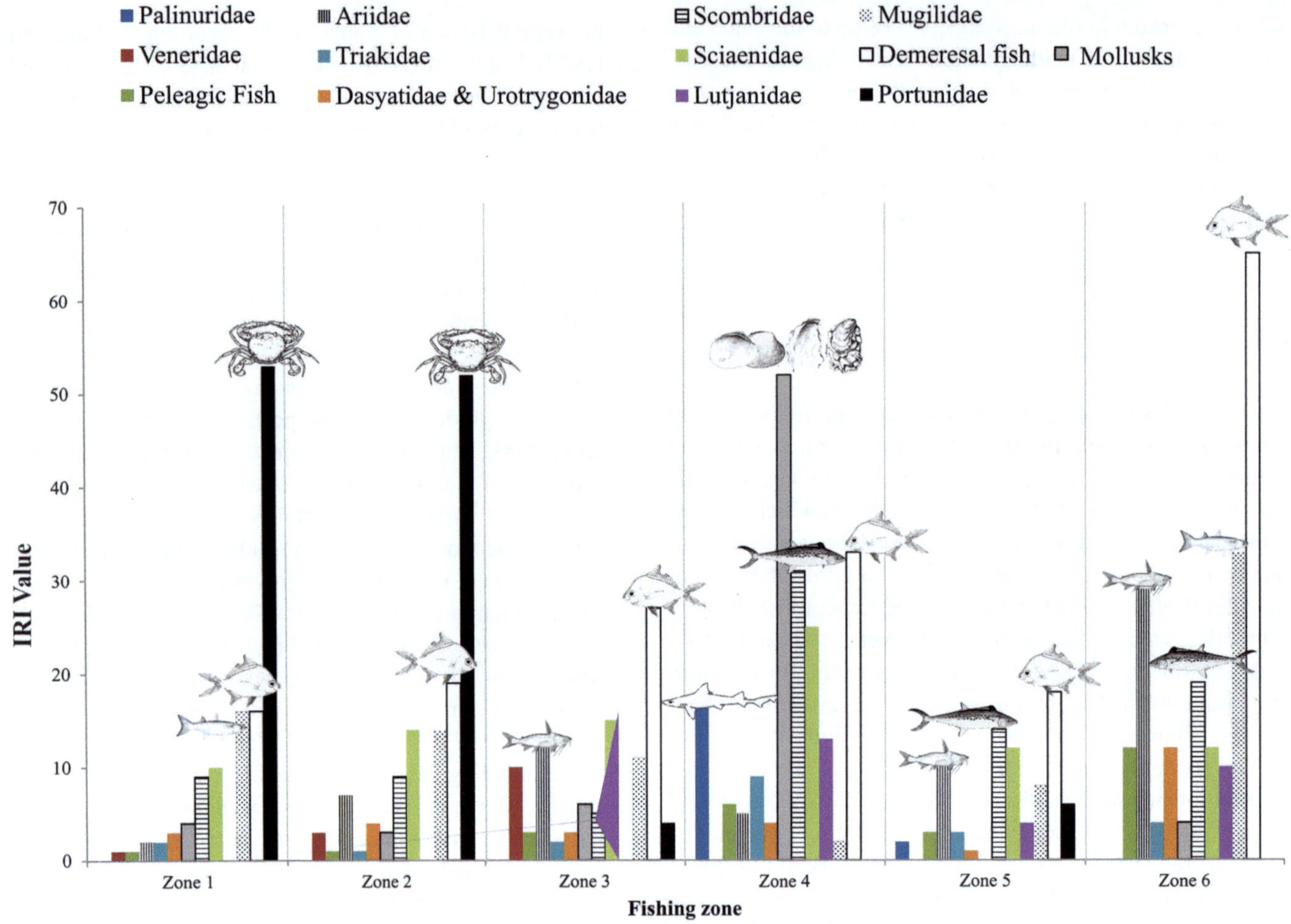

Fig. 8.3 IRI values of artisanal fisheries in Sinaloa organized by taxonomic family and fishing zone. Penaeidae is excluded from the figure, as it was the most important in all zones

The species catch was composed of at least 97 species belonging to 51 families. From these, 80 species were teleost fish and elasmobranchs, 11 were mollusks and 6 were crustaceans. Nine families represented 84.5 % of the total capture: Penaeidae (shrimps) (39 %), Portunidae (swimming crabs) (21 %), Veneridae (clams) (7 %), Mugilidae (mullets) (5 %), Scombridae (Pacific sierras) (4 %), Sciaenidae (weakfish) (3 %), Ariidae (sea catfish) (2 %), Lutjanidae (snappers) (2 %) and Triakidae (houndsharks) (1.5 %). Other species included oysters, squids, scallops, and snails (9 families); demersal fishes from at least 17 families that included puffer fish (Tetraodontidae), snooks (Centropomidae), trigger fish (Balistidae), mojarras (Gerreidae), grunts (Haemulidae), pompanos and jacks (Carangidae), flatfish and flounders (Paralichthyidae), groupers (Serranidae), cusk-eels (Ophidiidae), tilefish (Malacanthidae), bobos (Polynemidae), and croakers (Sciaenidae); and some pelagic fish such as sardines and herrings (Clupeidae), anchovies (Engraulidae), barracuda (Sphyraenidae), dolphin fish (Coryphaenidae), halfbeaks (Hemiramphidae), ladyfish (Elopidae), and milkfish (Chanidae). Approximately 27 % of the species reported in the landing log books were not identified because the common name was not recognized.

The most important artisanal fishery in this region of Mexico is the shrimp fishery based on IRI values, biomass and frequency of capture. Consequently, the total economic value of the shrimp fishery in this region is high, even though the shrimp fishery is closed for about 6 months each year. Other fisheries have a lower total captured biomass, but the frequency of capture throughout the year is relatively high. After the shrimp fishery, other important fisheries in the region are for demersal fishes, which include a large number of fish species. Swimming crabs, weakfish, sierra, mullet, mollusks, snappers, sea catfish, sharks, rays, small pelagic fish (such as anchovies and sardines), lobsters and clams are also important fisheries.

When the fisheries were analyzed by zone, the shrimp fishery was also the most important in all zones (Figs. 8.2 and 8.3). The importance of the non-shrimp fisheries varied by zone. The catch of swimming crabs was relatively high in zones 1 and 2, but the catch of crabs in the other zones was low. No species were dominantly captured in zones 3

Table 8.1 Fishing seasons and the months of highest capture of the most important commercial species of the artisanal fishery in Sinaloa, Mexico. Symbols denote open seasons (*), highest landings (*light grey* shading) and closed seasons for females only (*light blue* shading)

Fishery/month	Sep	Oct	Nov	Dec	Jan	Feb	Mar	Apr	May	Jun	Jul	Aug
Shrimp	*	*	*	*	*	*	*					
Swimming crab	*	*	*	*	*	*	*	*			*	*
Mullet	*	*	*		*	*	*	*	*	*	*	
Pacific sierra	*	*	*	*	*	*	*	*	*	*	*	*
Weakfish	*	*	*	*	*	*	*	*	*	*	*	*
Sea catfish	*	*	*	*	*	*	*	*	*	*	*	*
Snapper	*	*	*	*	*	*	*	*	*	*	*	*
Shark	*	*	*	*	*	*	*	*		*		
Clam	*	*	*	*	*	*	*	*	*	*	*	*

and 5; captures were relatively even among fisheries. Zone 4 was characterized by the large catch of Pacific sierra, sharks and lobsters. Zone 6 was characterized by large landings of fin fish; the primary species were mullets, sea catfish (Ariidae) and Pacific sierra. The demersal fishery was important in terms of frequency in all zones throughout the year, but was especially important in zones 3, 4 and 6. However, it is necessary to consider that the demersal fishery is composed of a large number of species and families that may overlap with other single-species fisheries.

The timing of fishing seasons did not vary significantly among zones due to government regulations. The shrimp fishery runs from September or October to March or April of each year. The swimming crab season is open following the closure of the shrimp fishery, but closes shortly from May to June, and reopens again in July; only males can be captured. The seasons of the highest catch of other important fisheries are February to June for the mullet, November to March for Pacific sierra, weakfish and sea catfish, November to May for snappers (Lutjanidae), and November to March for sharks. The demersal fishery operates through the year, but the composition changes both spatially and seasonally; however, because it consists of a large suite of fish species, usually grouped under the same common name, it was not possible to establish seasonal trends (Table 8.1).

8.4 Discussion

Indicators used to assess the state of small-scale fisheries are difficult to formulate because they must integrate biological, technological, economic and social information. However, the results of the current study represent a first step in managing the artisanal fisheries of Sinaloa. The artisanal fisheries of the SE Gulf of California use a diverse suite of fishing gears (e.g., gill nets, seine nets, trawl nets, cast nets, hook and line, traps) to capture at least 92 species including shrimp, sharks, rays, snails, clams, swimming crabs and several species of finfish. Spatial patterns in the complex artisanal fisheries of Sinaloa were identified that allowed the identification of distinct fishing zones. The relative importance and trend of each fishery varied in relation to the availability, economic value and frequency of use of this resource.

This suggested system of fishing zones allows the identification of the most important resources and strategies of the fishing operation for a given region; however, there are also some drawbacks of this approach. This system defines a fishery as a unit of boats that exploit similar resources, but this oversimplifies the fishery as a whole. For example, catch amounts are used as an indicator of the population dynamics and biology of the exploited populations without consideration of the basic biology of the exploited species or the differential use of fishing gears among zones. Additionally, because fishing effort was not recorded, the intensity of use of each species was based on the frequency of reports at a given fishing ground. Lastly, the economic component of each fishery was based solely on the economic revenue of the landed catch and there were no indicators related to fisher's organization and commercialization.

Despite the challenges of the method proposed, this system made it possible to identify seasonal changes of the artisanal fisheries which is needed information for proper fisheries regulation (Panayotou 1983). In all, three categories of target species were identified and could be used in setting broad-based management goals: (1) seasonal species with high value and abundance, some of which are already managed under an established management plan (e.g., shrimp, swimming crabs, oysters, clams, sharks, Pacific sierra), (2) low-abundance and high value species that are residents of estuarine and coastal areas (non-migratory) and do not have management plans (e.g., snappers, puffers (Tetraodontidae), groupers (Serranidae), snooks (Centropomidae), weakfish, flounders (Paralichthyidae)), and (3) high abundance, low value resident species that typically do not have management plans (e.g., mullets, sea catfish, mojarra (Gerreidae), grunt (Haemulidae), cusk-eels (Ophidiidae), jacks and pompanos(Carangidae), and approximately 100 additional species of fish.

The information generated in this study is comprehensive, and the use of different zones as well as the types of target species could be used as a mechanism of fisheries management and regulations. However, much more information is required to adequately assess the artisanal fisheries of Sinaloa. More information is needed about the merchandising and infrastructure of these fisheries, as well as the fishing operation itself including the total number of fishing permits, boats, fishermen and fishing gears used for each fishery and in each zone. Our results indicate that fishery resources are used in different ways in each zone, which is required information for the development of management strategies (Hilborn and Walters 1992). However, for a more proper assessment, the reasons and causes of these differences need to be more thoroughly understood through information about catch per unit effort and migratory patterns of the species. Further, there is also a lack of information about the selectivity of fishing gears (species composition and size structure of the catch), although it is known that trawl and gill nets employed for the shrimp fishery capture many juvenile fishes (Amezcua et al. 2006, 2009).

Correct identification of the exploited species is also essential. Currently, most captured species are labeled by common names, which prohibit the use of actual catch-specific composition that could include up to 200 species in some zones (Amezcua et al. 2006). Additionally, grouping landings by Family, or other grouping classifications such as demersal fishes, also masks the effects of fishing in certain populations because individual species are not denoted when the catch record data are analyzed.

At present, management policies have focused on maintaining single-species fisheries (i.e., shrimp and swimming crab) and few solutions have been discussed on how to manage a multi-species fishery. One typical option is the ecosystem-based management approach, which takes into account the effects of harvest on the community (Plagányi 2007); however, this requires a synthesis of a wide array of physical, biological and socioeconomic data—data that are incomplete for the artisanal fisheries of Sinaloa. Despite this lack of information, small-scale fisheries management should consider fisheries regionalization to apply zone-specific regulations.

Acknowledgments We thank C. Suarez and G. Ramirez for editing the figures. F. Amezcua thanks the project PAPIIT IN208911 for partially funding this research.

References

Accadia P, Franquesa R (2006) The operational units approach for fisheries management in the Mediterranean sea. General Fisheries Commission for the Mediterranean, FAO, Rome

Amezcua F, Madrid-Vera J, Aguirre-Villaseñor H (2006) Effect of the artisanal shrimp fishery on the ichthyofauna in the coastal lagoon of Santa Maria la Reforma, Gulf of California. Cienc Mar 32(1B):97–109

Amezcua F, Madrid-Vera J, Aguirre-Villaseñor H (2009) Incidental capture of juvenile fish from an artisanal fishery in a coastal lagoon in the Gulf of California. North Am J Fish Manag 29:245–255

Amezcua-Linares F (1977) Generalidades ictiológicas del sistema lagunar costero Huizache-Caimanero, Sinaloa. An Inst Cienc Mar y Limnol UNAM 4(1):1–25

Amezcua–Linares F (2009) Peces demersales del Pacifico de México. Universidad Nacional Autonoma de Mexico, Mexico

Aquino-Guzmán MA, Alonso-Rojo F, Millan-Garcia JG et al (1983) Composición y abundancia de las larvas de peces en el canal de agua-dulce del sistema lagunar Huizache-Caimanero. Dissertation, Universidad Autonoma de Sinaloa

Arreguín-Sánchez F (2006) Pesquerías de México. In: Guzmán-Amaya P, Fuentes-Castellanos D (eds) Pesca, acuacultura e investigación en México. Centro de Estudios para el Desarrollo Rural Sustentable y la Soberanía Alimentaria, Cámara de Diputados LIX Legislatura/Congreso de la Unión, Mexico, pp 13–36

Briggs JC (1974) Marine zoogeography. Mc Graw Hill, USA

Briggs JC (1995) Global biogeography. Elsevier, Amsterdam

Cano-Pérez FA (2000) Oceanografía física. In: de la Lanza-Espino G (ed) Oceanografía de los mares mexicanos. AGT Editor, Mexico, pp 453–511

Castro-Aguirre JL (1978) Catálogo sistemático de los peces marinos que penetran a las aguas continentales de México con aspectos zoogeográficos y ecológicos. Dep Pesca Dir Gen INP Ser Cient, Mexico

Castro-Aguirre JL, Espinosa HS, Schmitter-Soto J (1999) Ictiofauna estuario-lagunar y vicaria de México. Limusa/IPN, Mexico

Chapa-Saldaña H, Soto-López R (1969) Relación de algunos factores ecológicos en la producción camaronera de las lagunas litorales del sur de Sinaloa. In: Ayala-Castañares A, Pheger B (eds) Lagunas Costeras. UNAM-UNESCO, Mexico, pp 653–662

Contreras-Espinoza F, Castañeda-López O (2003) Los ecosistemas estuarinos del Estado de Sinaloa. In: Cifuentes-Lemus JL, Gaxiola-López J (eds) Atlas de los ecosistemas de Sinaloa. El Colegio de Sinaloa, Culiacán, pp 175–196

de la Cruz-Agüero J (2002) Peces marinos de Sinaloa: historia, distribución y diversidad. In: Cifuentes-Lemus JL, Gaxiola-López J (eds) Atlas de la biodiversidad de Sinaloa. El Colegio de Sinaloa, Culiacán, pp 301–309

Escobar–Fernández R, Siri M (1997) Nombres vernáculos y científicos de los peces del Pacífico Mexicano. UABC-SIMAC, México

Espino-Barr E, Cruz-Romero M (2006) Aspectos generales de la pesca ribereña en el Pacífico Mexicano. In: Guzmán-Amaya P, Fuentes-Castellanos D (eds) Pesca, acuacultura e investigación en México. Centro de Estudios para el Desarrollo Rural Sustentable y la Soberanía Alimentaria-Cámara de Diputados LIX Legislatura/Congreso de la Unión, Mexico, pp 37–47

Espino-Barr E, Cruz-Romero M, Garcia-Boa A (2003) Peces marinos con valor comercial de la costa de Colima, México. CONABIO-INP, Mexico

Espino-Barr E, Cabral-Solís EG, Garcia-Boa A et al (2004) Especies marinas con valor comercial de la costa de Jalisco, México. SAGARPA—INP, Mexico

Fisher W, Krupp F, Schneider W et al (1995) Guía FAO para la identificación de especies para los fines de la pesca. Pacífico centro-oriental, vol I. FAO, Roma

Hendrickx ME, Brusca RC (2002) Biodiversidad de los invertebrados marinos de Sinaloa. In: Cifuentes-Lemus JL, Gaxiola-López J (eds) Atlas de la biodiversidad de Sinaloa. El Colegio de Sinaloa, Culiacán, pp 141–163

Lankford R (1977) Coastal lagoons of Mexico: their origin and classification. In: Wiley M (ed) Estuarine processes. Academic Press, New York, pp 182–205

Méndez N (2002) Annelid assemblages in soft bottoms subjected to human impact in the Urías estuary (Sinaloa, Mexico). Oceanol Acta 25:139–147

Miller DJ, Lea RN (1972) Guide to the coastal marine fishes of California. Dept Fish and Game Bull 157, USA

Páez-Osuna F, Guerrero-Galván SR, Ruiz-Fernandez AC et al (1997) Fluxes and mass balances of nutrients in a semi-intensive shrimp farm in North-Western Mexico. Mar Pollut Bull 34:290–297

Panayotou T (1983) Conceptos de ordenación para las pesquerías en pequeña escala: aspectos economicos y sociales. FAO, Rome

Pitcher TJ, Hart PJB (1982) Fisheries ecology. Chapman and Hall, London

Plagányi ÉE (2007) Models for an ecosystem approach to fisheries. FAO Fisheries Technical Paper 477, Rome

Ramírez-Rodríguez M (2009) Pesquerías de pequeña escala en el noroeste de México. In: Urciaga-García J, Beltrán-Morales LF, Lluch-Belda D (eds) Recursos y servicios ambientales en el desarrollo regional. CIBNOR, UABCS, CICIMAR-IPN, Mexico, pp 157–176

Ramírez-Rodríguez M (2011) Data collection on the small-scale fisheries of México. ICES J Mar Sci 68:1611–1614. doi:10.1093/icesjms/fsr089

Ramírez-Rodríguez M, Ojeda-Ruíz MA (2012) Spatial management of small-scale fisheries on the west coast of Baja California Sur, Mexico. Mar Policy 36:108–112

Ramírez-Rodríguez M, López-Ferreira C, Hernández-Herrera A (2007) Atlas de localidades pesqueras de México, Libro cuatro: Sinaloa. IPN-CONAPESCA, Mexico

Robertson DR, Allen GR (2006) Shorefishes of the tropical eastern pacific: an information system. Smithsonian Tropical Research Institute, Balboa

Rubio-Rocha YG, Beltrán-Magallanes JA (2003) Biodiversidad y conservación en ambientes costeros. In: Cifuentes-Lemus JL, Gaxiola-López J (eds) Atlas de la biodiversidad de Sinaloa. El Colegio de Sinaloa, Culiacán, pp 99–109

SAGARPA (2007) Ley general de pesca y acuacultura sustentable. Diario Oficial de la Federación, Mexico

SAGARPA (2011) Anuario estadístico de acuacultura y pesca. Secretaría de Agricultura, Ganadería, Desarrollo Rural, Pesca y Alimentación, Mexico

SAGARPA (2012) Carta Nacional Pesquera. Diario Oficial de la Federación, Mexico

Thomson DA, Findley TL, Kerstiitch NA (1979) Reef fishes of the sea of Cortez: the rocky-shore fishes of the gulf of California. University of Arizona, Tucson

Van der Heiden AM, Findley LT (1988) Lista de peces marinos de Sinaloa. An Inst Cienc Mar y Limnol UNAM 15(2):209–224

Warburton K (1978) Community structure, abundance and diversity of fish in a Mexican coastal lagoon system. Estuar Coast Mar Sci 7:497–519

Part III
Socioeconomic Considerations

9 Ecosystem-Based Fisheries Management of a Biological Corridor Along the Northern Sonora Coastline (NE Gulf of California)

Peggy J. Turk-Boyer, Hem Nalini Morzaria-Luna, Iván Martinez-Tovar, Caroline Downton-Hoffmann and Adrian Munguia-Vega

Abstract

The coastline of Northern Sonora is dominated by hypersaline estuaries and vast rocky intertidal zones that are intermittently covered by the extreme tides characteristic of the Northern Gulf of California. Research on the spatial-temporal distribution of flora and fauna in wetland, sandy-muddy bottoms, the pelagic zone, subtidal rocky reefs and an offshore island offer an in depth characterization of the region's habitats and allow the definition of a unique biological Corridor for the coastal zone between Punta Borrascoso and Puerto Lobos, Sonora. Trophic studies and coupled oceanographic-biological models validated by larval dispersal and population genetic studies on commercial species highlight the connectivity between marine and coastal habitats and support the Corridor as a distinct management unit, especially for fisheries. Patterns of human use along the coast (fisheries, tourism and coastal development) have been documented and currently stakeholders in six communities are engaged in fisheries monitoring and management. The wealth of information available on this Corridor supports an ecosystem-based approach for fisheries management. The traditional hurdles to successful implementation of ecosystem-based fisheries management can be overcome for the coastal fisheries of the Peñasco Corridor by defining essential habitats for important target species, identifying trophic interactions, involving fishers and coastal communities in spatial planning and decision-making, and creating a positive incentive system.

Keywords

Ecosystem-based fisheries management (EBFM) · Connectivity · Marine spatial planning · Northern Gulf of California · Food web connectivity

P. J. Turk-Boyer (✉) · H. Morzaria-Luna · I. Martinez-Tovar · C. Downton-Hoffmann
Centro Intercultural de Estudios de Desiertos y Océanos (CEDO), Apartado Postal 53, CP 83550 Puerto Peñasco, SON, Mexico
e-mail: peggy@cedointercultural.org

H. Morzaria-Luna
e-mail: hmorzarialuna@gmail.com

I. Martinez-Tovar
e-mail: ivan.mar.tovar@gmail.com

C. Downton-Hoffmann
e-mail: caroline@cedointercultural.org

A. Munguia-Vega
Conservation Genetics Laboratory, School of Natural Resources and the Environment, University of Arizona, Tucson, AZ 85721, USA
e-mail: airdrian@email.arizona.edu

9.1 Introduction

Estuarine environments are known to be among the world's most important ecosystems for marine fisheries production (Houde and Rutherford 1993). In the Gulf of California, it has been estimated that 32 % of landings of small-scale fisheries depend on mangroves and coastal fringe environments (Aburto-Oropeza et al. 2008). The Northern Gulf, a distinct ecoregion of the Gulf of California, boasts of 2,116 km of coastline, three riparian systems, and 114,206 ha of coastal wetlands (Glenn et al. 2006). The region yields over 15 % of Mexico's total fisheries volume (Brusca and Bryner 2004). Fishing is the primary economic activity for the region's

F. Amezcua, B. Bellgraph (eds.), *Fisheries Management of Mexican and Central American Estuaries*, Estuaries of the World, DOI 10.1007/978-94-017-8917-2_9, © Springer Science+Business Media Dordrecht 2014

seventeen permanent fishing communities and 3500 small-scale fishers (Cudney-Bueno and Turk Boyer 1998; Moreno-Baez et al. 2012). Valuable shrimp, blue crab, clam, and scallop fisheries abound, with approximately 80 target species harvested by small-scale fishers (Cudney Bueno and Turk Boyer 1998; PANGAS 2012a, b, c, d, e).

Bound by the Midriff Island archipelago in the south and by the Colorado River Delta in the north, the Northern Gulf of California has a variety of topographic and bathymetric features and habitats that support the region's biodiversity (CONABIO 2004 AICA NO-34; Thomson et al. 2000). The shallow sloping shelf of the Northern Gulf is characterized by extreme tides, temperatures and salinities (Alvarez-Borrego et al. 1975) that fuel a seasonally reversing oceanographic gyre (Marinone 2003), distributing nutrients, larvae and food, and connecting populations and communities throughout the ecosystem (Marinone et al. 2008).

High biodiversity is another hallmark of the region. Forty-six percent of the Gulf's total ichthyofauna and a similar proportion of the Gulf's macro-invertebrates are present in the Northern Gulf (Hastings et al. 2010; Brusca and Hendrickx 2010). Cartilaginous fish are especially well represented, with 70 % of the Gulf's 87 species found in the region (Hastings et al. 2010). The Northern Gulf is also known for its large predatory fish of commercial and recreational importance: for example the families Serranidae and Scianidae are especially abundant (Rupnow 2008; Hernández-Velasco 2010; Aburto-Oropeza et al. 2008). Over 23 species of marine mammals (Urban 2010; Brusca et al. 2004) and five of the world's seven sea turtle species (Seminoff and Nichols 2007) use the area for feeding, breeding and nesting. The coastal wetlands offer feeding and resting spots for migratory birds of the Pacific Flyway as well as many resident and breeding species (Hinojosa-Huerta et al. 2007) and serve many other ecosystem functions.

Among the most renowned species in the Northern Gulf are the endemic vaquita porpoise (*Phocoena sinus*); the world's most endangered marine cetacean (Rojas-Bracho et al. 2006; Urban 2010), and the endemic giant croaker (*Totoaba macdonaldi*), one of the world's first marine fish to be listed as endangered (Barrera-Guevara 1990; Cisneros-Mata et al. 1995). Protection of these two species has been a primary driver for biodiversity conservation and fisheries management in the entire region, but especially in the Upper Gulf of California and Colorado River Delta Biosphere Reserve (Fig. 9.1). The reserve was established in 1993 (DOF 1993) largely to protect these two species and the spawning habitat of over 22 commercially important fish species within the Colorado River Delta (Barrera-Guevara et al. 2004). A Vaquita Refuge was created in 2005 (DOF 2005) overlapping the Reserve, offering an additional 1,263 km^2 to protect this species from entanglement in gillnets, the primary cause of incidental mortality (Rojas-Bracho et al. 2006).

The Northern Gulf of California was once a vast estuarine environment (Lavín and Sánchez 1999), fed by the mighty Colorado River and other riparian systems along the coast of Sonora, Mexico. The Colorado once contributed up to 50 % of all the freshwater entering the Gulf, with a freshwater lens extending up to 65 km south towards San Felipe, B.C. (Lavín and Sánchez 1999). Temperature and salinity profiles (Alvarez-Borrego et al. 1975), sedimentation patterns (Carriquiry and Sánchez 1999) and the perception of fishers (Cudney-Bueno and Turk-Boyer 1998) all indicate that the influence of this system was greatest along the northwestern coast of the Northern Gulf, where distinct fisheries are harvested. Today, only intermittent or subsurface flow enters the Gulf due primarily to upstream diversion for urban and agricultural use (Glenn et al. 2006; CONAGUA 2011). The impacts of the lack of freshwater from the Colorado River on the marine ecosystem have been well documented; these impacts include reductions in shrimp landings (Galindo-Bect et al. 2000), lower abundance and reduced distribution of freshwater clams (Rodríguez et al. 2001), riparian habitat loss (Glenn et al. 2001), reduction in avian breeding populations (Hinojosa-Huerta et al. 2004a, b), and alteration of reproduction and spawning habitat of the endemic totoaba population (Lercari-Bernier and Arreguín-Sánchez 2009). Incidental capture of juvenile totoaba in shrimp trawling activities (Barrera-Guevara 1990) and over-fishing (Flanagan and Hendrickson 1976) are contributing factors that led to the near extinction of this anadromous species (Rowell et al. 2008), underscoring the complexities inherent in management of coastal fisheries.

Despite the changes in freshwater input, the Upper Gulf continues to provide important ecosystem services reminiscent of estuarine conditions and primary productivity remains high (Alvarez-Borrego 2010). There is evidence, however, that fisheries populations in the region have declined (Cudney-Bueno and Turk-Boyer 1998; Sáenz-Arroyo et al. 2005; Erisman et al. 2012; Rupnow 2008). Open access, increased fishing effort, use of non-selective fishing gears, lack of management and insufficient enforcement are some of the primary causes for this decline (Cinti et al. 2010a, b). In the Northern Gulf, lack of information to advance fisheries management has been a major challenge, particularly amongst end users. In the last decade, several multi-disciplinary research efforts have advanced understanding of fisheries in the context of the Northern Gulf ecosystem and new strategies have emerged to solve this situation (PANGAS 2012a, b, c, d, e; Ainsworth et al. 2011).

In the Upper Gulf Reserve, where the survival of the world's most endangered marine mammal is threatened by

Fig. 9.1 Northern Gulf of California, Mexico, it's 17 fishing communities and 11 natural protected areas. The Upper Gulf of California/Colorado River Delta Biosphere Reserve is the northern-most red area

the region's most important fisheries, these various management challenges are compounded and complex. Though management structures have evolved, the process has been characterized by conflict and consumed the energy of environmental and fisheries authorities. In contrast, at the southeastern corner of the Reserve a different approach to fisheries management has emerged. Here, participatory research and management processes are unfolding with the collaboration of six coastal communities that exploit distinct fisheries in a diversity of habitats. A wealth of scientific information has been generated that characterizes the Peñasco Corridor, linking coastal and marine habitats and communities from Punta Borrascoso to Puerto Lobos in Northern Sonora as a distinct eco-region. Here, important advances have been made on key components for implementation of Ecosystem-based Fisheries Management (EBFM).

This chapter explores EBFM for the Peñasco Corridor. In this Corridor, fisheries production rivals that at the mouth of the Colorado River, supported by riparian, wetland, rocky and soft-bottom habitats. Especially in areas as diverse (at the genetic, species and environmental levels) as the Peñasco Corridor, effective fisheries management can be well served by an integrated approach that recognizes the connectivity between marine and coastal systems, habitats, species and human communities. We present a characterization of the diverse habitats of this Corridor, documenting physical characteristics, overall biodiversity, presence and abundance of fisheries species, and life stages present. We describe fishing effort, the fisheries exploited in each habitat, and present an analysis of catch composition and diversity by community. We describe studies on the connectivity between habitats and species by looking at trophic structure within the estuary system and offshore environments. We also summarize research on larval dispersal and genetic structure for three important commercial species using coupled biological-oceanographic models. Finally, this work highlights an emerging participatory process for ecosystem-based fisheries management in the rich biological Corridor from Punta Borrascoso to Punta Lobos, Sonora.

The information presented here is derived from research of an ecosystem-based management research consortium studying small-scale fisheries in the Northern Gulf of California-PANGAS (http://www.pangas.arizona.edu); and from the Atlantis Ecosystem-based modeling project for the Northern Gulf ecosystem developed by the Northwest Fisheries Science Center (NWFSC), National Oceanic and Atmospheric Administration (NOAA) in collaboration with PANGAS and CEDO. Methods and results for different components of these projects are available as theses, reports, or in some cases published papers, while other publications are in process. Additional unpublished data and analyses from these studies are presented in this chapter, such as the calculation of indices of richness, trophic structure, and diversity, generation of species lists, and site-specific comparisons.

A general description of the methods employed for surveys conducted in subtidal rocky reefs of the Peñasco Corridor and Northern Gulf from 2007 to 2012 can be found in Appendix 9.1. Methods for a 2-year community catch monitoring program in the Peñasco Corridor between September 2010 to 2012 and for data analysis are outlined in Appendix 9.2.

9.2 Ecosystem-Based Fisheries Management

Traditional fisheries management is slowly evolving from single species stock assessment models to multi-species approaches that incorporate management of biodiversity, habitat and ecosystem processes as well as socio-economic considerations (Koen-Alonso 2009; Crowder et al. 2008; Pikitch et al. 2004). This change has resulted from the recognition of the ineffectiveness of past approaches, as well as the importance of managing not only target species, but also considering food web dynamics where both predator and prey species abundance are secured (Crowder et al. 2008) and habitats are protected for maintenance of a healthy ecosystem (Pikitch et al. 2004). Ecosystem-based fisheries management addresses many of these issues. The overall objective of EBFM is to sustain healthy marine ecosystems and the fisheries they support (Pikitch et al. 2004). Ecosystem-based fisheries management can be defined in a variety of ways, but at its core, it recognizes: (1) that the use of all resources should be sustainable for the long-term (from both ecological and socio-economic perspectives); (2) ecosystem structure and function must be preserved or restored (if possible); and (3) uncertainties, weaknesses and knowledge gaps need to be explicitly recognized and the management structure needs to address unexpected changes in system conditions (Done and Reichelt 1998).

While the Gulf of California's sardine fishery incorporates ecosystem approaches (Bakum et al. 2010), as yet, no integrated EBFM exists for Gulf of California fisheries (Lluch-Cota et al. 2007), although the most recent fisheries law calls for more integrated management approaches. In 2012, Mexico's first fisheries refuges were established in Baja California Sur (DOF 2012), and a process was developed to define conditions for their establishment. Mexico has also established eleven Natural Protected Areas in the Northern Gulf (Fig. 9.1). These areas are managed by the National Commission of Natural Protected Areas (Comisión Nacional de Áreas Naturales Protegidas, CONANP), an agency of the Environmental Secretariat (Secretaría del Medio Ambiente y Recursos Naturales, SEMARNAT). A comprehensive Environmental Impact Assessment process, which is a broad environmental planning tool, is now required in all of Mexico's protected areas (DOF 2000); to date this instrument has only been enforced for fishers in the Upper Gulf of California Reserve and Vaquita Refuge.

Lack of basic life history information on individual target species, ecological and historical data on most fisheries, and limited characterization of the diverse habitats important for fisheries has made it all but impossible to move forward with an EBFM approach in the Gulf (Lluch-Cota et al. 2007). Ecosystem modeling is a tool that can support EBFM; several ecosystem models have been developed for the Northern Gulf, where a growing body of information has made this possible (Lercari-Bernier and Arreguín-Sánchez 2009; Ainsworth et al. 2011, Díaz-Uribe et al. 2012). Nonetheless, so far there is no formal pathway to incorporate model predictions into management. Given the enormity of the task required to implement EBFM practical approaches are being sought (Cowan et al. 2012, Crowder et al. 2008). Hilborn (2011) distinguishes between core and extended EBFM and proposes that at the most basic level EBFM should: (1) manage individual species by keeping fishing mortality below Maximum Sustainable Yields and fleet capacity within the resource limits; (2) prevent bycatch of non-target species through gear modification, offering incentives to avoid bycatch, or use area and seasonal closures; and (3) avoid fishing practices that modify habitats through spatial closures in sensitive areas.

The processes necessary to implement EBFM are already present in many fisheries systems throughout the world and important advances towards implementing them are being made within the Upper Gulf of California Biosphere Reserve and the Vaquita Refuge. A permit and gear buy-out program was implemented by the Mexican government in 2008 to reduce bycatch on the vaquita porpoise (Rodríguez-Ramírez 2010). Industrial fishers in the Upper Gulf Reserve were the first group required to implement an Environmental Impact Assessment; and in 2009 the same was required from small-scale fishers, setting an important precedent. Small-scale fishers have been active participants in conducting studies necessary for the assessment process, which includes onboard observations, fisheries monitoring, training programs, participation in decision-making processes for managing bycatch and improving compliance, based on socially accepted indicators (Pérez-Valencia et al. 2013a).

Finally, fisheries are part of a complex socio-ecological system and fishers must be incorporated into management (Gutiérrez et al. 2010; Hilborn 2011). Stakeholders need to be involved in processes to understand their fisheries, make decisions about management, and commit to comply with agreements; these aspects are as important as top-down approaches and legal tools (Cudney-Bueno et al. 2009; Espinoza-Romero et al. 2011; Cinti 2010a). Where co-management processes exist, leadership, social capital and incentives have been shown to promote success (Gutiérrez et al. 2010). The need for cross-scale forums to bring different parties together has been identified (Cudney-Bueno and Basurto 2009); many such forums are emerging in the Upper Gulf where progress is being made, such as in the Vaquita Refuge (Turk-Boyer and Barrera 2012). A balanced and comprehensive incentive system is needed to keep all parties motivated and working together towards developing functional fisheries systems (Jones et al. 2011). Socio-economic and market incentives will be the ultimate driver to maintain sustainable fisheries. Catch share systems with harvest quotas or turfs are actively being promoted in the region as one such incentive. These management tools can help prevent overfishing, promote stability and ecological stewardship (Gutiérrez et al. 2010; Fujita et al. 2012).

9.3 The Peñasco Biological-Fisheries Corridor

The Peñasco Biological-Fisheries Corridor, hereafter Peñasco Corridor, is located along the northeastern coast of the Northern Gulf of California between Punta Borrascoso and Puerto Lobos, Sonora (Fig. 9.2). Stakeholders from six coastal communities who engage in small-scale coastal fisheries are an integral part of this Corridor. With the exception of Puerto Peñasco, fishing is the primary economic activity for all of these communities. Puerto Peñasco has a more diversified economy with tourism, industrial (shrimp and finfish) and recreational fisheries in addition to small-scale fishing. The Corridor's small-scale fishers engage in a variety of fisheries, predominantly using small skiffs, called "pangas". The fisheries use a variety of gears, including commercial diving, gillnets and traps in the shallow subtidal zones up to 30 m depth, and longline and deepwater nets used in waters up to 150 m depth. Fishing for the star-studded grouper (*Hyporthodus niphobles*) occurs occasionally in depths of up to 300 m using long-line, but this depth is not considered in our definition of the Corridor. An average fishing day might take a fisherman 45 km from the coast. Deeper water is closer to the shore at the southern part of the Corridor than at Puerto Peñasco.

The southern limit of this Corridor, Puerto Lobos, is the approximate latitude where the summer (May to September) counterclockwise current (Marinone 2003, 2012) turns northward along the Sonoran coast, transporting larvae along the way (Soria et al. 2012). Puerto Lobos also is the southern limit of the Puerto Peñasco regional fisheries office (Oficina de Pesca y Acuacultura, Puerto Peñasco, Sonora) which manages all fisheries issues for the communities of the Corridor as well as for the community of Golfo de Santa Clara. Given that the biological, ecological and fisheries context for Golfo de Santa Clara is distinct, we have excluded this community from the definition of the Peñasco Corridor.

9.4 Characterization of Habitats of the Peñasco Corridor

The variety of coastal habitats in the Peñasco Corridor, from riparian to saltwater wetlands, sandy and muddy bottom environments, intertidal and offshore rocky reefs, an island archipelago and a rich pelagic environment give sustenance to marine fisheries and a diversity of marine life.

9.4.1 Pelagic Environment

The offshore pelagic environment has lower salinity than other Upper Gulf sites, defining a distinct hydrogeographic zone (Sánchez-Velasco et al. 2012). Here a physical-chemical

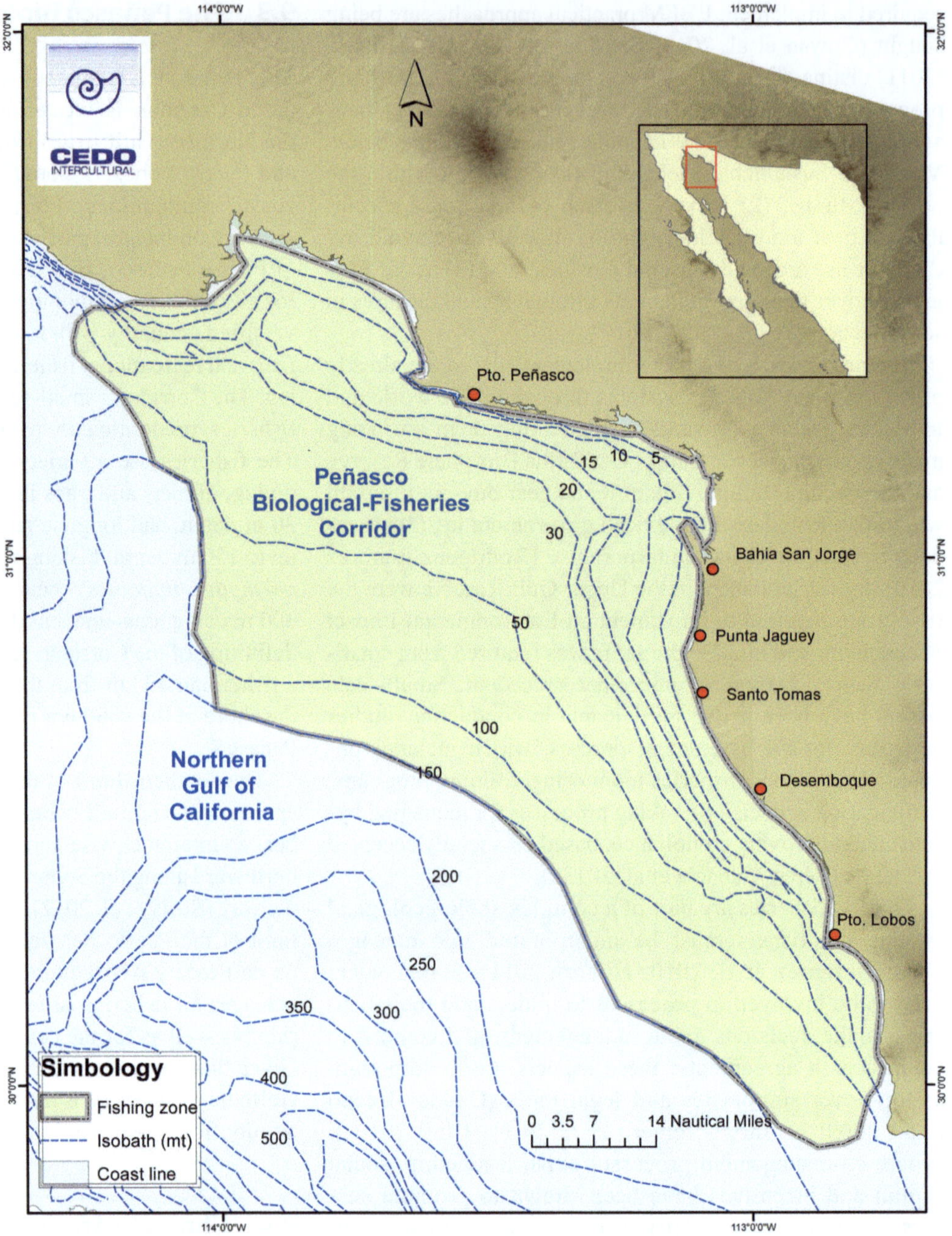

Fig. 9.2 Peñasco Biological-Fisheries Corridor in the Northern Gulf of California, Sonora, Mexico and the six coastal communities it sustains

front separates the shallow sloping coastal Corridor from the deeper waters to the south, roughly corresponding to the area we describe as the "Peñasco Corridor", up to the 40-m isobath.

Both in this pelagic zone and in adjacent wetlands of the Peñasco Corridor the abundance and richness of fish larvae are high, which underscores its relevance as nursery habitat (Iris-Maldonado 2011; Sánchez-Velasco et al. 2012). *Anchoa* spp., an important prey for targeted commercial species, nesting birds and sea lions populations, is the dominant species group in these waters (Iris-Maldonado 2011; Sánchez-Velasco et al. 2012). A significant portion of the nesting population of the least tern, *Sternula antillarum*, uses the wetlands in the Corridor (Palacios and Mellink 1996; Rosemartin 2008). Here, the terns feed and nourish their young with the abundant *Anchoa* spp., *Leuristhes sardina*, and *Colpichthys regis* found in the pelagic environment (Zuria and Mellink 2005; Iris-Maldonado 2011; Sánchez-Velasco et al. 2012).

9.4.2 Riparian Habitats

Riparian vegetation gives evidence of fresh water input at three sites along the Sonora coast (Felger 2000) (Fig. 9.3). North of Puerto Peñasco at Bahía Adair in the high intertidal zone, inside the coastal dunes, artesian springs create

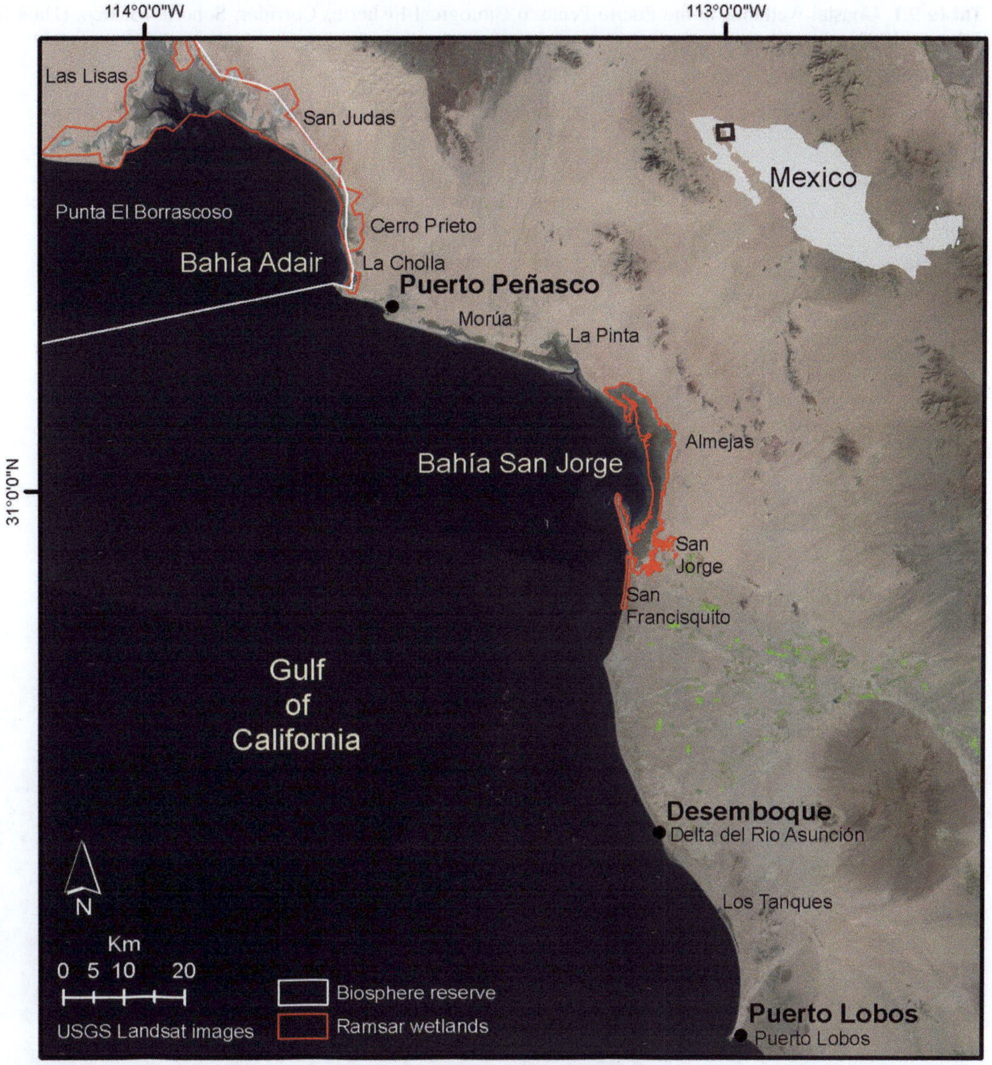

Fig. 9.3 Coastal wetlands in the Puerto Peñasco Biological-Fisheries Corridor, Sonora, Mexico are hypersaline negative estuaries that cover 114,206 ha. (Glenn et al. 2006)

freshwater oases that harbor relict vegetation from the Colorado River (Felger 2000; Ezcurra et al. 1988). Riparian vegetation is also found along the now dry Sonoyta River channel (Felger 2000), which has its terminus between Estero Morúa and Estero La Pinta, once a single estuarine system (Turk-Boyer 1998). At Desemboque, the Asunción River terminates inland of the coastal dunes forming a small wetland with both mesquite trees and halophytic vegetation (Glenn et al. 2006). The marine zones offshore from these riparian areas are recognized as key spawning habitat for shrimp, blue crabs and sharks, species generally known as estuarine spawners (Calderón-Aguilera et al. 2003; Loaiza-Villanueva and Downton- Hoffmann 2011).

9.4.3 Wetland Habitats

In the Puerto Peñasco Corridor, there are eight estuaries that range in size between 235 and 98,740 ha and cover a total of 114,206 ha (Table 9.1; Fig. 9.3) (Glenn et al. 2006). These are negative estuaries with higher salinities at the head than at the mouth, resulting from high evaporation rates and lack of freshwater input (Brusca et al. 2006; Glenn et al. 2006). Environmental conditions within these estuaries are extreme and highly dependent on depth, solar heating, and tidal exchange (Morzaria-Luna et al. 2010), with tidal fluctuations of up to 8 m (Brusca and Bryner 2004). Such extreme environmental conditions determine faunal community structure. Fish community structure is determined by tidal cycle, with higher richness and abundance during neap tide (Iris-Maldonado 2011); while fish larvae are found in zones defined by salinity and temperature gradients, with lower diversity in hypersaline waters (Sánchez-Velasco et al. 2012).

Physically, the estuaries are a complex of intertidal mudflats, tidal channels, marshes, saltpans, coastal dunes, beaches and the wetland-terrestrial ecotone (Fig. 9.4) (Ezcurra et al. 1987; Felger 2000; Johnson 1982). The Peñasco Corridor is beyond the northern limit for mangrove distribution except for a small stand of black mangroves at Puerto Lobos (Brusca et al. 2006; Glenn et al. 2006). The marsh is

Table 9.1 Coastal wetlands in the Puerto Peñasco Biological-Fisheries Corridor, Sonora, Mexico. (Data from Glenn et al. 2006; Felger 2000; Johnson 1982)

Wetland	Area	Main habitats	Tidal regime
Estuaries of Bahía Adair	29226	Halophyte marsh, mud flats, coastal dunes, salt flats, fresh-water springs	Macrotidal, mixed semidiurnal
Cerro Prieto	1987	Halophyte marsh, mud flats	Macrotidal, mixed semidiurnal
La Cholla	235	Halophyte marsh, mud flats, coquina reefs (consolidated shells)	Macrotidal, mixed semidiurnal
Morúa	1097	Halophyte marsh, mudflats	Macrotidal, mixed semidiurnal
La Pinta	3338	Halophyte marsh, mudflats	Macrotidal, mixed semidiurnal
Estuaries of Bahía San Jorge	9874	Halophyte marsh, mudflats, coastal dunes, salt flats	Macrotidal, mixed semidiurnal
San Francisquito	543	Halophyte marsh	Macrotidal, mixed semidiurnal
Delta del Rio Asunción	9233	Brackish marsh (vegetated by mesquites, *Prosopis* spp.) and halophyte marsh	Mixed semidiurnal
Los Tanques	543	Halophyte marsh	Mixed semidiurnal
Puerto Lobos	4	Halophyte marsh, mangrove (black mangrove, *Avicennia germinans*)	Mixed semidiurnal

Fig. 9.4 Estero Cerro Prieto, Sonora, Mexico. Estuaries are a complex of intertidal mudflats, tidal channels, marshes, salt pans, coastal dunes, beaches and the wetland-terrestrial ecotone

vegetated by halophytes and surrounded by Sonoran Desert vegetation characteristic of the Lower Colorado River Valley subdivision (Shreve and Wig-gins 1964). This is the most arid type of Sonoran Desert, because of the combination of high temperature and low precipitation (Turner and Brown 1994).

Wetland-dependent fisheries represent a large proportion of the commercial harvest in the Northern Gulf of California. For example, in 2008 shrimp landings in Sonora represented over 8000 metric t in live weight or 16% of national production (CONAPESCA 2008). The Corridor's most important commercial fish use these wetlands in their adult phases (Table 9.2). Fish and invertebrates of commercial importance such as shrimp and blue crab use estuaries as nursery sites. Blue shrimp (*Litopenaeus stylirostris*) larvae recruit within coastal wetlands and the coastal area outside estuaries in the Corridor has the highest concentration of mature females along the coast of Sonora (Calderón-Aguilera et al. 2003). A majority of estuary fish are juveniles; an analysis of fish community structure in Estero Morúa and Estero La Salina found that 60 to ~80% of fish found throughout the year were immature (Iris-Maldonado 2011).

Table 9.2 Commercial fish species that use estuaries of the Peñasco Corridor, Sonora, Mexico. (Cudney-Bueno and Turk-Boyer 1998; Hastings and Findley 2007; Iris-Maldonado 2011)

Species	Common name
Paralabrax maculatofasciatus	Spotted sand bass
Orthopristis reddingi	Bronze-striped grunt
Lutjanus argentiventris	Yellow snapper
Mycteroperca jordani	Gulf grouper
Mustelus lunulatus; Mustelus henlei	Smooth-hound sharks
Dasyatis brevis	Whiptail stingray
Rhinobatos productus	Shovelnose guitarfish
Paralichthys aestuarius	Cortez flounder
Menticirrhus nasus	Highfin king croaker
Mugil cephalus, Mugil curema	Mullets
Lobotes pacificus	Pacific tripletail
Hyporthodus niphobles	Star-studded grouper
Hyporthodus acanthistius	Gulf coney

9.4.4 Sandy and Muddy Bottom Habitats

The vast majority of the seafloor in the Puerto Peñasco Corridor is either sandy or muddy bottom habitat (Carriquiry and Sánchez 1999) (Fig. 9.5). The Colorado River once contributed over 160 million t of sediment per year to the Gulf (Van Andel 1964). Combined with river deposits from mainland Sonora, these sediments now cover the entire continental shelf of the Northern Gulf and fill two deep water basins with sand and mud (Carriquiry and Sanchez 1999). With no additional input today, tidal and wind driven currents continually rework these sediments (Carriquiry et al. 2011). Particle size and associated oxygen content are primary factors that determine the distribution of both infauna and epifauna in these habitats.

Sandy and muddy bottom environments in the Gulf of California are known for their high diversity of invertebrate species. Considering all depths, sandy and muddy habitats account for 41.2 and 26.7% of the total macro invertebrate species in the Gulf, respectively (Brusca and Hendrickx 2010).

Sharks and rays (Chondrichthys) are particularly diverse in soft-bottom habitats, with 52% of the Gulf's species present in the Upper Gulf, probably in part related to the prevalence of soft substrates (Hastings and Findley 2007). Other species groups typically found in soft bottoms in the Corridor include: anchovies (Engraulidae), herrings (Clupeidae), new world silversides (Atherinopteridae), a variety of perciform fish such as corvinas and croakers (Sciaenidae), grunts (Haemulidae), flounders and flatfishes (Pleuronectiformes) (Hastings and Findley 2007).

Sandy bottoms are the primary habitat for the region's commercial shrimp and the valuable geoduck clam (*Panopea globosa*) (Pérez-Valencia and Aragón–Noriega 2012), as well as other invertebrates of commercial importance (Appendix 9.3).

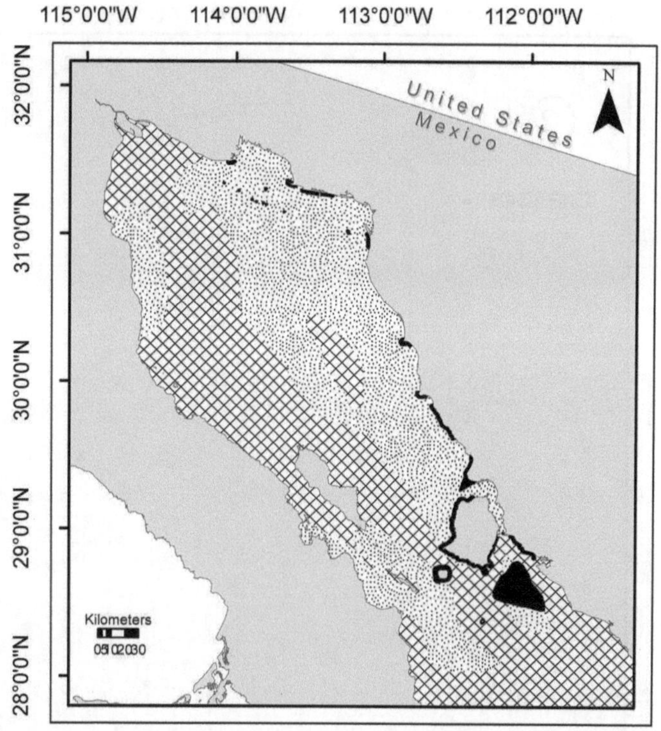

Fig. 9.5 Substrate map showing relative distribution of major habitat types of the northern Gulf of California, Mexico: Rock (*solid black*); mud (*cross-hatch*); sand (*fine dots*). (From Ainsworth et al. 2011)

9.4.5 Rocky Reefs

Intertidal and subtidal rocky reefs are also found in the Northern Gulf of California. Intertidal reefs are limited to only a few sites, and three of them are in the Peñasco Corridor (Borrascoso, Puerto Peñasco, and Puerto Lobos). Subtidal rocky outcrops appear in the Corridor as a series of patch reefs that parallel the shore along the continental shelf between Bahía Adair and Bahía San Jorge, in both shallow water (2–25 m depth) and deep water (>25 m depth) (Fig. 9.6). Composed predominantly of beach rock (limestone), the reefs appear as small exposed islands of hard substrate in a predominantly sandy and muddy bottom. These reefs were probably once shoreline intertidal habitats inundated by Pleistocene sea level rise (Thomson and Gilligan 1983). Recreational fishers identify 17 reefs they fish regularly (Pérez-Valencia et al. 2007; Rupnow 2008); while industrial trawlers have located many more rocky outcrops as targets to avoid during their trawling activities (Turk-Boyer et al. 2009).

Rocky reefs vary in size and rugosity. Some create substrate for habitat forming species, such as *Sargassum* spp. (Aburto-Oropeza et al. 2007) and rhodoliths (Steller et al. 2003). They offer shelter, spawning and feeding sites for large predatory fish and support coastal fisheries. Though reefs represent an essential habitat for many species, they are a patchy resource, limited in geographical coverage and

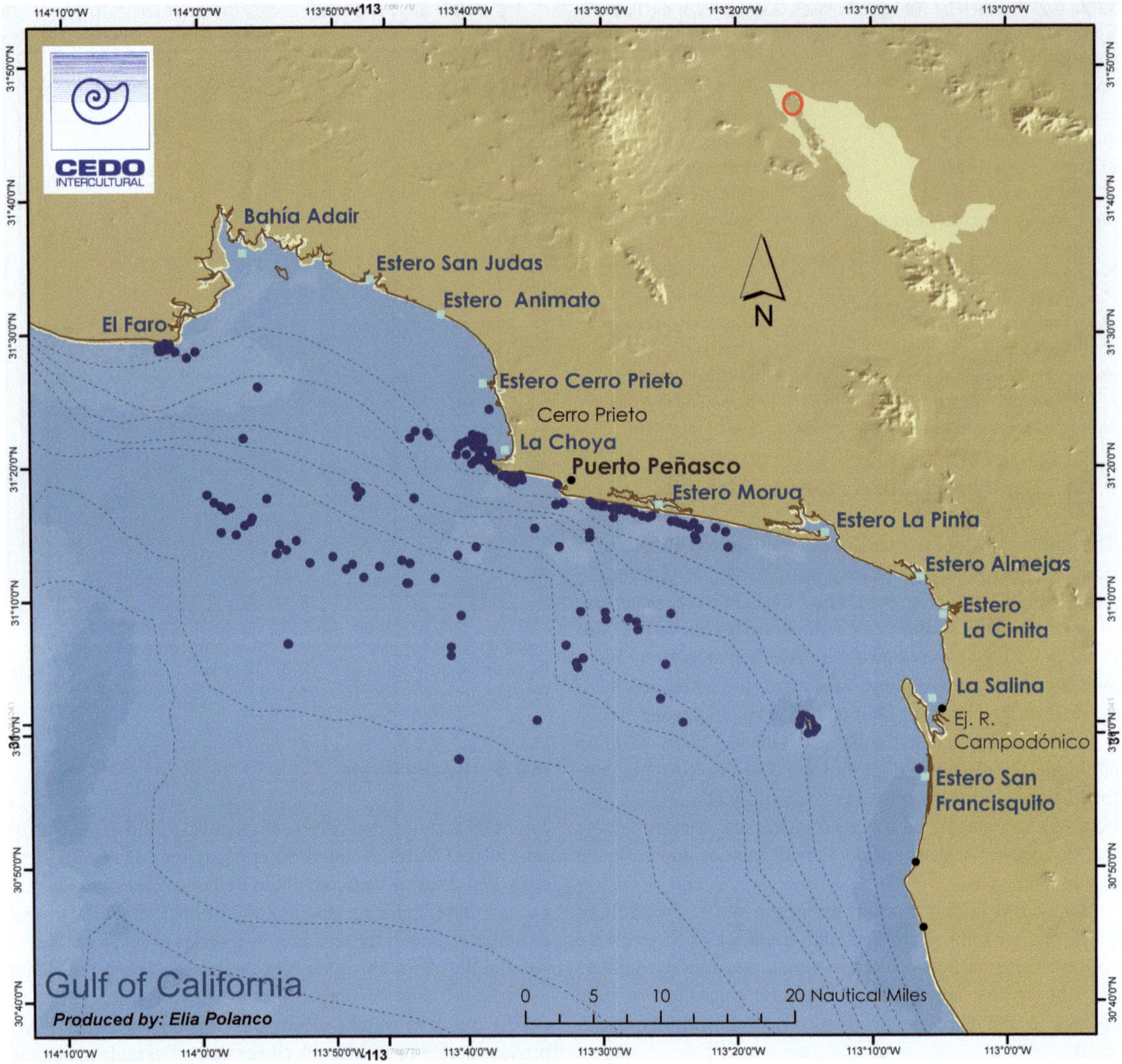

Fig. 9.6 Subtidal reefs and bathymetry offshore of Puerto Peñasco, Sonora, Northern Gulf of California. Data from research expert workshop, commercial divers who fish these areas and industrial shrimp trawlers who avoid these rocks

physically separated from other reefs. Despite their limited abundance, rocky habitats harbor 33.4% of the total Gulf macroinvertebrate faunal species (Brusca and Hendrickx 2010), with mollusks and echinoderms commonly associated with these habitats (Keen 1971; Brusca 1980; Maluf 1988; Cintra-Buenrostro 2001; Solís-Marín et al. 2005).

9.4.6 Offshore Islands

The San Jorge Island reefs are the most important offshore rocky habitats in the Peñasco Corridor in terms of diversity (Martinez 2010). San Jorge is the northern-most rocky island in the Gulf, located approximately 41 km southeast of Puerto Peñasco. With one main island and four islets, San Jorge is an island archipelago that extends 1,097 m in length. The islands are composed of granite and are arid and steep, with terrestrial and intertidal habitats that decline precipitously into subtidal rocks. San Jorge Island is a protected area under the CONANP category of Special Area for Protection of Flora and Fauna (Area Especial de Protección de Flora y Fauna: Islas del Golfo de California, Sonora), but the marine portion of this island archipelago is not included in this designation and is under no special management or protection category.

Table 9.3 Number of small skiffs (pangas), permits, commercial species fished and fishing gear used by three communities of the Upper Gulf of California Biosphere Reserve in 2011, and five communities of the Peñasco Corridor in the Northern Gulf of California, Sonora, and capture in tons

Site	Number of pangas	Number of permits	Capture in tons	Number of species fished	Fishing gear
San Felipe	315[a]	603	7,848.6[b]	65	Hooka, gillnet, longline, cane and traps
Golfo de Santa Clara	457[a]	925	11,619[b]	59	Hooka, gillnet, longline and traps
Puerto Peñasco	350[a]	278	19,045[b]	46	Hooka, gillnet, longline, pole and traps
Subtotal	1,122	1,806	38,512.6		
Bahia San Jorge	30	124	205.21	13	Gillnet, traps
Punta Jagüey	48	19	128.41	17	Gillnet, pots and traps
Santo Tomás	37	0	72.34	24	Gillnet, longline, traps
Desemboque	194	65	291.05	30	Hooka, gillnet, longline, pole, traps
Puerto Lobos	42	8	74.91	36	Hooka, gillnet, longline, cane, pots and traps
Subtotal	351	216	771.92		
Total	1473	2022	31435.92		

[a] Data from Upper Gulf small-scale fishers environmental impact study (Pérez-Valencia et al. 2012)
[b] Data from OEIDRUS (2011)

The San Jorge Island reefs represent one of the Northern Gulf's biodiversity hot spots. Diversity indices for fish show the island to be among the highest of the Northern Gulf (Martinez 2010). Large predatory fish are abundant; the island presented the highest average trophic level for fish of all surveyed sites in the Northern Gulf in 2010 and 2011. Subtidal surveys in 2007 and 2011 show commercially important invertebrates were most abundant at San Jorge Island when compared to other sites in the Northern Gulf. Rock scallop (*Spondylus limbatus*), black murex snail (*Hexaplex nigritus*), pen shells, the pearl oyster (*Pinctada margaritensis*), and *Octopus* spp. (Martinez 2010) were the most abundant at San Jorge Island. The island also had the second greatest abundance of the brown sea cucumber (*Isostichopus fuscus*). Winged oyster (*Pteria sterna*), an important fishery in the Corridor, was more abundant at Angel de la Guarda and Tiburon Island.

San Jorge has no terrestrial plants but supports an important seabird colony (Everett and Anderson 1991; Velarde and Anderson 1994). Reported nesting species include brown boobies (*Sula leucogaster*), blue-footed boobies (*Sula sula*), double-crested cormorants (*Phalacrocorax auritius*), Heerman's gulls (*Larus heermanni*), and red-billed tropicbirds (*Phaethon aetherus*) (Velarde and Ander-son 1994). The island is the Gulf's second most important breeding site for the California sea lion (*Zalophus californianus californianus*); sea lion censuses conducted in 2004 found 3,822 individuals with 968 pups (Szteren et al. 2006), but in July 1998, an ENSO year that number was almost double (6,717 sea lions and 793 were pups) (Mellink and Romero-Saavedra 2005).

9.5 Fishing Communities

The Peñasco Corridor is comprised of six fishing communities along a 200 km stretch of coast (Fig. 9.2). The largest community is Puerto Peñasco (pop. 57,342 in 2010; INEGI 2011). Puerto Peñasco also has the most diversified economy with small-scale, industrial, and recreational fisheries as well as commerce, tourism, other service industries and related activities; the municipal government is also based here. The other communities of the Corridor are much smaller and have varying degrees of infrastructure and access to basic services and education. Three of these coastal communities are fishing camps, associated with inland ejidos as follows: Bahia San Jorge (BSJ)—Ejido Rodolfo Campodónico; Santo Tomas (STO)—Ejido Álvaro Obregón; Punta Jagüey (PJA)—Ejido 15 de Septiembre. In this work we use the names of the coastal villages. The local fisheries office that services these communities is a dependency of the National Commission for Fisheries and Aquaculture (Comisión Nacional de Acuacultura y Pesca-CONAPESCA). The CONAPESCA office is located in Puerto Peñasco, where all fisheries catch is officially landed. Because of the distance, most of the catch from communities south of Puerto Peñasco was largely unreported until a community logbook program was initiated in September 2010–2012 (Downton-Hoffmann et al. 2013c). Due to its size, Puerto Peñasco has the largest number of pangas (Table 9.3). In contrast, Santo Tomás has 37 total (20 active) pangas, and no fishing permits. Bahia San Jorge, the fishing community that is most organized, has only one cooperative fishing unit and all the members of the community belong to it.

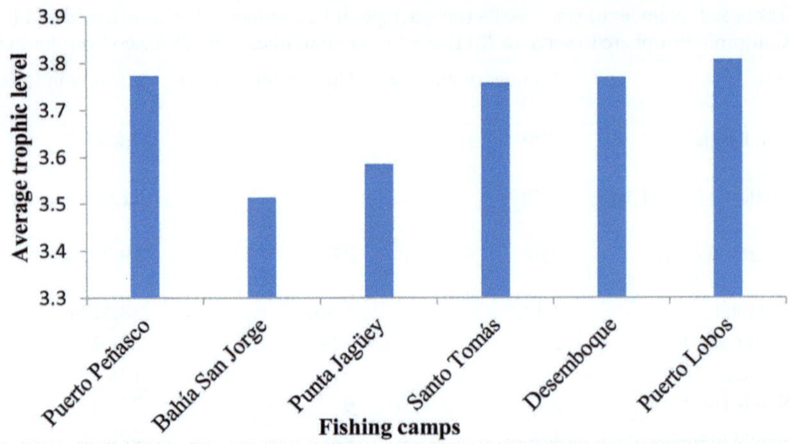

Fig. 9.7 Average trophic level of catch from the six communities of the Puerto Peñasco Corridor, Northern Gulf of California, Sonora. Based on catch data from September 2010 to September 2012. (Downton-Hoffmann et al. 2013c)

Puerto Peñasco is the community that targets the most species (46) and Bahía San Jorge the least (13). Only four target species are shared by all six communities: Gulf drum (chano), guitarfish, blue crab and flounder (Downton-Hoffmann et al. 2013c). A cluster analysis of catch composition separates Bahía San Jorge and Punta Jagüey, communities close to each other that fish in nearby sites and have similar catch (Downton-Hoffmann et al. 2013c; Appendix 9.2). The second cluster groups the two communities that are furthest apart from each other, Puerto Lobos and Puerto Peñasco. Both of these communities harvest resources in rocky habitats, which are likely connected via larval dispersal (see section on marine connectivity). Desemboque also has catch in common with Puerto Lobos, but the latter also fishes in other types of habitats.

An analysis of the trophic level of catch by community shows in general the communities of the Corridor are impacting top predators, with average trophic level ranging between 3.5 and 3.8 (Fig. 9.7). The exceptions are the communities of Bahía San Jorge and Punta Jagüey, which fish predominantly in Bahía San Jorge and lower down the food chain.

Habitat characterizations and fishery-use patterns define a distinct coastal zone and oceanographic region for the Corridor. The offshore limits for the Corridor definition are based on depth and/or distance from the shore. Fishing with traps, hooka and gillnets takes place in shallow areas. In Puerto Peñasco and Bahía San Jorge most areas used by small-scale fisheries do not exceed 30 m depth; while at Desemboque, Puerto Lobos and Santo Tomas fishing takes place in waters up to 150 m deep (Fig. 9.8), which is relatively close to shore. In the deepest areas (300 m), the star-studded grouper is caught using long line and some sharks are also captured, but most catch comes from below the 100 m isobath. At depths between 60 to 100 m longline and gillnets are used to catch angel shark, smoothhound shark and large flounders. Distance from shore might be the most realistic way to define the Corridor, as logistically and economically this distance is what limits daily trips.

9.6 Fisheries in the Peñasco Corridor

A total of 75 species are captured, either as target species or secondary catch by small-scale fishers from the six communities of the Peñasco Corridor. For communities other than Puerto Peñasco the most important commercial fish species include blue crab, winged oyster, squid (occasionally), murex snail, and guitarfish, flounder, angel shark, smoothhound shark and drum (Fig. 9.9) and at Puerto Peñasco geoduck clam, blue crab, and black and pink murex snail are the most important fisheries. Most of these are soft-bottom species or species that use both soft bottom and hard substrates.

Though it does not register as a dominant species anymore, shrimping is an important fishery occurring in sandy and muddy habitats in the Peñasco Corridor, though it is not all backed by permits. Shrimp represented 10 % of the total catch landed in Puerto Peñasco in 2011 (OEIDRUS 2011), including landings from small-scale and industrial fishermen. Small-scale fishers use 2 ¾ inch gillnets from small boats with outboard motors called pangas, while industrial otter trawlers target over nine shrimp species. Approximately 80 % of the catch from trawlers is composed of blue shrimp (*Litopenaeus stylirostris*) and most of the remaining 20 % are brown shrimp (*Farfantepenaeus californiensis*) and white shrimp (*Litopenaeus vannamei*), all of high economic value.

Trawling for finfish takes place after the shrimp season is closed; this fishery is directed towards big fishes such as flounder (families Paralichthyidae and Pleuronectidae), trigger fish (cochito, *Balistes* spp), gold-spotted sand bass (extranjero, *Paralabrax* spp.), corvinas (*Cynoscion* spp.), Pacific hake (*Merluccius* spp.) and several species of elasmobranchs (rays and sharks). Occasionally, when trawls are done close to reefs, predatory sea bass and groupers are caught as well.

The extensive subtidal sandy and muddy plains of the Bahía Adair and Bahía San Jorge Corridor are key habitats for commercially important benthic mollusks harvested by

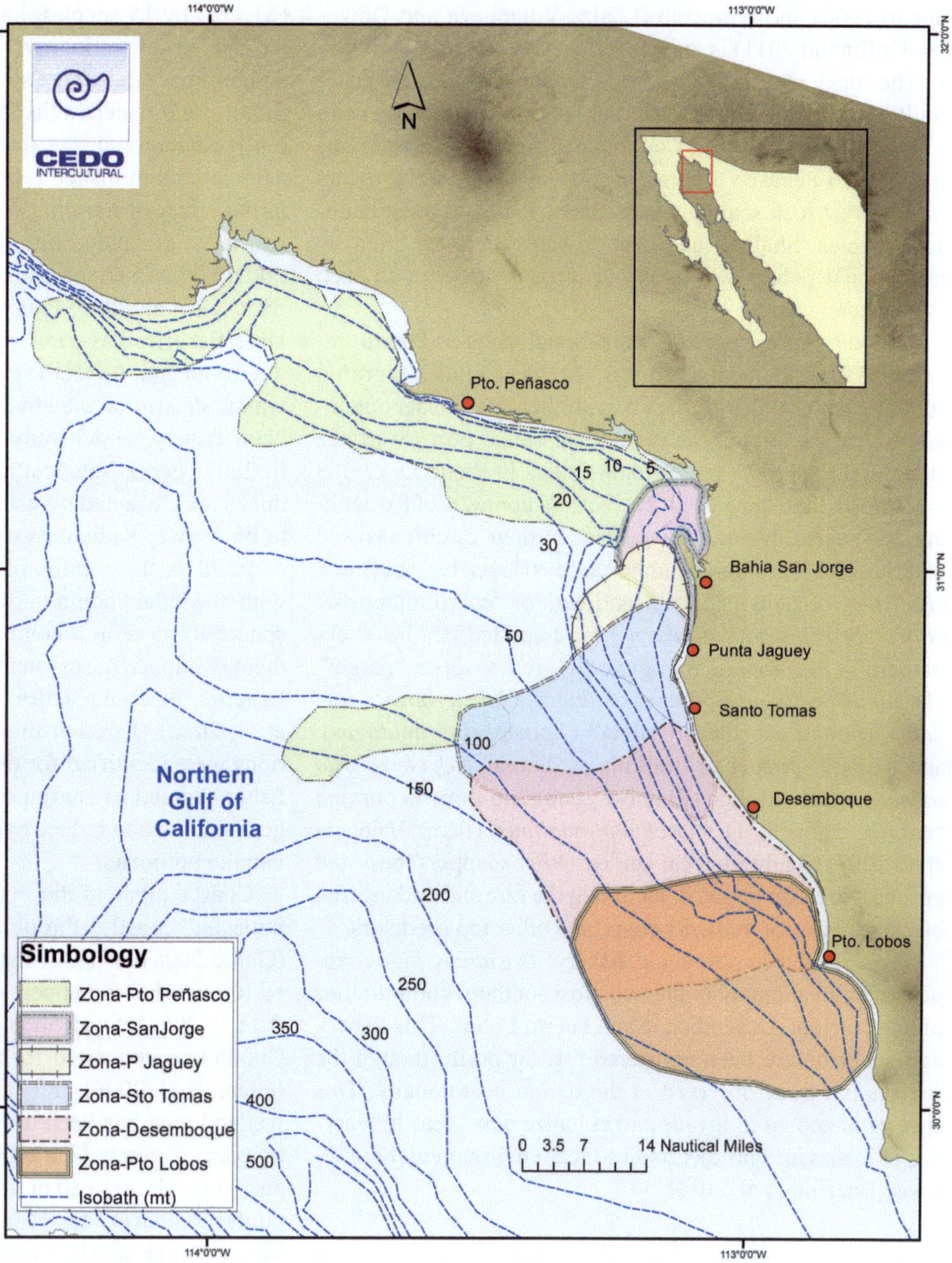

Fig. 9.8 Fishing zones for communities of the Peñasco Corridor in the Northern Gulf of California, Sonora. (Downton-Hoffmann et al. 2013c, Appendix 9.3)

commercial divers, based mainly in Puerto Peñasco. Diving takes place in water up to 30 m, depending on the species, generally with the use of surface supplied air (hookah). Geoduck clam (*Panopea globosa*), white clam (*Dosinia ponderosa*), *Chione* spp. clams and speckled scallop (*Argopecten circularis*) as well as other bivalve species; pen shells (*Pinna* spp.) and winged oyster are the most important of these fisheries (Cudney Bueno and Turk Boyer 1998; Downton-Hoffman et al. 2013a). The spatial distribution of geoduck clam beds (Perez-Valencia and Aragón-Noriega 2012), and winged oyster beds (INAPESCA 2011; Downton-Hoffmann 2012) have been delineated in this region. Also the density of black murex snail (Loaiza-Villanueva et al. 2012; Cudney-

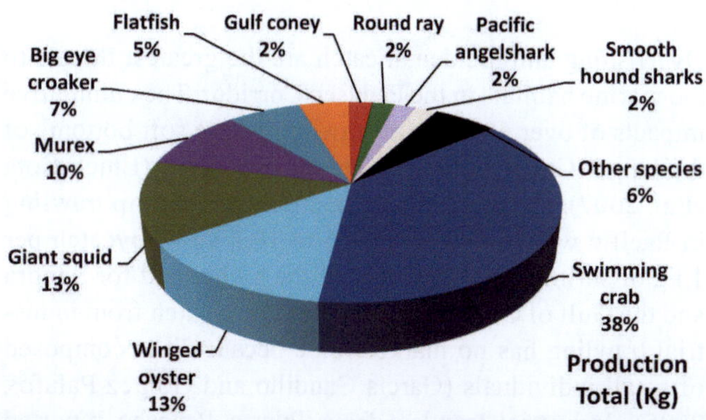

Fig. 9.9 Composition of small-scale fishing catch (in percentages) from five communities of the Peñasco Corridor, Sonora, Mexico in 2011, excluding Puerto Peñasco. (Downton-Hoffman et al. 2013a)

Bueno 2000) and blue crab (Loaiza-Villanueva and Downton-Hoffmann 2011) is known from some areas.

The rocky reef habitats of the Peñasco Corridor are especially important sites for artisanal fisheries (Cudney-Bueno and Turk-Boyer 1998). The shallow water reefs (<25 m) near Puerto Peñasco are used mostly by commercial divers who harvest rock scallop, black murex snail and other abundant species. Shallow and deeper water reefs are used by recreational fishers who seek out large predatory fish such as groupers.

In 2009, there were 207 recreational users at Puerto Peñasco (35 small local operators, 32 local guides working aboard yachts, and 140 tourists with their own boats operating out of La Cholla, the main recreational port (Martinez et al. 2011). In 2003, recreational fishers in Bahía La Cholla mentioned that they used 27 reefs (Rupnow 2008), landing 26 species. Ninety-five percent of their catch consisted of 8 species, in order of importance: triggerfish "cochito" (*Balistes polylepis*), spotted sand bass or "cabrilla arenera" (*Paralabrax maculatofasciatus*), gold-spotted sandbass "extranjero" (*Paralabrax auroguttatus*), red snapper "pargo", (*Lutjanus peru*), gulf grouper or "baya" (*Mycteroperca jordani*), spotted cabrilla or "pinta" (*Epinephelus analogus*), and leopard grouper or "cabrilla sardinera" (*Mycteroperca rosacea*), Gulf corvina (Rupnow 2008) and shortfin corvina "curvina de orilla" (*Cynoscion parvipinnis*) (Pérez-Valencia et al. 2007) caught by local guides. Other snappers, bass and groupers are also targeted, including the rare and endangered black sea bass, *Stereolepis gigas*, and other top predators.

In 2011, the jumbo squid fishery, *Dosidicus gigas*, appeared as an anomaly in the two most southern communities of the Corridor, Desemboque and Puerto Lobos. This fishery had never before been registered this far north, though the species has been observed in the region occasionally. This species is known to invade and colonize new areas by varying its behavior and diet to suit local environmental conditions (Bazzino et al. 2010).

9.7 Human Impacts

Overfishing and incidental catch are the greatest threats to the marine habitats in the Peñasco Corridor. The cumulative impacts of over 60 years of trawling on the soft bottoms of the Upper Gulf have raised the most concern (Lluch-Cota et al. 2007). Bycatch volume and ratio for shrimp trawling in Pacific waters ranges from 5 to 10.2 kg of bycatch per 1 kg of shrimp, with the ratio on the higher end for Sonora and the Gulf of California. Much of the bycatch from industrial trawling has no market value because it is composed of small individuals (García Caudillo and Gómez Palafóx 2005). Industrial trawlers from Puerto Peñasco reported 110 bycatch species in the 2004–2005 season, dominated (61.1 %) by 15 species, including big eye croaker, bronze-striped grunt, stingray, lizardfish, weak-fish, flatfish, swimming crab, robust swimming crab, sardine and shark. Small-scale fisheries that target the same species once they reach commercial size are impacted by trawler bycatch and associated diminished recruitment. Additionally, the cumulative effect of trawling on overall biodiversity and on the resilience of sandy bottoms can be high (Lluch-Cota et al. 2007; Higuera-Ciapara and Mayorga-Ríos 2006; Arreguín-Sánchez et al. 2005; Pauly et al. 1998; Rogers et al. 1999). Dive fisheries have low incidental catch and their overall environmental impact is considered low. Studies are underway to determine the environmental impact of the geoduck clam fishery, which only started in the Peñasco Corridor in 2007 (Pérez-Valencia and Aragon-Noriega 2012). Geoducks are extracted by using high powered air compressors to blow away sediments over the deeply-embedded clam.

In 2010, the community of Puerto Peñasco participated with the other communities within the Upper Gulf Biosphere Reserve in a comprehensive participatory Environmental Impact Assessment for the nine major small-scale fisheries that take place in the Reserve (Pérez-Valencia et al. 2012). Through this process, specific mitigation actions were identified for their most important fisheries, and fishers agreed to continue to develop spatial-temporal and gear solutions to reduce bycatch of the critically endangered vaquita porpoise.

Coastal areas in the Puerto Peñasco Corridor have been impacted mostly through tourism-related developments (Guido 2006). In 1972, Estero Puerto Peñasco was converted to a marina to support the large industrial fleet (Munro-Palacio 2007). Over 50 % of the original area of Estero La Cholla was converted for a residential tourist development (Glenn et al. 2006). In Estero La Pinta, 20 % of the original wetland area has been affected by dredging and filling for tourism activities (Turk-Boyer 2008). In Mexico, beaches and wetlands up to 20 m inland from the highest tide line are public domain (Federal Maritime-Terrestrial Zone) (Cortina-Segovia et al. 2007). Long-term use and exploitation of the Federal Zone requires an environmental review and permits from the Environmental Secretariat, which can define use and in some cases limit conversion of coastal areas to other uses (Morzaria-Luna et al. 2013).

In 2012, the Puerto Peñasco Tourist Homeport project, proposed to receive 50,000 tourists and 48 cruise ships per year, was approved for construction along Sandy Beach, northeast of Puerto Peñasco (SEMARNAT 2012). With the construction of a 3,000 m long and 51 m wide sea wall, the project is poised to change sediment dynamics along the coast and alter key soft bottom and rocky habitats used by commercial divers for harvest of the valuable geoduck clam and rock scallop fisheries (Cudney-Bueno et al. 2009; Pérez-Valencia and Aragón–Noriega 2012).

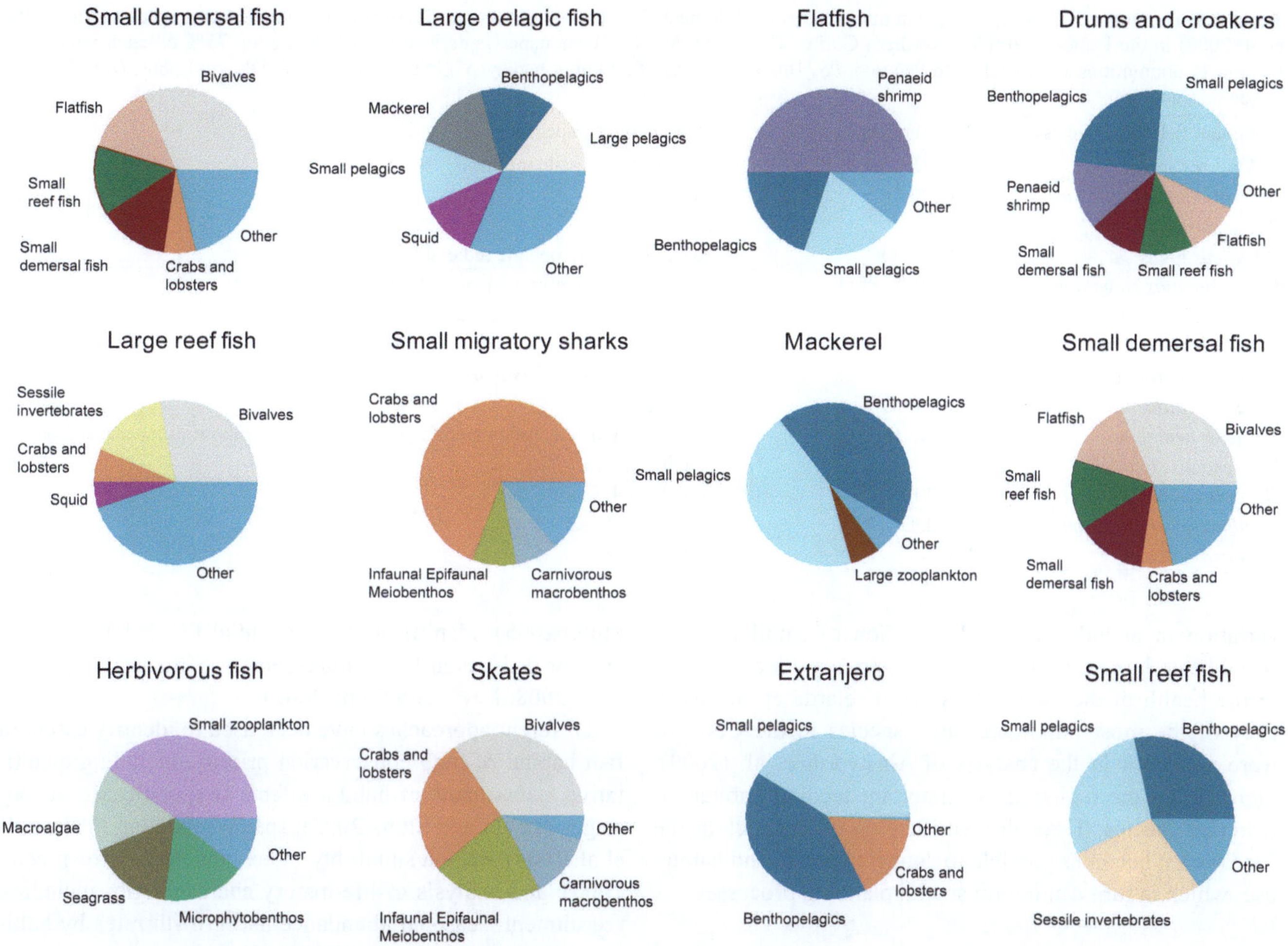

Fig. 9.10 Diet composition of functional fisheries groups of the Northern Gulf of California as defined for the Atlantis ecosystem model, derived from stomach content analysis of fish in the Peñasco Corridor (Ainsworth et al. 2009)

9.8 Food Web Linkages Between Fisheries Species in the Peñasco Corridor

Food web considerations are important in EBFM; there is increasing emphasis on the importance of understanding food web linkages and in maintaining the diversity of such linkages (Marasco et al. 2007). In the most simplistic view, assuring an adequate food supply for fisheries species is key for sustaining fisheries populations. The first step for management is to identify the diet of the primary fisheries species and to understand trophic linkages because changes in one component of the food web can have cascading effects on other components of the food web and the ecosystem overall (Crowder et al. 2008). Overharvesting low on the food chain can have disproportionately larger impacts on the top of the food chain, and harvesting top predators leads to simplified community structure and lack of stability or resilience in the system (Pauly et al. 2002; Essington et al. 2006).

Diet composition of functional fisheries groups of the Northern Gulf was determined through stomach content analysis (Ainsworth et al. 2009). Figure 9.10 shows the relative importance of different prey to different fisheries groups identified for the Northern Gulf of California, which included samples collected from the Peñasco Corridor and in the Guaymas region.

The abundant bivalves and mollusks that inhabit the Corridor's rocky reefs and sandy-muddy bottoms and support Puerto Peñasco commercial dive fisheries, are part of the diet of blue crab, guitarfish, smoothhound shark and predatory murex snails (Table 9.4). Other non-commercial species, mussels for example (*Modiolus capax*), are a key component of the diet of the black murex snail and other species. Small pelagic fishes sustain many of the region's fisheries species: Gulf croaker, gold-spotted sand bass, flatfish and shrimp (Table 9.4). There is currently no formal fishery for the small pelagic species in this region of the Gulf.

Many species are opportunistic feeders and change diet as food availability varies. This shows the importance of linking food abundance to oceanographic changes and climate conditions (Marasco et al. 2007). For the Corridor, the

Table 9.4 Dominant fisheries species, communities that exploit them (Downton-Hoffmann et al. 2013c) and dominant prey species (Ainsworth et al. 2009) in the Peñasco Corridor, Northern Gulf of California, Mexico. Dominance is defined to include the top 75 % of catch volume or income. (Communities are: *PPE* Puerto Peñasco, *BSJ* Bahía San Jorge, *PJA* Punta Jagüey, *STO* Santo Tómas, *PLO* Puerto Lobos, *DDC* Desemboque de Caborca)

Dominant fisheries species	Communities	Dominant prey species
Callinectes spp.	BSJ, PJA, STO, PPE	Crustaceans, sessile mollusks
Hyporthodus acanthistius	STO, PLO	Blue crab
Pteria sterna	DDC	Filter feeder
Dosidicus gigas	DDC, PLO	Hake, lizardfish, pelagic red crab*
Micropogonias megalops	STO, PJA, PLO	Hake, small pelagics, pennaeid shrimp, flatfish
Paralabrax auroguttatus	STO, PLO	Hake, small pelagics, crabs and lobsters
Pleuronectidae	PJA, PLO	Pennaeid shrimp, small pelagics, hake
Rhinobatos productus	BSJ	Crabs and lobsters, bivalves
Squatina californica	DDC, PLO	Bivalves, infauna
Musteles henlei	STO, PLO	Crabs and lobsters, infauna/epifauna meibenthos, carnivorous macrobenthos
Litopenaeus stylirostris	BSJ	Small pelagics
Hexaplex nigritus	BSJ, PPE	Sessile mollusks
Phyllonotus erythrostomus	BSJ, DDC, PPE	Sessile mollusks
Panopea globosa	PPE	Filter feeder

variation in abundance and distribution of small pelagics could be an important indicator of fisheries productivity and of the health of the entire ecosystem (Velarde et al. 2004). Prey of the most important target species in the Corridor were abundant in the analysis of Ainsworth et al. (2009), highlighting the region as an important feeding habitat for fisheries species. Through identification of linkages in the food web it becomes possible to connect species and habitat use, which in turn can inform spatial planning processes.

9.9 Identification of Essential Fish Habitat for Different Life Stages

Essential fish habitat includes areas that are most favorable for fish populations to spawn, feed and mature (Levin and Stunz 2005). Fishery management plans in the US and the European Union take into consideration the essential fish habitat of exploited species (Bellido et al. 2008; Levin and Stunz 2005). Fishery managers are required to identify threats to essential fish habitats and design steps to ameliorate those threats (Lindeman et al. 2000). As with protected area initiatives, the essential fish habitat designation can be used to assure the sustainable use of exploited coastal resources (Lindeman et al. 2000). Although Mexico does not consider essential fish habitat within the current regulatory framework, identification of these key habitats for exploited species can nonetheless be used to support biodiversity conservation and fisheries management. Understanding which habitats and sites in the Puerto Peñasco Corridor serve as essential fish habitat could help target conservation efforts to protect a variety of species, lead to better regulatory decisions for fisheries management, and allow efficient use of limited conservation resources (Beck et al. 2001; Levin and Stunz 2005). Identification of essential fish habitats is crucial for ecosystem-based management of fisheries (Bellido et al. 2008; Levin and Stunz 2005).

Different approaches have been used to identify essential fish habitat of high conservation priority, including quantitative assessments of habitat effects at specific life history stages (Levin and Stunz 2005), spatial modeling (Valavanis et al. 2004), habitat suitability index modeling (Brown et al. 2000), and analysis of life-history characteristics including recruitment, seasonal abundance and growth rates by habitat (Gallagher and Heppell 2010). These approaches require key information for managed groups, including potential spawning aggregation sites, larval duration patterns and inshore and offshore use areas for different life stages (Lindeman et al. 2000). Often, commercial fishing activities take place over essential fish habitat (i.e. spawning aggregations); thus mapping fishing grounds can also provide relevant information on areas of biological importance (Erisman et al. 2012). Within the Puerto Peñasco coastal Corridor, emphasis has been placed in identifying essential fish habitat for fish larvae and juvenile fish (Iris-Maldonado 2011; Peguero-Icaza et al. 2008, 2011), and predatory reef fishes (Aragón-Noriega et al. 2009; Erisman et al. 2007, 2012). The spatial distribution of fishing activities has also been assessed for fisheries within the coastal Corridor (Aragón-Noriega et al. 2012; Cudney-Bueno and Turk-Boyer 1998; Erisman et al. 2012; Moreno-Báez 2010; Rodríguez-Quiroz et al. 2010) (Fig. 9.8).

Table 9.5 summarizes the primary habitats of the dominant fisheries species in the Peñasco Corridor, generated from catch data records collected during 2011 (Downton-Hoffmann et al. 2013a). For this analysis dominant fisheries species are defined as the top 75 % of the catch volume and/or species that generated 75 % of the income for that

Table 9.5 Dominant fisheries species of the Peñasco Corridor, Northern Gulf of California, Sonora for different communities, their habitat occurrence with life history stage, when known (*A* Adult , *G* Gravid , *J* Juvenile , *L* Larval). Dominance is defined as the top 75 % of catch volume or income for a given community. (Communities are: *PPE* Puerto Peñasco, *BSJ* Bahía San Jorge, *PJA* Punta Jagüey, *STO* Santo Tómas, *PLO* Puerto Lobos, *DDC* Desemboque de Caborca)

Dominant fisheries species	Communities	Estuaries	Sandy-muddy bottoms	Rocky reefs	San Jorge island	Pelagic zone	River mouth
Callinectes spp.	BSJ, PJA, STO, PPE	A	A		A	A	A
Hyporthodus acanthistius	STO, PLO	A		A, J	A, J	A	
Pteria sterna	DDC		A				A
Dosidicus gigas	DDC, PLO					A	
Micropogonias megalops	STO, PJA, PLO		A			A	
Paralabrax auroguttatus	STO, PLO		A	A	A	A	
Pleuronectidae	PJA, PLO	A	A				A, J
Rhinobatis productus	BSJ	A, J	A				
Squatina californica	DDC, PLO	A	A				A
Musteles henlei	STO, PLO	A	A			A	
Litopenaeus stylirostris	BSJ	A, J	A		A, J, L		A, J
Hexaplex nigritus	BSJ, PPE	A	A	A	A		A
Phyllonotus erythrostomus	BSJ, DDC, PPE	A	A	A	A		A
Panopea globosa	PPE		A, J, L				

community. Presence or absence of these species in a given habitat is indicated in Table 9.5. Quantitative assessments of abundance of each of these species are available at some sites and for some life stages, but not all. In the case of some of the habitats data is available at multiple sites and for multiple seasons, which adds another dimension for assessing critical sites, key for spatially explicit management considerations.

9.10 Wetland Connectivity

Wetlands along the Puerto Peñasco Corridor provide a connection between the marine system and the surrounding Sonoran Desert (Glenn et al. 2006); they export organic nutrients (i.e. from bird guano), detritus (i.e. from halophyte plants), prey (i.e. invertebrates), and consumers (through foraging across habitats and migrations) (Polis et al. 1996; Rose and Polis 1998; Spackeen 2009). Thus, coastal wetlands in the Northern Gulf represent true "desert oases" that concentrate resources in the middle of an arid landscape (Glenn et al. 2006). Trophic connectivity, the transfer of food resources which can occur through the movement of either consumers or resources across habitats (Talley et al. 2006), has just begun to be studied for estuaries and surrounding environments (Alvirde-López 2012; Spackeen 2009); however linkages can be inferred from similar systems.

Estuaries and the near shore marine environment are likely linked functionally by flows of nutrients and organic matter moved through the tidal cycle (Polis et al. 1997). Salt marshes contribute to secondary productivity in nearby coastal waters through in situ consumption of marsh-derived sources, the export of dissolved and detrital organic carbon and nutrients, and the movement of fish and shellfish (Weinstein et al. 2000). The degree of contribution by salt marsh outwelling to coastal productivity is unclear, as primary production in estuaries in the coastal Corridor has not been assessed. However, stable isotope analysis of the food web in Northern Gulf estuaries suggests that phytoplankton and macrophyte detritus maintain secondary estuarine productivity (Alvirde-López 2012; Spackeen 2009).

The role of large consumers in transporting materials into the marine environment likely plays an important role for estuarine trophic connectivity. Within salt marshes, cyclic and ontogenetic movements provide trophic connectivity among habitats and with the marine environment (Rountree and Able 2007). Many fish and invertebrate species use estuaries during larval and juvenile stages, then move out as adults where they inhabit nearby subtidal habitats (Cudney-Bueno and Turk-Boyer 1998; Iris-Maldonado 2011). In this way, movements of fish and zooplankton driven by the macrotidal regime may facilitate rapid nutrient translocation (Polis et al. 1997). Adult fish, entering estuaries during spring tides (Iris-Maldonado 2011) feed and then transport materials from marshes to coastal environments. Coastal fisheries are supported by the export of juvenile fish and invertebrates from marsh channels (Calderón-Aguilera et al. 2003; Cudney-Bueno and Turk-Boyer 1998; Loaiza-Villanueva et al. 2009). Waterbirds feeding on fish and invertebrates also concentrate and transport large amounts of nutrients in their guano (Polis et al. 1997). Many migratory water birds, moving through the Pacific Flyway

(Vega et al. 2006) use estuaries as wintering and stopover sites feeding on invertebrates in the mudflats (Glenn et al. 2006). Guano can then increase nutrient status and primary production in the near shore marine environment (Stapp and Polis 2003).

Marine productivity can also subsidize estuaries, through input of phytoplankton. The importance of phytoplankton can be inferred from its high palatability and the efficiency of algae-derived food webs (Deegan and Garritt 1997). The isotopic signatures of wetland dependent species from Northern Gulf estuaries, such as blue crab, suggest that marine phytoplankton is an important carbon source (Alvirde-López 2012; Spackeen 2009). Tidal exchange also deposits macroalgal wrack onshore; wrack can then be broken down by invertebrates which are food for shore birds (Polis et al. 1997). Estuaries may also provide a link between the Sonoran Desert and the marine ecosystem of the Gulf of California. For example, terrestrial reptiles and mammals, such as the coyote (*Canis latrans*) subsidize their diets extensively with marine resources (Rose and Polis 1998). Trophic connectivity of estuaries should be taken into account when considering habitat protection mechanisms. However, there is little information regarding the distance offshore reached by estuarine resources or the distance that marine resources travel to subsidize estuaries. Thus, it is unclear how far is it necessary to protect marine areas in order to ensure adequate connectivity (Talley et al. 2006).

9.11 Marine Connectivity

Demographic connectivity between populations is characterized by relatively large amounts of propagules that have measurable effects on fisheries over ecological timescales. Genetic connectivity on the other hand usually involves fewer propagules that are key for biodiversity conservation (including genetic diversity and structure) over evolutionary time scales and affect the evolutionary potential for adaptation (Lowe and Allendorf 2010). The seasonally reversing oceanographic gyre that characterizes the predominant ocean circulation of the Northern Gulf (Marinone 2003) is the primary engine that drives marine connectivity (both demographic and genetic) in the region (Marinone et al. 2008; Marinone 2012). Since most marine organisms pass through larval dispersal phases, ocean currents are key for their transport between populations. Currents also move passive particles such as nutrients and contaminants to and from estuarine environments (Marinone 2012). By coupling biological information, such as spawning time, habitat, and larval behavior, with oceanographic models that predict ocean flow (Coupled-Biological Oceanographic models or CBOM) it is possible to begin to understand how populations are connected (Metaxas and Saunders 2009). Such information is key for understanding the flow of matter and energy within the ecosystem and for effective marine spatial planning and can significantly improve management and conservation strategies, particularly when currents are strongly asymmetric (Gaines et al. 2003; Beger et al. 2010; Kininmonth et al. 2011) as in the Northern Gulf. Cyclonic (counter-clockwise) circulation from May-September and anticyclonic circulation (clockwise) from October-April (Marinone 2012) create an asymmetrical system in the Northern Gulf. With some exceptions (e.g., geoduck clam, catarina scallop), most commercial species in the Northern Gulf spawn during May to September (Soria et al. 2014).

A coupled biological oceanographic model developed for the Northern Gulf of California (Marinone et al. 2008) was used to analyze connectivity in three fisheries species that spawn during May-July and that are of prime economic importance in the Northern Gulf ecosystem with special emphasis on the Peñasco Corridor (Soria et al. 2012; Soria et al. 2014). These three species are broadcast spawners with distinct life histories: Blue crab is a muddy bottom species with long pelagic larval duration (PLD) of 70 days, the leopard grouper is a rocky reef species with intermediate PLD (30 days) and the rock scallop is a rocky reef species of shorter larval duration (PLD=21 days) (Soria et al. 2012). CBOM models of all species were validated using empirical data from multiple hypervariable genetic markers and, in the case of the rock scallop, data from larval spat abundance on artificial collectors at various sites where suitable habitat occurs was used as well (Soria et al. 2012) (Fig. 9.11).

Larval dispersal within the Peñasco-Lobos Corridor is analogous to a river that flows in a single direction (i.e., anticlockwise for summer spawners, Fig. 9.11). This implies that upstream sources (e.g., Puerto Lobos) are key for sustaining fisheries and biodiversity at downstream sites (e.g., Bahia San Jorge) (Beger et al. 2010; Kininmonth et al. 2011). In such an asymmetric system, the spatial scale of connectivity scales-up with the length of the PLD. However, connectivity does not decrease with increasing distance. Instead, connectivity is low between some adjacent populations, and high between many non-adjacent localities, in a pattern described as diagonal connectivity. This points to potential challenges where local communities might not benefit from conservation and management actions within their nearby fishing sites, but may benefit from what happens at distant upstream sites.

In general, high levels of gene flow are found in the Northern Gulf of California for blue crab (*Callinectes bellicosus*), leopard grouper and rock scallop (*Spondylus limbatus*), which prevents local differentiation at most sites. However, modeling and empirical evidence indicates that Bahía San Jorge in the Upper Gulf, shows higher levels of genetic structure, larval retention and genetic relatedness,

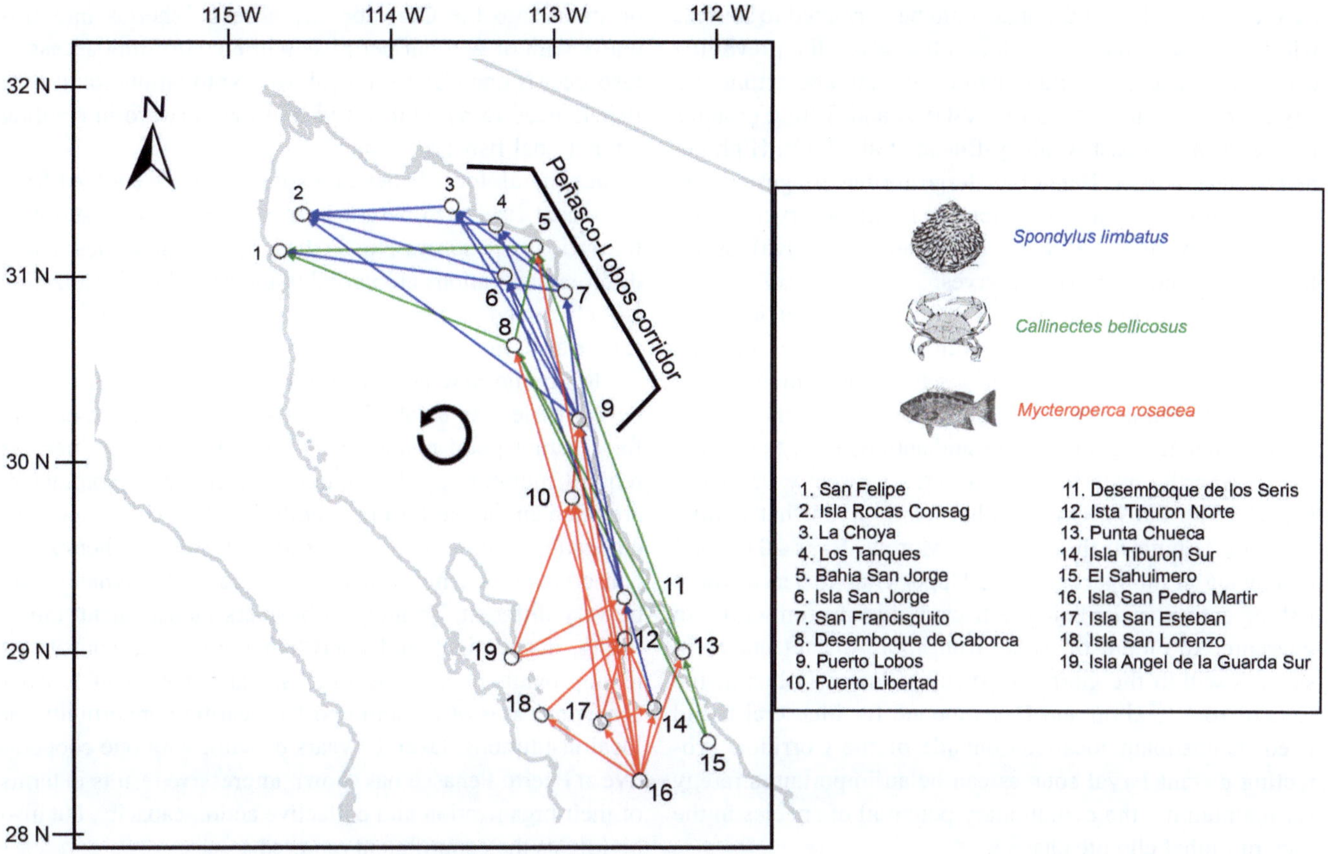

Fig. 9.11 Marine connectivity for three commercial species that spawn during May-July in the Northern Gulf of California, Mexico. Sites (nodes) are connected via larval dispersal (*links*) according to oceanographic models validated with population genetics. *Arrows* show the direction of the dispersal events that follow the anti-cyclonic circulation (Soria et al. 2012; Soria et al. 2014). For clarity, some sites are not situated on their actual coastal location

suggesting high isolation within this bay for multiple species (Soria et al. 2012). Despite large tidal fluxes, slow residual currents seem to explain why larvae get caught up in this bay. It seems that local circulation patterns within this bay likely contribute to isolating populations of other species as well, which is corroborated by the evolution in-situ of micro-endemic fish (e.g., Huang and Bernardi 2001).

Additional results for rock scallop showed connectivity (genetic and demographic) between the fished subpopulations of the Peñasco Corridor and southern sites (Soria et al. 2012). For the rock scallop, Puerto Lobos and Desemboque de Caborca which are located 150 km south of Puerto Peñasco, are potentially important upstream larval sources at the ecological scale, but Puerto Libertad, Las Cuevitas (~175 km away) and Desemboque de los Seris (~200 km away), showed weak connectivity links that imply connectivity with the Corridor at the evolutionary and genetic scales (Soria et al. 2012). On the other hand, sites as distant as Ángel de la Guarda could also be genetically linked to San Jorge Island (Soria et al. 2012). For blue crab, besides sources within the Corridor, other sites that provide significant amounts of larvae at the ecological scale include the Guaymas region south of Tiburon Island, separated by 325 km. During June, when currents reach their highest speed, marine reserves located as far away as San Pedro Martir Island (350 km distant) are key sources of larvae that could affect the demography of commercial fish populations in the Peñasco-Lobos Corridor in the case of species with a PLD of 30 days, such as leopard grouper (Soria et al. 2014). These results highlight that EBFM within the Corridor depends on multiple upstream sources that are distant and not intuitive.

San Jorge Island had the highest rock scallop spat recruitment (35–45 %) in the Corridor (Soria et al. 2012), and of all the islands surveyed throughout the Northern Gulf; it had the greatest adult densities of rock scallop (Martinez et al. 2011).

In 2002 a network of experimental marine reserves was established in the Puerto Peñasco Corridor, including a reserve around San Jorge Island, through a community-based process with a group of local divers concerned about declining populations of rock scallop and other benthic fisheries (Cudney-Bueno et al. 2009). Authors hypothesized that increases in rock scallop abundance documented at the northern-most sites of the area could be attributed to sources within the rocky reef network and that given the prevailing currents, San Jorge Island in particular, could be acting as a key source for larval export to reserves and fishing grounds on the Peñasco coast (Cudney-Bueno et al. 2009). High genetic relatedness at Bahía San Jorge further suggested that local retention was also contributing to the observed recovery at nearby sites (i.e. Sandy Beach and La Choya), established as voluntary marine reserves.

The high level of connectivity along the Corridor and the high potential for local recruitment within Bahía San Jorge have important implications for management of the Corridor. In Bahía San Jorge, local communities that participate in management decisions and actions have the potential to reap the benefits of their actions directly (Cudney-Bueno et al. 2009). On the other hand, given that fishing intensity is high in the Corridor (Moreno-Baez 2010) and increasing at the margin of the Upper Gulf Reserve where fishing activities are being restricted, it is also important to guarantee the input of larvae from more distant upstream sources within the Corridor (such as Puerto Lobos in the case of rock scallop and Desemboque for blue crab), and even in upstream locations outside of the Corridor. Protecting distant larval sources can be an important strategy for maintaining the evolutionary potential of species in the face of global climate change.

By coupling information on biological connectivity with species distribution and location of essential habitats for key species within the Corridor with patterns of use of local communities, we can begin to create management scenarios that make sense considering ecosystem trends and offer acceptable tradeoffs for the benefit of conservation, management and economic sustainability of local communities. For example, the decreasing degree of overlap of fishing zones used by communities of the Corridor from south to north (Fig. 9.8) and the grouping of Puerto Peñasco and Puerto Lobos according to the commercial species captured at each site might relate to these oceanographic dynamics.

9.12 Discussion: Management of the Puerto Peñasco Corridor

Until the economic boom of the mid 2000s, the Peñasco Corridor was an isolated stretch of beach, where fishers went about their business with little knowledge of or concern for regulations, except at Puerto Peñasco. As natural as it is to pick wild berries, coastal inhabitants would go out to fish. There was little local understanding of how to obtain fishing permits or why it was necessary. The end result is that even today approximately 75 % of the Corridor's fishers are irregular; with inadequate legal access to all the resources they fish. Fishers also come from nearby towns or states into the Corridor and access fisheries intermittently with or without permits. While controlling access to resources is one way to control overexploitation, traditional fishers need to be legitimized to move forward in creating a functional fisheries system.

In general, local fisheries institutions have evolved from individual fishers to permit holders, cooperatives and then federations of cooperatives in the larger communities. These different institutions are varied in their level of organization and effectiveness for collective action. The level of organization also varies by community. For some fisheries, such as shrimp, no new permits are available, while permits for new fisheries can be obtained. This dynamic offers potential for creating good management structures, as new permits require monitoring of resources. Independent contractors are often employed for this monitoring, but when a local organization with a commitment to sustainable fisheries and conservation is employed in the process, the dynamic can be very different. Training in business management, monitoring, negotiations and many other processes contributes to improving internal operations. Participation in broader scale processes offers an important learning opportunity for local institutions. Over 15 years of work with one cooperative at Puerto Peñasco has shown impressive results in terms of their organization and collective action capacity, but also highlights the commitment required.

At its peak in the 1980s the number of industrial boats operating in the Upper Gulf reached 240 (Lluch-Cota et al. 2007; Meltzer et al. 2012; Galindo-Bect et al. 2000), the majority of these working out of Puerto Peñasco. Beginning in 2000, fishing activities with incidental catch greater than 50 % were officially prohibited in all Natural Protected Areas in Mexico, and the industrial trawlers in the Upper Gulf, with a home base at Puerto Peñasco, were the first group required to present a formal impact assessment. Fish excluders, turtle excluders, and other gear modifications are now mandatory. By 2012, only 58 boats were authorized at Puerto Peñasco, though another fisher group has a request for authorization pending to fish in the Reserve. A new impact study presented by trawlers in 2012 proposed spatial-temporal restrictions that in combination with reductions in fleet size would effectively reduce the area trawled in the Reserve by 43 % compared to previous estimates. While the season for trawling in the Reserve is only 3 months, outside the Reserve, in the Peñasco Corridor, for example, it extends through March.

In 2009, small-scale fishers of the Upper Gulf Reserve began working under mitigation measures mandated as part of the Environmental Impact Assessment (EIA) process, which includes a comprehensive monitoring program to document reduction in impacts (Pérez-Valencia et al. 2012). Through this process specific mitigation actions were identified for their most important fisheries, and they agreed to continue to develop spatial-temporal approaches to reduce impacts on the critically endangered vaquita porpoise. The EIA program requires fishers to develop training, logbook monitoring, onboard observation and participatory programs including creating a decision-making body and indicators to measure their compliance with the programs they outlined. This type of process is unprecedented in Mexico and throughout the world for traditional economic activities such as fishing. The fishers hired a local conservation NGO (Non Governmental Organization) to partner with them to develop these programs and have indicated their willingness to find the appropriate trade-offs to protect this porpoise, which is the primary driver for management in the region. It is too early to see ecosystem-wide results from these efforts. Ample evidence exists, however, to raise concerns about increased effort in areas outside the Reserve's boundaries, which are often more heavily targeted as a result of fishing restrictions within protected areas (Kellner et al. 2007). This process is being observed at Puerto Peñasco, at the southeast margin of the Reserve, where fishers have the option to fish within or outside the Reserve, and prefer to fish outside rather than participate in the Environmental Impact process and associated costs. Some fishers from San Felipe, B.C. have expressed similar approaches to new regulations in the Reserve. At Puerto Peñasco this has the effect of increasing effort and impacts on the Corridor.

A new shrimp law (NOM-002-PESC-2013, in DOF 2013) is poised to change these dynamics even further. Through a 3 year process beginning in 2013 shrimp fishing with gillnets will be phased out in the Upper Gulf Reserve and Vaquita Refuge and replaced with a light trawl, known as RS-INP-MX to be pulled from a skiff, or other gear types as they are developed. Fishers, especially at Puerto Peñasco, are resistant to adopting this new gear, but incentives to switch are being offered by the government.

The Upper Gulf small-scale fishers' environmental impact study is being managed at the level of communities and federations, but it is also working through cooperatives to reach individual fishermen. Decision-making committees were formalized in each community in May and June 2013 to move the impact study process forward and increase compliance. The cross-scale management groups that have formed in association with the Upper Gulf Reserve and Vaquita Refuge have mostly focused on forwarding vaquita conservation. In the Peñasco Corridor, on the other hand, collective action has taken place around individual species. Permit holders of black murex snail, for example, have gathered to discuss management options; Blue crab Subcommittees of Aquaculture and Fisheries, which link to state committees of the Secretary of Aquaculture and Fisheries, have formed to manage this resource across the Corridor. In some cases more formal integration of groups for better management of the entire chain of custody of a resource has been promoted. The Productive System is a model promoted by CONAPESCA for improving management and marketing of a product. Operated as a civil association, a Productive System organization was created for blue crab in the state of Sonora and one was initiated for the geoduck clam. These organizations are meeting with varying degrees of success depending on the leadership. An Integrator, which collects a group of producers for formalizing their commercialization of a product, is an-other type of institution and one has been created for geoduck clam, but many challenges to successful operation still exist.

A variety of management actions have been implemented on fisheries and coastal habitats of the Corridor. The Upper Gulf environment is generally known to be dynamic and unpredictable (Cudney-Bueno and Turk Boyer 1998). One year a species is abundant, the next it is not, and then new species such as the giant squid appear. The communities in the region have learned to adapt to these conditions and opportunities. The fisheries regulatory systems, however, are less adaptable. Obtaining permits for a new species is a slow process. Economic security and food security are considerations. There are few legitimate economic opportunities for communities in the region and buffers against this are needed for the most vulnerable communities (Morzaria-Luna et al. 2013). Management should consider creating sustainable development alternatives.

Management plans have been developed for rock scallop and geoduck clam), fisheries that are being managed with turfs and quotas for individual beds, while management regulations for other species are still being developed (blue crab, octopus) (PANGAS 2012c, d). Feasibility permits for the winged oyster fishery have been granted to two cooperatives at Desemboque. Permits regulate access to some of these fisheries by controlling the number of boats, gear, and sometimes dictating a quota and season.

Rock scallop is the only commercial species harvested from rocky reefs which has a management plan. This species is listed in Mexico as protected (NOM-059-ECOL), based on its population status in southern Mexico (DOF 2010). As such it cannot be exploited with commercial fishing permits, but can be harvested under a strict management program that is supervised by Mexico's Wildlife Department (Cudney-Bueno 2007). Rock scallop has been harvested sustainably by a group of commercial divers since 2008 when they received a permit and a quota to harvest the species in an exclusively-designated area of the Peñasco Corridor. The management plan requires annual censuses of different fishing beds, estimation of biomass, and designation of site-specific quotas, season closures and size limitations (Martinez and Turk-Boyer 2011). Under this management regime, San Jorge Island was maintained as an area of high density of the species until 2011–2012, when densities began to drop. New highly valued fisheries (geoduck clam, Mexican bay scallop, *Argopecten circularis*, and concave scallop, *Euvola vogdesi*) have attracted many into diving fisheries, increasing pressure on the benthic resources (Martinez and Turk-Boyer 2011) such as rock scallop.

Commercial species such as black and pink murex snail, rock scallop and winged oyster inhabit the patch reefs that are interspersed throughout the Corridor, but the snail is also associated with the sandy bottoms where they may find prey nearby. For such benthic species, delineation of beds and management of quotas offer a good option to maintain viable densities. Such management tools are being used for geoduck clam, winged oyster and rock scallop. Use of turfs with clearly defined property rights and quotas by bed can be an important means for managing soft-bottom habitats, patch reefs and associated species. Turf rights are given in a variety of ways. The CONAPESCA permitting process gives access to a resource in a particular area and may delimit particular beds within the area for extraction by a resource user. The permit process might also designate some beds for conservation within a fisherman's designated area, as has been done with geoduck clam. Fishermen might also be granted property rights through the Secretary of the Environment's Wildlife Department, which gives exclusive use of an area for protected species that have management plans, such as rock scallop. Finally, long-term guarantees for the use of a resource can be obtained with a concession, which gives exclusive use of an area for extraction for an extended period of time. The process for obtaining a concession is more complicated and requires a Technical-Economic Study that evaluates the capacity of the solicitor to administer the technical, administrative and financial components of the resource adequately. As geoduck fishers obtain commercial permits, they are looking to guarantee their rights for this valuable resource through concessions.

Turfs can be effective for protecting habitats and species, only if there is compliance on the part of fishers backed by official enforcement. The number of divers that have access to diving resources in the Corridor has increased, and with this the overlap in areas of use for different species has also increased. If concessions for multiple species were possible, such that specific areas could be exclusive to a defined group of fishers, it would help resolve this issue and strengthen the sense of ownership for protection of the resources and habitat. This situation has evolved almost naturally for the San Jorge community, which has unified into one cooperative. The geographic extent of access given through fishing permits has traditionally been very extensive, however, allowing overlap between fishers and conflicts to ensue. If property rights (permits or concessions) are matched to actual fishing zones in each community the chances of management success can be increased. Implementation of such a system would be challenging, as it would limit access that fishermen currently have to larger areas, but information is available to assist in taking these steps. Local fisher knowledge has been used in the Corridor to generate a large scale perspective on spatial and temporal use of fishing zones and overlap (Cudney Bueno and Turk Boyer 1998; Moreno-Baez et al. 2012). By Coupling this information with the finer scale spatial temporal assessment obtained from 2010–2012 daily catch records (Fig. 9.8), we have a multi-dimensional perspective that can inform management decisions.

San Jorge Island stands out among the rocky habitats of the Corridor for its biodiversity value and importance for commercial species. Until recently use and impact of fishing on the island has been low. Marine connectivity studies highlight the island as a potential link with more southerly rocky habitats (Midriff islands), and as a larval source for more northerly sites, such as the rocky reefs at La Choya, Sandy Beach and Borrascoso. These studies indicate that the offshore waters at San Jorge Island represent a good site to protect, either through marine conservation tools or as fisheries refuges. This strategy could guarantee dispersal of larvae to the coastal sites along Puerto Peñasco, and could conserve the evolutionary potential of the island by maintaining connectivity with sites further south. San Jorge Island might also be an important larval source for sites as distant as Rocas Consag within the Upper Gulf Reserve, which is already a protected site and possibly a source for more southerly sites along the Baja coast. Trophic structure of catch has been calculated and related to standing trophic structure at San Jorge Island. The high trophic level of fish species at the island suggests special management for the large predatory species that live there might be in order. These are targeted by commercial fishers as well as recreational fishers, who have indicated an understanding of the importance of protecting the island (Rupnow 2008). This data on trophic level offers an important baseline for measuring changes in ecosystem health.

Mexico's first fisheries refuges were formalized in Baja California Sur in 2012 (DOF 2012); this designation offers an alternative for protecting the important biodiversity of San Jorge Island that might be socially acceptable. The establishment of fisheries refuges in Mexico in 2012 and definition of a process for their creation sets the stage with fisheries authorities for approaching management in the Corridor from an ecosystem perspective. The new law allows for establishment of networks of refuges as well.

During 2002–2003, commercial divers voluntarily ceased fishing activities at the island, creating a no-take area around its perimeter (Cudney et al. 2009). Closed areas in coastal reefs (where new development projects have been approved) were paired with fishing areas and 2 years of data yielded important results showing the benefits of marine reserves for two dive fisheries (Cudney et al. 2009). The benefits were seen most clearly in reefs at the northern end of the study area within the Peñasco Corridor. The rapid recovery of rock scallop and black murex snail populations was attributed to local larval recruitment by dispersion from the more southerly sites, including San Jorge Island (Cudney et al. 2009).

Subsequent studies of larval recruitment for rock scallop (Soria 2010; Marinone et al. 2008), one of the primary target species studied, confirmed the importance of local recruitment for this species and give a wider view of recruitment throughout the Northern Gulf (Soria et al. 2012) (see Sect. 9.11 on marine connectivity).

A time series of on-site density by bed available for rock scallop shows a clear decrease in density at San Jorge Island compared to other sites, a reflection of increased fishing pressure on diving fisheries in general (Martinez and Turk Boyer 2011). Though a good management plan exists for this species, quotas are determined annually by bed, legal access is defined, and the users are committed to complying with regulations, a widespread lack of enforcement of laws, turfs, and other regulations has caused a disintegration of good management practices. Until there is a strong enforcement process for these benthic fisheries and others, it will be difficult to maintain fisher interest in responsible fishing. Some changes have been observed in the Upper Gulf Reserve where cross-scale forums have fishers and government authorities at the table together and fishers make their pleas for enforcement clear.

Puerto Lobos represents another potential rocky site for special management. Here populations of rock scallop, winged oyster, and possibly other species (black murex snail) disperse to sites near San Francisquito and Puerto Peñasco, where they are accessed by a large number of divers. Diving at Puerto Lobos is a new occupation. There are no permits for rock scallop, but the winged oyster population has shown a resurgence here and access was opened up for two cooperatives in 2010 and 2011. Dive fisheries are increasing in this community as an economic option. New banks of the winged oyster fishery have also appeared north of San Jorge Island. When fishers try to obtain access to new fisheries, it offers an opportunity to establish limits to areas that might be important for recruitment.

Studies on marine connectivity highlight the importance of Desemboque, the mouth of the Asunción River, for recruitment of blue crab. We also know that this is a key spawning site for shrimp, shark and blue crab. Shark fishing is not permitted at the mouth of a river, and all three species have closed seasons during summer reproduction. Enforcement activities are minimal in the area. Members of this community recognize that they are extracting resources during reproduction and that resources congregate in front of their community for this purpose. The children in this community have actively promoted that fishers respect the laws; offering promise that this kind of social pressure might influence compliance. Identification of viable alternatives for this fishing community should be investigated, while also seeking a more enforceable type of closure for this key habitat through the use of fisheries refuges. The social capital that has been built in this community through education programs and other activities has potential for being harnessed in support of tangible management action such as refuges. Little is known about the opportunities that might exist to secure upstream water sources from the Asunción River for environmental purposes, but it should be considered for maintenance of this important spawning site.

Most of the region's important fisheries species (except squid, Gulf croaker and gold-spotted sandbass) use wetland habitats as nursery and/or feeding grounds. For this reason, management of wetland habitats directly affects fishing activities. Several wetlands in the Corridor are partially protected by several tools and management processes. The Bahía Adair wetland complex and offshore sandy and muddy habitats are located within the Upper Gulf of California and Colorado River Delta Biosphere Reserve (CONANP 2007) and were declared a Ramsar Wetland of International Importance in 2009 CONANP (CONANP 2011). The Biosphere Reserve is managed through a management program and other policy instruments, such as the Environmental Impact Assessment. In 2010 the San Jorge wetland complex to the southeast of the Reserve was also designated a Ramsar Wetland of International Importance (CONANP 2011). A participatory management plan has been produced for Bahía Adair under the Ramsar designation, while processes have initiated for developing a management program for the San Jorge wetlands (see Chap. 11 in this series). At Bahía Adair federal zone transfer agreements were obtained for conservation, with 306 ha registered (DOF 2012). At Estero Morúa a federal zone transfer agreement was obtained for the municipality of Puerto Peñasco for 68,257 m^2 (DOF 2012) for protection of nesting sites of the Least tern. Estero La Salina in the San Jorge wetland complex is part of Mexico's Important Bird Areas Program (CONABIO 2004).

Federal and municipal concessions, the Reserve's management plan, and Ramsar designations all are potential deterrents to development pressure, and help maintain the integrity of these coastal systems, but don't directly address fisheries management. A more active strategy, however, is underway through participatory processes being promoted as part of the Ramsar sites at Bahía Adair and Bahía San Jorge. In Chap. 11 these processes are described in detail; landowners and fishers of these wetlands have engaged actively in management of resources, zoning, and development of sustainable, economic alternatives, such as monitoring and ecotourism. The social capital developed in these sessions will be leveraged for formalizing a management group for this larger concept of a Biological-Fisheries Corridor. In fact landowners in both bay regions have formalized a network for sharing experiences.

Additional evidence discussed in the section on connectivity suggests a degree of local larval retention in the Peñasco Corridor for multiple species, which lends itself to implementing local management actions in which the users will benefit directly, as was the case with the temporary reserves implemented in 2002–2003 (Cudney-Bueno and Basurto 2009; Cudney-Bueno et al. 2009). In that case, the socio-ecological feedback that resulted from fishers involvement in monitoring helped fishers maintain interest in the reserves until outsiders began to benefit from their management actions (Cudney-Bueno et al. 2009). Multiple processes are currently underway with local fishers to engage them to understand and make decisions about the management of different habitats and species of the Corridor.

Understanding of trophic interactions in the Corridor has advanced significantly. The abundance and importance of small pelagics for sustenance of the Gulf croaker, gold-spotted sand bass, fish and shrimp, all commercially important species, is rivaled only by the importance of bivalves, some of commercial value, which support crabs, snails and shark fisheries species. Several commercial species are a key part of the diet of others: shrimp is an important prey for flounder for example and crabs sustain Gulf coney, sharks, and gold-spotted sand bass. Care must be taken when managing these species for maximum sustainable yield that there is biomass available to support the diet of other targeted species. As yet there is no fishery for the small pelagic species in this region, but diving fisheries for bivalves have increased, with a potential impact on targeted populations that depend on these as prey. Whatever management scenarios and targets are proposed, they must recognize these important interactions. Controlling access to small pelagics, or mussels (which are not yet actively fished), through the permitting process for new fisheries, might help limit new pressures on these important prey resources. Especially for benthic resources, management of key habitat or areas through fisheries refuges or marine protected areas with spatial or season closures might offer some protection if clearly identified.

Though no formal integrated management initiative is yet underway in the Peñasco Corridor, many of the steps outlined by the U.S. National Marine Fisheries Service (1999) for implementing EBFM have advanced significantly. By leveraging fisher participation, government and foundation funding, collaboration with two consortiums of researchers and NGOs, and over 30 years of community experience and trust, progress has been made in collecting baseline data for forwarding EBFM in the Peñasco Corridor as discussed in this Chapter. Much of the data was collected through collaborative processes with fishers, and so the resulting co-produced data sets the stage for dialogue about potential management scenarios.

To promote an ecosystem approach to management in the Corridor will require consolidation of a management group that shares an ecosystem vision. An analysis of catch data (Downton-Hoffmann et al. 2013c) suggests that it would be appropriate to start with the communities of Puerto Peñasco and Puerto Lobos based on their similarity of catch and shared habitats. Connectivity studies also suggest the logic in promoting shared processes in resource management of these two communities as well as other groupings to initiate the collaborative integrated process. The next step for moving forward will be to survey the local communities about their objectives with fisheries management, collaborative processes and socio-economic goals. This will be followed with the generation of a specific set of management scenarios that address community interests, ecosystem processes and fisheries management needs and presenting these to the management group. The tremendous amount of data generated in this project will be overwhelming to most of the stakeholders that are involved. We will need to develop some alternative scenarios that illustrate clear tradeoffs to allow communities and other stakeholders to collectively evaluate the direction for management of the Corridor. Atlantis is a spatially explicit ecosystem model built for the Northern Gulf ecosystem that might be used to evaluate tradeoffs and operationalize decision-making, although it is unclear whether it can manage data at the scale of the Corridor.

Most of the management scenarios that emerged through analysis of this ecosystem suggest the use of spatial planning instruments. Spatial planning might incorporate fisheries refuges, clear definition of boundaries and access rights through permits or concessions, while at the same time address fisheries and ecosystem goals. Definition of access rights through the use of areas and/or quotas can be an important incentive for fishers in this region, where such basic rights as permits are lacking. In return for legal access to resources, fishers might be open to restrictions in spatial or temporal use that would otherwise be unacceptable. It will be important to evaluate stakeholders' interest before structuring a set of viable management scenarios.

It will be necessary to assess where uncertainty lies in this system and incorporate buffers in management and conservation actions. Changes in the dynamics of the oceanographic system due to climate change are one such uncertainty that impacts larval recruitment, trophic linkages and fisheries dynamics. Connectivity studies illustrate the importance of considering both distant and local source populations to manage species in both ecological and evolutionary time frames. Though we propose the Peñasco Corridor as an ecosystem-level management unit, we also recognize the need to relate it to Northern Gulf ecosystem dynamics and the opportunity to create a network of managed areas that offer insurance against such uncertainties.

We developed a proposed set of indicators to measure ecosystem health and fisheries equilibrium, based on the data from studies on biodiversity and fisheries within the Corridor (Table 9.6). These indicators cover measures of ecosystem structure, biodiversity conservation and equilibrium of the potential resources. These indicators, if developed,

Table 9.6 Proposed indicators for use in the Peñasco Corridor, Northern Gulf of California, Mexico to measure ecosystem and fisheries health. Baseline information is available for most of these parameters

Indicators	Measure	State/Tendency	Management objective
Average size	Size of fish	S, T	ES[a]
Trophic level of catch	TL	S, T	ES[a]
Proportion of predatory species	% Predators	S, T	BC[b]
Total sampled biomass	Biomass	T	PR[c]
Species richness	Number of species	S	BC[b]
Shannon's biodiversity index	H'	S	BC[b]

[a] Ecosystem structure (ES)
[b] Biodiversity Conservation (BC)
[c] Equilibrium of Potential Resource (PR) (Pauly et al. 2000)

monitored and evaluated through collaborative processes, can serve as a key communication tool for implementing EBFM and adapting management for the benefit of society, fishers and the ecosystem.

An important next step is to determine socio-economic management objectives and simple indicators that track economic health and well-being with input and for the use of local communities. The final set of indicators will be realized through a participatory process with the appropriate stakeholder groupings. We propose establishing a set of management scenarios, using the information generated about the ecosystem and the preliminary lists of both ecological, fisheries and socio-economic indicators as a guide.

We have described various monitoring processes that have been implemented for monitoring targeted catch, bycatch, size structure and trophic level of catch, biomass, and in situ density of various species. Through both the PANGAS initiative and the Environmental Impact Study process, procedural manuals have been developed that present standardized monitoring methods that can be used over time and throughout the ecosystem (Downton-Hoffmann et al. 2013a, b, c; Pérez-Valencia et al. 2013a, b). The ecosystem initiative supported by the David and Lucile Packard Foundation provided the funding to initiate this monitoring process. Today fishers and government agencies are sharing in this responsibility as well. To access a new species fishers must work with a research organization to conduct the needed monitoring. Fishermen also are financially responsible for implementing the programs associated with the Environmental Impact Study in the Upper Gulf Reserve, and have contracted third parties to develop the programs and conduct the needed monitoring. Within the Reserve CONANP has offered financial support to fishers through their PROCODES (Programa de Conservación para el Desarrollo Sostenible) Program and state fisheries agencies in both Baja California and Sonora have contributed together with NGOs to support fishers' financial obligations for this process. State and national fisheries authorities are giving priority to ordering fisheries in the Upper Gulf. To insure success a clear and transparent plan of co-responsibility for financing and implementing monitoring will be needed. Involvement of fishers in monitoring has the advantage of reducing costs, but also is an important mechanism of socio-ecological feedback on the state of the resources. Precedents are already established with fishers in the region to participate in monitoring and financing such processes. In general there is a forward momentum for implementing many components of Ecosystem-based fisheries management in the Peñasco Corridor.

9.13 Conclusions

While there are many ways to approach managing fisheries, marine systems and, most importantly, their users are space-based. Many authors contend that place-based management and marine spatial planning can provide the most promising approach to implementing ecosystem-based fisheries management (Crowder et. al 2008; McLeod et al. 2005). Some consider this should be the second tier in implementation of EBFM, simply because of the amount of data required. Place-based management will require a hierarchy of management practices, beginning with a collective vision for ecosystem-based management and moving toward an integrated approach that contemplates multiple spatial and temporal scales. The key will lie in developing governance systems that align stakeholder incentives with management objectives.

Meaningful involvement of government agencies is also crucial. The Peñasco Corridor is not a region of high priority: the populations it serves are small communities, marginalized in many ways. Until recently the region was isolated by poor quality roads and other basic services are minimal. It is just such communities, however, that need help in sustaining their livelihoods that depend almost exclusively on resource extraction. Our experience has shown that such communities are open to dialogue, interested in connection and willing to participate in such processes, setting the stage for success.

Acknowledgments This project would not have been possible without the vision and financial support of the David and Lucile Packard Foundation's ecosystem-based management initiative. Too many researchers, partners and co-workers have been involved to give mention to all, but staff members that have been key to conducting this work and deserve special thanks for their dedication to different components of this long-term process, include Dr. Sergio Perez-Valencia, Rene Loaiza-Villanueva, Elia Polanco, Angeles Yazmin Sánchez Cruz, Ema Elizabeth González Plascencia, Marlene Anaid Luquin Covarrubias, David Buitrago, Verónica Castañeda, Paloma Valdivia, Hitandehui Tovar Vázquez and Alejandro Castillo. Most importantly and significantly we give thanks to the fishers of the Peñasco Corridor, who have openly shared their lives, their fisheries and aspirations to move towards creating a more functional fisheries system.

References

Aburto-Oropeza O, Sala E, Paredes G et al (2007) Predictability of reef fish recruitment in a highly variable nursery habitat. Ecology 88(9):2220–2228

Aburto-Oropeza O, Ezcurra E, Danemann G et al (2008) Mangroves in the Gulf of California increase fishery yields. Proc Natl Acad Sci U S A 105:10456–10459. doi:10.1073/pnas.0804601105

Ainsworth CH, Kaplan IC, Levin PS et al (2009) Diet compositions of fish species in the Northern Gulf of California. In: Morzaria-Luna HN (ed) A handout of the integrative marine ecology. Ecosystem Sciences Program, Northwest Fisheries Science Center, NOAA, Seattle, p 11

Ainsworth CH, Kaplan IC, Levin PS et al (2011) Atlantis model development for the Northern Gulf of California. US Dept Commer NOAA Tech Memo NMFS-NWFSC-110

Alvarez-Borrego S (2010) Physical, chemical and biological oceanography of the Gulf of California. In: Brusca R (ed) The Gulf of California: biodiversity and conservation. University of Arizona and The Arizona-Sonora Desert Museum, Tucson, pp 24–48

Alvarez-Borrego S, Flores-Báez BP, Galindo-Bect LA (1975) Hidrología del Alto Golfo de California II. Condiciones durante invierno, primavera y verano. Cien Mar 2(1):21–36

Alvirde-López SB (2012) Estimación de la red trófica en el estero Morúa, Sonora, México, un humedal costero del Golfo de California. Universidad Autónoma de Baja California Sur

Aragón-Noriega AE, Valenzuela-Quiñones W, Esparza-Leal HM et al (2009) Analysis of management options for artisanal fishing of the Bigeye Croaker *Micropogonias megalops* (Gilbert 1890) in the upper Gulf of California. Int J Biodivers Sci Ecosyst Serv Manage 5:208–214. doi:10.1080/17451591003709371

Aragón-Noriega EA, Cisneros-Mata MA, Rodríguez-Quiroz G (2012) Artisanal fishery of elasmobranches within two marine protected areas in the upper Gulf of California. Indian J Fish 59:1–6

Arreguín-Sánchez F, Lercari D, Galván-Piña VH et al (2005) Modelación de ecosistemas como soporte para manejo de recursos marinos: Casos de estudio en el Pacífico Mexicano. XI Congreso Latinoamericano de Ciencias

Bakum A, Babcock EA, Lluch-Cota SE et al (2010) Issues of ecosystem-based management of forage fisheries in "open" non-stationary ecosystems: the example of the sardine fishery in the Gulf of California. Rev Fish Biol Fish 20:9–29. doi:10.1007/s11160-009-9118-1

Barrera-Guevara JC (1990) The conservation of *Totoaba macdonaldi* (Gilbert), (Pisces: Scianidae), in the Gulf of California, Mexico. J Fish Biol 37(Suppl. A):201–202

Barrera-Guevara JC, Parra Salazar I, Albar-Reynoso MP et al (2004) El Golfo de California: Prioridades de Conservación. World Wildlife Fund Mexico

Bazzino G, Gilly WF, Markaida U et al (2010) Horizontal movements, vertical-habitat utilization and diet of the jumbo squid (*Dosidicus gigas*) in the Pacific Ocean off Baja California Sur, Mexico. Prog Oceanogr 86:59–71

Beck MW, Heck KL, Able KW et al (2001) The identification, conservation, and management of estuarine and marine nurseries for fish and invertebrates. BioScience 51(8):633–641

Beger M, Linke S, Watts M et al (2010) Incorporating asymmetric connectivity into spatial decision making for conservation. Conserv Lett 3(5):359–368. doi:10.1111/j.1755-263X.2010.00123.x

Bellido JM, Brown AM, Valavanis VD et al (2008) Identifying essential fish habitat for small pelagic species in Spanish mediterranean waters. Hydrobiologia 612:171–184. doi:10.1007/s10750-008-9481-2

Brown SK, Buja KR, Jury SH et al (2000) Habitat suitability index models for eight fish and invertebrate species in Casco and Sheepscot bays, Maine. North Am J Fish Manage 20:408–435. doi:10.1577/1548-8675(2000) 020<0408:HSIMFE>2.3.CO;2

Brusca RC (1980) Common intertidal invertebrates of the Gulf of California. The University of Arizona Press, Tucson

Brusca RC, Bryner GC (2004) A case study of two Mexican biosphere reserves: the upper Gulf of California and Colorado river delta and the El Pinacate and Gran Desierto de Altar biosphere reserves. In: Harrison NE, Bryner GC (eds) Science and politics in the international environment. Rowman and Littlefield Publishers, Lanham, pp 28–64

Brusca R, Hendrickx M (2010) Invertebrate biodiversity and conservation in the Gulf of California. In: Brusca R (ed) The Gulf of California: biodiversity and conservation. University of Arizona and The Arizona-Sonora Desert Museum, Tucson, pp 72–95

Brusca RC, Findley LT, Hastings PA et al (2004) Macrofaunal biodiversity in the Gulf of California. In: Cartron JLE, Ceballos G, Felger R (eds) Biodiversity, ecosystems, and conservation in Northern Mexico. Oxford University Press, New York, pp 179–203

Brusca RC, Cudney-Bueno R, Moreno-Báez M (2006) Gulf of California esteros and estuaries. Analysis, state of knowledge, and conservation and priority recommendations. Final Report to the David and Lucile Packard Foundation by the Arizona-Sonora Desert Museum

Calderón-Aguilera LE, Marinone SG, Aragon-Noriega EA (2003) Influence of oceanographic processes on the early life stages of the blue shrimp (*Litopenaeus stylirostris*) in the upper Gulf of California. J Mar Syst 39:117–128

Carriquiry JD, Sánchez A (1999) Sedimentation in the Colorado river delta and upper Gulf of California after nearly century of discharge loss. Mar Geol 158:125–145

Carriquiry J, Villaescusa J, Camacho-Ibar V et al (2011) The effects of damming on the materials flux in the Colorado river delta. Environ Earth Sci 62:1407–1418. doi:10.1007/s12665-010-0626-z

Cinti A, Shaw W, Cudney-Bueno R, Rojo M (2010a) The unintended consequences of formal fisheries policies: social disparities and resource overuse in a major fishing community in the Gulf of California, Mexico. Mar Policy 34:328–339

Cinti A, Shaw W, Torre J (2010b) Insights from the users to improve fisheries performance: fishers' knowledge and attitudes on fisheries policies in Bahia de Kino, Gulf of California, Mexico. Mar Policy 34:1322–1334

Cintra-Buenrostro CE (2001) Los Asteroideos (Echinodermata: Asteroidea) de aguas someras del Golfo de California, México. Oceánides 16:49–90

Cisneros-Mata MA, Montemayor-Lopez G, Roman-Rodriquez MJ (1995) Life history and conservation of *Totoaba macdonaldi*. Conserv Biol 9(4):806–814

CONABIO (2004) AICA NO-34. Bahía e Islas de San Jorge. http://conabioweb.conabio.gob.mx/aicas/doctos/NO-59.html. Accessed 29 June 2009

CONAGUA (2011) Estadísticas del agua en México. Comisión Nacional del Agua. Secretaría de Medio Ambiente y Recursos Naturales, Comisión Nacional del Agua, Mexico

CONANP (2007) Programa de conservación y manejo de la Reserva de la Biosfera Alto Golfo de California y Delta del Río Colorado. Comision Nacional de Areas Naturales Protegidas, Mexico

CONANP (2011) Listados de Sitios Ramsar. http://ramsar.conanp.gob.mx/lsr.php. Accessed 1 Oct 2011

CONAPESCA (2008) Anuario estadístico de acuacultura y pesca 2008. CONAPESCA, Mexico

Cortina-Segovia S, Brachet-Barro G, Ibañez de la Calle M, Quiñones-Valades L (2007) Océanos y costas. Análisis del marco jurídico e instrumentos de política ambiental en México. Secretaría de Medio Ambiente y Recursos Naturales. Instituto Nacional de Ecología, Mexico

Cowan JH Jr, Rice JC, Walters CJ et al (2012) Challenges for implementing an ecosystem approach to fisheries management. Mar Coast Fish 4(1):496–510

Crowder LB, Hazen EL, Avissar N et al (2008) The impacts of fisheries on marine ecosystems and the transition to ecosystem-based management. Annu Rev Ecol Evol Syst 39:259–278

Cudney-Bueno R (2000) Management and conservation of benthic resources harvested bysmall-scale hookah divers in the Northern Gulf of California, Mexico: the black murex snail fishery. MS thesis. University of Arizona

Cudney-Bueno R (2007) Plan de manejo regional para el aprovechamiento y la recuperación del callo de escarlopa, *Spondylus calcifer*. Documento preparado para la Dirección General de Aprovechamiento de Vida Silvestre. Centro Intercultural de Estudios de Desierto y Océanos, Puerto Peñasco

Cudney-Bueno R, Basurto X (2009) Lack of cross-scale linkages reduces robustness of community-based fisheries management. PLoS One 4(7):6253. doi:10.1371/journal.pone.0006253

Cudney-Bueno R, Turk-Boyer PJ (1998) Pescando Entre Mareas del Alto Golfo de California: Una Guía Sobre la Pesca Artesanal, Su Gente, y Sus Propuestas de Manejo. Serie Técnica # 1. Centro Intercultural de Estudios de Desiertos y Océanos, Puerto Peñasco

Cudney-Bueno R, Bourillón L, Sáenz-Arroyo A et al (2009) Governance and effects of marine reserves in the Gulf of California, Mexico. Ocean Coast Manage 52:207–218

Cudney-Bueno R, Lavin M, Marinone S, Raimondi P, and Shaw W (2009) Rapid effects of marine reserves via larval dispersal PLoS One 4:1 e4140. doi:10.1371/journal.pone.0004140

Deegan LA, Garritt RH (1997) Evidence for spatial variability in estuarine food webs. Mar Ecol Prog Ser 147:31–47. doi:10.3354/meps147031

Díaz-Uribe JG, Arreguín-Sánchez F, Lercari-Bernier D et al (2012) An integrated ecosystem trophic model for the north and central Gulf of California: an alternative view for endemic species conservation. Ecol Model 230:73–91. doi:10.1016/j.ecolmodel.2012.01.009

DOF (1993) Decreto por el que se declara área natural protegida con el carácter de Reserva de la Biosfera, la región conocida como alto Golfo de California y delta del Río Colorado, ubicada en aguas del Golfo de California y los municipios de Mexicali, BC, del Puerto Peñasco y San Luis Río Colorado, Sonora. Diario Oficial de la Federación, Secretaría de Gobernación, México

DOF (2000) Reglamento de la Ley General del Equilibrio Ecológico y la Protección al Ambiente en Materia de Evaluación del Impacto Ambiental. Artículo 2, Capítulo 5, Inciso S. Diario Oficial de la Federación, Secretaría de Gobernación, México

DOF (2005) Programa de protección de la Vaquita dentro de área de Refugio ubicada en la porción occidental del Alto Golfo de California. Diario Oficial de la Federación, Secretaría de Gobernación, México

DOF (2010) Protección ambiental-especies nativas de México flora y fauna silvestres-categorías de riesgo y especificaciones para su inclusión, exclusión o cambio-lista de especies en riesgo. Diario Oficial de la Federación, Secretaría de Medio Ambiente y Recursos Naturales, Mexico

DOF (2012) Acuerdo por el cual se establece una red de zonas de refugio en aguas marinas de jurisdicción federal frente a la costa oriental del Estado de Baja California Sur, en el corredor marino San Cosme a Punta Coyote. Diario Oficial de la Federación, Secretaría de Gobernación, Mexico

DOF (2013) Respuesta a los comentarios efectuados al Proyecto de Modificación a la Norma Oficial Mexicana NOM-002-PESC-1993, Para ordenar el aprovechamiento de las especies de camarón en aguas de jurisdicción federal de los Estados Unidos Mexicanos, publicada el 31 de diciembre de 1993 y sus modificaciones publicadas los días 30 de julio de 1997 y 28 de noviembre de 2006, publicado el 22 de febrero de 2013. Diario Oficial de la Federación, Secretaría de Gobernación, Mexico

Done TJ, Reichelt RE (1998) Integrated coastal zone and fisheries ecosystem management: generic goals and performance indices. Ecol Appl 8:110–118

Downton-Hoffmann CA (2012) Reporte final de actividades perteneciente al permiso de pesca de fomento No. DGOPA.00695.080211.0200. No. PESQ-0612-07, Centro Intercultural de Estudios de Desiertos y Océanos AC. Puerto Peñasco

Downton-Hoffmann CA, Turk-Boyer PJ, Thaxton B et al (2013a) Corredor biológico pesquero Puerto Peñasco-Puerto Lobos, Sonora. Reporte Interno CEDO. Centro Intercultural de Estudios de Desiertos y Océanos AC, Puerto Peñasco

Downton-Hoffmann CA, Loaiza-Villanueva RD, Turk-Boyer PJ et al (2013b) Protocolo de monitoreo biológico-pesquero de peces de importancia comercial en la región del norte del golfo de California. Programa de pesca sustentable. Centro Intercultural de Estudios de Desiertos y Océanos, A.C. Puerto Peñasco

Downton-Hoffmann CA, Martinez-Tovar I, Sánchez-Cruz AY et al (2013c) Protocolo de Monitoreo del Corredor Biológico Puerto Peñasco-Puerto Lobos, Sonora. Programa de Pesca Sustentable. Centro Intercultural de Estudios de Desiertos y Océanos, A.C. Puerto Peñasco

Erisman BE, Buckhorn ML, Hastings PA (2007) Spawning patterns in the leopard grouper, *Mycteroperca rosacea*, in comparison with other aggregating groupers. Mar Biol 151:1849–1861

Erisman BE, Aburto-Oropeza O, Gonzalez-Abraham C et al (2012) Spatio-temporal dynamics of a fish spawning aggregation and its fishery in the Gulf of California. Sci Rep 2:284. doi:10.1038/srep00284

Espinoza-Romero MJ, Chan KMA, McDaniels T, Dalmer DM (2011) Structuring decision-making for ecosystem-based management. Mar Policy 35:575–583

Essington TE, Beaudreau AH, Wiedenmann J (2006) Fishing through marine food webs. PNAS 103(9):3171–3175

Everett WT, Anderson DE (1991) Status and conservation of the breeding seabirds on offshore Pacific islands of Baja California and the Gulf of California. Seabird status and conservation: a supplement. In: Croxall JP (ed) ICBP Technical Publication No. 11. International Council for Bird Preservation, Cambridge

Ezcurra E, Equihua M, Lopez-Portillo J (1987) The desert vegetation of El-Pinacate, Sonora, Mexico. Vegetatio 71:49–60

Ezcurra E, Felger RS, Russell A, Equihua M (1988) Freshwater islands in a desert sand sea: the hydrology, flora and phytogeography of the Gran Desierto oases of Northwestern Mexico. Desert Plant 9:35–44

Felger RS (2000) Flora of the Gran Desierto and Rio Colorado of Northwestern Mexico. University of Arizona Press, Tucson

Flanagan CA, Hendrickson JR (1976) Observations on the commercial fishery and reproductive biology of the totoaba, *Cynosaon macdonaldi*, in the Northern Gulf of California. Fish Bull 74:531–544

Fujita R, Lynham J, Micheli F et al (2012) Ecomarkets for conservation and sustainable development in the coastal zone. Biol Rev Camb Philos Soc 88:273–286. doi:10.1111/j.1469-185X.2012.00251.x

Gaines SD, Gaylord B, Largier JL (2003) Avoiding current oversights in marine reserve design. Ecol Appl 13:32–46. doi:10.1890/1051-0761 013[0032:ACOIMR]2.0.CO;2

Galindo-Bect M, Glenn EP, Page HM et al (2000) Penaeid shrimp landings in the upper Gulf of California in relation to Colorado river freshwater discharge. Fish Bull 98(1):222–225

Gallagher MB, Heppell SS (2010) Essential habitat identification for age-0 rockfish along the central Oregon coast. Mar Coast Fish 2:60–72. doi:10.1577/C09-032.1

García-Caudillo JM, Gómez-Palafóx JV (2005) La pesca industrial de camarón en el golfo de California: Situación Económico-Financiera

e Impactos Socio-ambientales Conservación Internacional México. AC, Mexico
Glenn EP, Zamora-Arroyo F, Nagler PL et al (2001) Ecology and conservation biology of the Colorado river delta, Mexico. J Arid Environ 49:5–15
Glenn EP, Nagler PL, Brusca RC, Hinojosa-Huerta O (2006) Coastal wetlands of the Northern Gulf of California: inventory and conservation status. Aquat Conserv 16:5–28
Guido S (2006) Red de monitoreo de los desarrollos turísticos del Noroeste costero. ALCOSTA-CIAD, Mazatlan
Gutiérrez NL, Hilborn R, Defeo O (2010) Leadership, social capital and incentives promote successful fisheries. Nature 470(1):386–389. doi:10.1038/nature09689
Hastings PA, Findley LT (2007) Marine fishes of the upper gulf biosphere reserve, Northern Gulf of California. Dry borders: great natural areas of the Gran Desierto and upper Gulf of California. University of Utah Press, Salt Lake City
Hastings P, Findley L, Van Der Heiden AM (2010) Fishes of the Gulf of California. In: Brusca R (ed) The Gulf of California: biodiversity and conservation. University of Arizona and The Arizona-Sonora Desert Museum, Tucson, pp 96–118
Hernández-Velasco A (2010) Efecto de la pesca artesanal en la estructura comunitaria de peces de arrecife rocosa del Noreste del Golfo de California. Undergraduate Thesis, Universidad Autónoma de Baja California Sur, La Paz
Higuera-Ciapara I, Mayorga-Ríos A (2006) Investigación científica y tecnológica: Conacyt. In: Guzmán PG, Fuentes DF (eds) Pesca, acuacultura e investigación en México, Comisión de Pesca, pp 273–284
Hilborn R (2011) Future directions in ecosystem based fisheries management: a personal perspective. Fish Res 108:235–239
Hinojosa-Huerta O, Carillo-Guerrero Y, DeStefano S et al (2004a) Waterbird communities and associated wetlands of the Colorado river delta, Mexico. Stud Avian Biol 27:52–69
Hinojosa-Huerta O, Iturribarría-Rojas H, Carrillo-Guerrero Y et al (2004b) Bird conservation plan for the Colorado river delta. Pronatura Noroeste, San Luis Río Colorado
Hinojosa-Huerta O, Garcia-Hernandez J, Carillo-Guerrero Y, Zamora-Hernandez E (2007) Hovering over the Alto Golfo: the status and conservation of birds from the Rio Colorado to the Gran Desierto. In: Felger RS, Broyles B (eds) Dry borders: great natural reserves of the Sonoran desert. The University of Utah Press, Salt Lake City, pp 383–407
Houde ED, Rutherford ES (1993) Recent trends in estuarine fisheries: predictions of fish production and yield. Estuaries 16(2):161–176
Huang D, Bernardi G (2001) Disjunct sea of cortez-pacific ocean *Gillichthys mirabilis* populations and the evolutionary origin of their sea of Cortez endemic relative, *Gillichthys seta*. Mar Biol 138:421–428
INAPESCA (2011) Evaluación de un banco de callo de árbol, El Desemboque, Sonora. Instituto Nacional de Pesca, Guaymas
INEGI (2011) Información de México para niños. http://cuentame.inegi.org.mx/monografias/informacion/son/poblacion. Accessed 4 Feb 2013
Iris-Maldonado CA (2011) Estructura de la comunidad de peces de dos esteros en el norte del Golfo de California. Dissertation, Centro de Investigación Científica y de Educación Superior de Ensenada
Johnson AF (1982) Dune vegetation along the eastern shore of the Gulf of California. J Biogeogr 9:317–330
Jones PJS, Qiu W, De Santo EM (2011) Governing marine protected areas—getting the balance right. United Nations Environment Program, Nairobi
Keen AM (1971) Sea shells of tropical west America: marine mollusks from Baja California to Peru. Stanford University Press, California
Kellner JB, Tetreault I, Gaines SD, Nisbet RM (2007) Fishing the line near marine reserves in single and multispecies fisheries. Ecol Appl 17(4):1039–1054
Kininmonth S, Beger M, Bode M et al (2011) Dispersal connectivity and reserve selection for marine conservation. Ecol Model 222:1272–1282
Koen-Alonso M (2009) Ecosystem approach to fisheries. In: Beamish RJ, Rothschild BJ (eds) The future of fisheries science in North America. Fish Fish Ser 31:185–207
Lavín MF, Sánchez S (1999) On how the Colorado river affected the hydrography of the upper Gulf of California. Cont Shelf Res 19:1545–1560
Lercari-Bernier D, Arreguín-Sánchez F (2009) An ecosystem modeling approach to deriving viable harvest strategies for multispecies management of the Northern Gulf of California. Aquat Conserv Mar Fresh Ecosyst 19:384–397
Levin PS, Stunz GW (2005) Habitat triage for exploited fishes: can we identify essential "Essential Fish Habitat?" Estuar Coast Shelf Sci 64:70–78. doi:10.1016/j.ecss.2005.02.007
Lindeman KC, Pugliese R, Waugh GT, Ault JS (2000) Developmental patterns within a multispecies reef fishery: management applications for essential fish habitats and protected areas. Bull Mar Sci 66:929–956
Lluch-Cota SE, Aragón-Noriega EA, Arreguín-Sánchez F et al (2007) The Gulf of California: review of ecosystem status and sustainability challenges. Prog Oceanogr 73:1–26
Loaiza-Villanueva R, Downton-Hoffmann CA (2011) Informe de monitoreo poblacional de Jaiba *Callinectes* spp. en la Costa Norte de Sonora en el Corredor Bahía Adhair-Desemboque de Caborca 2009–2011, Región Puerto Peñasco, Sonora. Centro Intercultural de Estudios de Desiertos y Océanos, A.C. Puerto Peñasco
Loaiza-Villanueva R, Castañeda-Fernández de Lara V, Pérez-Valencia S (2009) Monitoreo poblacional de jaiba *Callinectes* spp. en la costa norte de Sonora en el corredor Bahía Adair-Desemboque de Caborca, 2008–2009. Centro de Estudios de Desiertos y Océanos, Puerto Peñasco
Loaiza-Villanueva R, Downton-Hoffmann CA, Martínez-Tovar I et al (2012) Recomendaciones para el manejo regional de las especies de caracol chino (*Hexaplex nigritus* y *Phyllonotus* (*Hexaplex*) *erythrostomus* en el Norte del Golfo de California, región Puerto Peñasco, Sonora. Informe. Centro Intercultural de Estudios de Desiertos y Océanos AC. Puerto Peñasco
Lowe WH, Allendorf FW (2010) What can genetics tell us about population connectivity? Mol Ecol 19:3038–3051. doi:10.1111/j.1365-294X.2010.04688.x
Maluf LY (1988) Composition and distribution of the Central Eastern Pacific echinoderms. Nat Hist Mus Los Angeles County. Tech Rep 2:1–242
Marasco RJ, Goodman D, Grimes CB et al (2007) Ecosystem-based fisheries management: some practical suggestions. Can J Fish Aquat Sci 64:928–939
Marinone SG (2003) A three-dimensional model of the mean and seasonal circulation of the Gulf of California. J Geophys Res 108(C10). doi:10.1029/2002JC001720
Marinone SG (2012) Seasonal surface connectivity in the Gulf of California. Estuar Coast Shelf Sci 100:133–141
Marinone SG, Ulloa MJ, Parés-Sierra A et al (2008) Connectivity in the Northern Gulf of California from particle tracking in a three-dimensional numerical model. J Mar Syst 71:149–158. doi:10.1016/j.marsys.2007.06.005
Martinez I (2010) Monitoreo participativo de especies de importancia ecológica y comercial en los arrecifes rocosos de la RB Alto Golfo de California y Delta del Río Colorado y su zona de influencia en Sonora. Centro Intercultural de Estudios de Desiertos y Océanos AC, Puerto Peñasco
Martinez I, Turk-Boyer P (2011) Actualización del Plan de Manejo de callo escarlopa (*Spondylus calcifer*) en el Golfo de California. Centro Intercultural de Estudios de Desiertos y Océanos, A.C. Puerto Peñasco
Martínez I, Turk-Boyer P, Sanchez A, Downton-Hoffmann CA (2011) Corredor Biológico-Pesquero Peñasco, Punta Borrascoso a Puerto Lobos, Sonora, Norte del Golfo de California. Una Propuesta de Ordenamiento Pesquero de la Pesca Ribereña. Proyecto PANGAS. Centro Intercultural de Estudios de Desiertos. Puerto Peñasco

McLeod KL, Lubchenco J, Palumbi SR, Rosenberg AA (2005) Scientific consensus statement on marine ecosystem-based management. Communication Partnership for Science and the Sea, Corvallis

Mellink BE, Romero-Saavedra AL (2005) Dieta del lobo marino de California, *Zalophus californianus* Lesson, 1828, en la Isla San Jorge en el norte del Golfo de California, México, 1998–1999. Cien Mar 31(2):369–377

Meltzer L, Blinick NS, Fleishman AB (2012) Management implications of the biodiversity and socio-economic impacts of shrimp trawler by-catch in Bahía de Kino, Sonora, México. PLoS One 7(6):e35609. doi:10.1371/journal.pone.0035609

Metaxas A, Saunders M (2009) Quantifying the "bio" components in biophysical models of larval transport in marine benthic invertebrates: advances and pitfalls. Biol Bull 216:257–272

Moreno-Báez M (2010) Mapping human dimensions of small-scale fisheries in the Northern Gulf of California, Mexico. Ph.D Thesis, School of Natural Resources and the Environment, University of Arizona, Tucson

Moreno-Báez M, Cudney-Bueno R, Orr B et al (2012) Integrating the spatial and temporal dimensions of fishing activities for management in the Northern Gulf of California, Mexico. Ocean Coast Manage 55:111–127. doi:10.1016/j.ocecoaman. 2011.10.001

Morzaria-Luna H, Iris-Maldonado A, Valdivia-Jiménez P (2010) Physico-chemical characteristics of negative estuaries in the Northern Gulf of California, Mexico. In: Crane JR, Solomon AE (eds.) Estuaries: Types, movement patterns and climatical impacts. Nova Science Publishers Inc. pp 201–223. ISBN: 978-1-60876-859-2

Morzaria-Luna H, Castillo-López A, Danemann GD, Turk-Boyer P (2013) Conservation strategies for coastal wetlands in the Gulf of California, Mexico Wetlands Ecol Manage doi: 10.1007/s11273-013-9328-0

Morzaria-Luna H, Turk-Boyer P, Moreno-Baez M (2013) Social indicators of vulnerability for fishing communities in the Northern Gul fof California, Mexico: Implications for climate change. Mar Policy (2013), http://dx.doi.org/10.1016/j.marpol.2013.10.013i

Munro-Palacio G (2007) Breve historia de Puerto Peñasco. De Cierto Mar Editores, Puerto Peñasco

U.S. National Marine Fisheries Service (1999) Ecosystem-based fisheries management: a report to Congress by the Ecosystem Principles Advisory Panel (online). United States Department of Commerce, National Oceanic and Atmosphereic Administration, National Marine Fisheries service, Silver Spring, MD. www.st.nmfs.gov/st7/document/epap_report.pdf. Accessed 15 Jan 2006

OEIDRUS (2011) Oficina Estatal de Información del Desarrollo Rural Sustentable del Estado de Sonora. http://www.oeidrus-sonora.gob.mx/. Accessed 26 March 2013

Palacios E, Mellink E (1996) Status of the least tern in the Gulf of California. J Field Ornithol 67:48–58

PANGAS (2012a) Callo de Escarlopa *Spondylus limbatus*. Ficha Informativa de la Pesca Ribereña del Norte del Golfo de California. Centro Intercultural de Estudios de Desiertos y Océanos, A.C. Puerto Peñasco

PANGAS (2012b) Caracol chino (Familia Muricidae). Ficha Informativa de la Pesca Ribereña del Norte del Golfo de California. Centro Intercultural de Estudios de Desiertos y Océanos, A.C. Puerto Peñasco

PANGAS (2012c) Pulpo *Octopus bimaculatus*. Ficha Informativa de la Pesca Ribereña del Norte del Golfo de California. Centro Intercultural de Estudios de Desiertos y Océanos, A.C. Puerto Peñasco

PANGAS (2012d) Jaiba *Callinectes bellicosus*. Ficha Informativa de la Pesca Ribereña del Norte del Golfo de California. Centro Intercultural de Estudios de Desiertos y Océanos, A.C. Puerto Peñasco

PANGAS (2012e) Baqueta *Hyporthodus acanthistius*. Ficha Informativa de la Pesca Ribereña del Norte del Golfo de California. Centro Intercultural de Estudios de Desiertos y Océanos, A.C. Puerto Peñasco

Pauly D, Christensen V, Dalsgaard J et al (1998) Fishing down marine food webs. Science 279:860–863. www.sciencemag.org

Pauly D, Christensen V, Walters C (2000) Ecopath, ecosim, and ecospace as tools for evaluating ecosystem impact of fisheries. ICES J Mar Sci 57:697–706

Pauly D, Christensen V, Guenette S et al (2002) Towards sustainability in world fisheries. Nature 418:689–695

Peguero-Icaza M, Sánchez-Velasco L, Lavín MF, Marinone SG (2008) Larval fish assemblages, environment and circulation in a semienclosed sea (Gulf of California, Mexico). Estuar Coast Shelf Sci 79:277–288. doi:10.1016/j.ecss.2008.04.008

Peguero-Icaza M, Sánchez-Velasco L, Lavín MF et al (2011) Seasonal changes in connectivity routes among larval fish assemblages in a semi-enclosed sea (Gulf of California, Mexico). J Plankton Res 33:517–533. doi:10.1093/plankt/fbq107

Pérez-Valencia SA, Aragón–Noriega EA (2012) Determinación de biomasa y bancos de la almeja de sifón *Panopea globosa* (Dall 1898) en una zona virgen del Alto Golfo de California. Interciencia 37(9):651–656

Pérez-Valencia S, Loaiza-Villanueva RD, Sánchez-Cruz AY (2007) Pesca Deportiva-Recreativa en Puerto Peñasco, Sonora; México. Informe Interno. Centro Intercultural de estudios de Desiertos y Océanos AC, Puerto Peñasco

Pérez-Valencia SA, Gorostieta-Monjaraz M, Castañeda-Fernández de Lara V et al (2012) MIA-R para la Pesca Ribereña Responsable en la Reserva de la Biosfera Alto Golfo de California y Delta del Río Colorado: Costa Oeste. Centro Intercultural de Estudios de Desiertos y Océanos, A.C. Puerto Peñasco

Pérez-Valencia SA, Polanco-Mizquez EI, Rodríguez-Uceda A et al (2013a) Programa de monitoreo abordo: manifestación de impacto ambiental regional para la pesca ribereña responsable en el Alto Golfo de California y Delta del Río Colorado, V.I. Centro Intercultural de Estudios de Desiertos y Océanos, A.C., Puerto Peñasco

Pérez-Valencia SA, Polanco-Mizquez EI, Rodríguez-Uceda A et al (2013b) Programa de monitoreo pesquero: manifestación de impacto ambiental regional para la pesca ribereña responsable en el Alto Golfo de California y Delta del Río Colorado, V.I. Centro Intercultural de Estudios de Desiertos y Océanos, A.C., Puerto Peñasco

Pikitch EK, Santora C, Babcock EA et al (2004) Ecology: ecosystem-based fishery management. Science 305:346–347

Polis GA, Holt RD, Menge BA, Winemiller K (1996) Time, space, and life history: Influences on food webs. 435–460 In: Polis GA, Winemiller K (eds) Food webs: integration of patterns and dynamics. Chapman and Hall, New York.

Polis GA, Anderson WB, Holt RD (1997) Toward an integration of landscape and food web ecology: the dynamics of spatially subsidized food webs. Annu Rev Ecol Syst 28:289–316

Rodríguez CA, Flessa KW, Dettman D (2001) Effects of upstream diversion of Colorado river water on the estuarine bivalve mollusc *Mulinia coloradoensis*. Conserv Biol 15:249–258

Rodríguez-Quiroz G, Aragón-Noriega A, Valenzuela-Quiñónez W, Esparza-Leal HM (2010) Pesca artesanal en las zonas de conservación del Alto Golfo de California. Rev Biol Mar y Oceanogr 45:89–98. doi:10.4067/S0718-19572010000100008

Rodríguez-Ramírez R (2010) Política ambiental, conservación y desarrollo regional sustentable: Una aproximación a través de los "Buy-backs" internacionales y el Pace Vaquita en la región del Alto Golfo de California. MSc Thesis, Centro de Investigación en Alimentación y Desarrollo A.C., Hermosillo

Rogers SI, Clarke RK, Reynolds JD (1999) The taxonomic distinctness of coastal bottom-dwelling fish communities of the North-east Atlantic. J Anim Ecol 68:769–782

Rojas-Bracho L, Reeves RR, Jaramillo-Legoretta A (2006) Conservation of the vaquita *Phocoena sinus*. Mammal Rev 36:179–216

Rose MD, Polis GA (1998) The distribution and abundance of coyotes: the effects of allochthonous food subsidies from the sea. Ecology 79:998–1007

Rosemartin A (2008) Biological dimensions of tern management: a case study of the least tern in Sonora, Mexico and a comparative analysis of reproductive investments in terns. MSc Thesis, School of Natural Resources, University of Arizona, Tucson
Rountree R, Able K (2007) Spatial and temporal habitat use patterns for salt marsh nekton: implications for ecological functions. Aquat Ecol 41:25–45. doi:10.1007/s10452-006-9052-4
Rowell K, Flessa KW, Dettman DL et al (2008) Diverting the Colorado river leads to a dramatic life history shift in an endangered marine fish. Biol Conserv 141(4):1140–1150
Rupnow J (2008) Recreational fisheries and the ecology and management of targeted resources in the Northern Gulf of California, Mexico. MSc Thesis, Arizona State University, Tucson
Sáenz-Arroyo A, Roberts CM, Torre J, Carino-Olvera M (2005) Using list naturalists' observations and grey literature to reassess marine species at the Gulf grouper in the Gulf of California, Mexico. Fish Fish 6:121–133
Sánchez-Velasco L, Lavín MF, Jiménez-Rosenberg SPA et al (2012) Larval fish habitats and hydrography in the biosphere reserve of the upper Gulf of California. Cont Shelf Res 33:89–99. doi:10.1016/j.csr.2011.11.009
SEMARNAT (2012) Resolutivos emitidos en el PEIA en la DGIRA. Proyecto "Home Port Turístico Puerto Peñasco" en la Gaceta Ecológica Año X Separata No. DGIRA/055/12 del 21 de noviembre de 2012. SEMARNAT/DGIRA, Mexico
Seminoff JA, Nichols WJ (2007) Sea turtles of the Alto Golfo. A struggle for survival. In: Felger RS, Broyles B (eds) Dry borders: great natural reserves of the Sonoran desert. University of Utah Press, Salt Lake City, pp 505–518
Shreve F, Wiggins IL (1964) Vegetation and flora of the Sonoran desert. Stanford University Press, Stanford
Solís-Marín FA, Laguarda-Figueras A, Durán-González A et al (2005) Equinodermos (Echinodermata) del Golfo de California, México. Rev Biol Trop 53(3):123–137
Soria G (2010) Connectivity of marine bivalve species in the Northern Gulf of California: implications for fisheries management and conservation. PhD Thesis, University of Arizona, Tucson
Soria G, Munguía-Vega A, Marinone SG et al (2012) Linking bio-oceanography and population genetics to assess larval connectivity. Mar Ecol Prog Ser 463:159–175
Soria G, Torre-Cosio J, Munguia-Vega A, Marinone SG, Lavín MF, Cinti A, Moreno-Báez M (2014) Dynamic connectivity patterns from an insular marine protected area in the Gulf of California. Journal of Marine Systems 129:248–258
Spackeen J (2009) Analysis of food web dynamics in the Northern Gulf of California using stable isotopes. Undergraduate degree, University of Miami, Miami
Stapp P, Polis GA (2003) Influence of pulsed resources and marine subsidies on insular rodent populations. Oikos 102:111–123
Steller DL, Riosmena-Rodriguez R, Foster MS, Roberts CA (2003) Rhodolith bed diversity in the Gulf of California: the importance of rhodolith structure and consequences of disturbance. Aquat Conserv Mar Fresh Ecosyst 13:S5–S20
Szteren D, Aurioles-Gamboa D, Gerber LR (2006) Population status and trends of the California sea lion (Zalophus californianus californianus) in the Gulf of California, Mexico. In: Trites A, Atkinson S, DeMaster D et al (eds) Sea lions of the world. Alaska Sea Grant College Program/University of Alaska Fairbanks, Fairbanks, pp 369–384
Talley DM, Huxel GR, Holyoak M (2006) Habitat connectivity at the land-water interface. In: Crooks KR, Sanjayan MA (eds) Connectivity conservation. Cambridge University Press, Cambridge, pp 97–129
Thomson DA, Gilligan M (1983) The rocky shore fishes. In: Case TJ, Cody ML (eds) Island biogeography in the sea of Cortez. University of California Press, Berkeley, pp 98–129
Thomson DA, Findley LT, Kerstitch AN (2000) Reef fishes of the sea of Cortez: the rocky shore fishes of the Gulf of California. University of Texas, Austin
Turk-Boyer PJ (1998) Estero Morua: the formation of an estuary. Centro Intercultural de Estudios de Desiertos y Océanos, A.C. Puerto Peñasco
Turk-Boyer PJ (2008) Integrating science and community participation: conservation and management of coastal wetlands in the Northern Gulf of California. Centro Intercultural de Estudios de Desiertos y Océanos, A.C. Puerto Peñasco
Turk-Boyer PJ, Barrera JC (2012) Indicadores y mecanismos para la evaluación de la efectividad del programa de protección de la vaquita dentro del área de refugio ubicada en la porción occidental del Alto Golfo de California. Centro Intercultural de Estudios de Desiertos y Océanos, A.C. Puerto Peñasco
Turk Boyer PJ, Castañeda-Fernández de Lara V, Morzaria-Luna HN et al (2009) Sustainable fisheries in the upper Gulf of California: characterization of fisheries and essential fish habitats, and community management interests. Final Grant Report 06-0357. Centro Intercultural de Estudios de Desiertos y Océanos, A.C. Puerto Peñasco
Turner R, Brown D (1994) Sonoran desertscrub. In: Brown D (ed) Biotic communities: Southwestern United States and Northwestern Mexico. University of Utah Press, Salt Lake City, pp 181–221
Urban J (2010) Marine mammals of the Gulf of California: an overview of diversity and conservation status. In: Brusca R (ed) The Gulf of California: biodiversity and conservation. University of Arizona and The Arizona-Sonora Desert Museum, Tucson, pp 188–209
Valavanis VD, Georgakarakos S, Kapantagakis A et al (2004) A GIS environmental modeling approach to essential fish habitat designation. Ecol Model 178:417–427. doi:10.1016/j.ecolmodel.2004.02.015
Van Andel TH (1964) Recent marine sediments of the Gulf of California. In: van Andel TJ, Shor GG (eds) Marine geology of the Gulf of California: a symposium. Am Assoc Pet Geol Mem 3:216–310
Vega X, González MA, Muñoz del Viejo A (2006) Potential new Ramsar sites in northwest Mexico: strategic importance for migratory waterbirds and threats to conservation. In: Boere GC, Galbraith CA, Stroud DA (eds) Waterbirds around the world. The Stationery Office, Edinburgh, pp 158–160
Velarde E, Anderson DW (1994) Conservation and management of seabird islands in the Gulf of California: setbacks and successes. In: Nettleship DN, Burger J, Gochfeld M (eds) Seabirds on islands: threats, case studies and action plans. Birdlife International, Cambridge, pp 229–243
Velarde E, Ezcurra E, Cisneros-Mata MA, Lavin MF (2004) Seabird ecology, El Niño anomalies, and prediction of sardine fisheries in the Gulf of California. Ecol Appl 14:607–615
Weinstein MP, Litvin SY, Bosley KL et al (2000) The role of tidal salt marsh as an energy source for marine transient and resident finfishes: a stable isotope approach. T Am Fish Soc 129:797–810. doi:10.1577/1548-8659(2000) 129<0797:TROTSM>2.3.CO;2
Zuria I, Mellink E (2005) Fish abundance and the 1995 nesting season of the least tern at Bahía San Jorge, Northern Gulf of California, Mexico. Waterbirds 28:172–180

"Marine Areas of Responsible Fishing": A Path Toward Small-Scale Fisheries Co-Management in Costa Rica? Perspectives from Golfo Dulce

10

Luc Fargier, Hans J. Hartmann and Helena Molina-Ureña

Abstract

This chapter analyzes participatory management processes of small-scale fisheries in two Pacific embayments of Costa Rica, a centralized state of Central America where fisheries management is traditionally "top-down", data deficient, and poorly adapted to local biological and socio-economic conditions. We provide an historical overview of coastal activities governance and fisheries national context, and describe different participative approaches to small-scale fishery management. The Marine Area of Responsible Fishing (Área Marina de Pesca Responsable, or AMPR), created in 2008, is a management tool developed by the Costa Rican government to effectively involve fishers organizations in small-scale fisheries management. In this paper, we compare participative management initiatives associated with AMPRs in the Golfo Dulce and Golfo de Nicoya (Palito and Tárcoles), and Marine Protected Areas (MPAs) in Cahuita and Marino Ballena National Parks. Based on our analysis, we recommend ten measures to improve the small-scale fisheries co-management process. Among these, five recommendations stand out: (1) increase the participation of artisanal fishers in the development of collective choice rules; (2) allocate costs and benefits of management measures among artisanal fishers; (3) improve local leadership; (4) improve understanding and transparency of the management process; and (5) formalize and implement strategic fisheries management plans.

Keywords

Small-scale fisheries · Marine area of responsible fishing · Co-management · Costa Rica · Tropical Eastern Pacific fisheries

10.1 Introduction

There are no universal definitions of small-scale or artisanal fisheries (hereby considered synonymous, for practical purposes), despite the fact that they are thought to represent over 90% of all fishers and maritime workers across the globe and responsible for roughly half of the total catch for human consumption (FAO 2012; World Bank et al. 2010). The meanings of the terms vary according to chrono-geographical context and conceptual currents (Berkes et al. 2001b; Carvalho et al. 2011; Durand et al. 1991; FAO and World Fish Center 2008; Johnson 2006; Kurien 2003; Salas et al. 2007; Smith 1979; Staples et al. 2004; World Bank et al. 2010).

L. Fargier (✉) · H. J. Hartmann
UMR 7266 LIENSs (Research Center for Coasts, Environment and Societies), Université de La Rochelle (ULR), 2 rue Olympe de Gouges, 17000 La Rochelle, France
e-mail: luc.fargier@univ-lr.fr

H. J. Hartmann
e-mail: hhartman@univ-lr.fr

H. Molina-Ureña
CIMAR (Research Center in Marine Sciences and Limnology), Universidad de Costa Rica (UCR), San Pedro de Montes de Oca, 2060-11501 San José, Costa Rica
e-mail: hmolina@rsmas.miami.edu

F. Amezcua, B. Bellgraph (eds.), *Fisheries Management of Mexican and Central American Estuaries*, Estuaries of the World, DOI 10.1007/978-94-017-8917-2_10, © Springer Science+Business Media Dordrecht 2014

After being neglected by decision makers for many decades, artisanal fisheries are increasingly recognized for their contribution to food security, poverty reduction, and local and national economies, especially in developing countries (Andrew et al. 2007; Béné et al. 2007; Evans et al. 2011; FAO 2007, 2009, 2011a, b, c, d; FAO and World Fish Center 2008).

In Central America, artisanal fisheries are especially difficult to encapsulate due to their heterogeneous and complex structure. Agüero (1992) and Salas et al. (2007, 2011) characterize them as follows: (1) a multi-species, multi-gear fishery that changes seasonally, targeting high-value commercial species for consumption, local sale, and/or export using low-technology gear requiring intensive labor and modest capital investment; (2) a poorly organized and marginalized professional sector, highly dependent on intermediaries in the supply chain and marketing component; (3) a population with limited access to capital, credit, education, health care, and other social benefits; (4) operating in many small, dispersed, and often isolated landing sites; and (5) a low-cost source of animal protein, a buffer against rural unemployment, and an economic stimulant, due in part to multiplier effects that accrue to fishing. Multiplier effects arise because fishing activities use inputs from other industries/businesses to generate their own products, which in turn become inputs to other economic sectors; this situation acknowledges the interdependency among sectors (FAO 2005). Overall, the multi-faceted nature of small-scale fisheries in Central America reflects the complex geophysical, bio-ecological, socio-economic and political realities of the region, rendering artisanal fisheries management intricate and delicate (Salas et al. 2011).

Following the international debt crisis of the 1980s and many Central American civil wars, fishing represents a "last frontier" for the rural unemployed sector in Central America (Elizondo Mora 2005; FAO 2011d; González Álvarez et al. 1993). The number of artisanal fishers in the region more than doubled between 1970 and 2000 (Agüero 1992; Chuenpagdee et al. 2006; FAO 1999; OSPESCA 2010). According to OSPESCA (2010), artisanal fisheries landings in all Central American countries, except Panama, are more important than industrial landings. However, these data are difficult to validate to date, as available publications neither specify census methodology, nor the type of fishing. Nevertheless, they confirm the economic importance of artisanal fishing in the region.

Gréboval (2007) identified six natural and anthropogenic factors that negatively affect coastal fisheries in Central America: (1) the absence of strong governance structures; (2) a poor understanding of coastal fisheries operations; (3) excessive fishing capacity; (4) a downward trend and increased variability in resource abundance due to overfishing, habitat degradation, and El Niño Southern Oscillation (ENSO) events; (5) increasing demand in the face of limited resources; and (6) poverty and lack of development alternatives in coastal areas.

The small-scale fisheries sector has generally been excluded from fisheries management due to its geographic, socio-economic, and political marginalization (Béné 2003, 2004; Jacquet and Pauly 2008; Jacquet et al. 2010; Pauly 1997, 2006; Pauly and Agüero 1992; Teh et al. 2011). Local traditions and socio-economic aspects of coastal fisheries are often not considered in management decisions (Martin 2001; Ruddle 2011; Ruddle and Hickey 2008; Ruddle and Satria 2010), despite evidence that the structure and dynamics of small-scale coastal fisheries in tropical developing countries are profoundly shaped by non-biological factors (Chauveau and Jul-Larsen 2000).

Evidence suggests that fisheries regulations and practices are much more likely to be successful if fishers and other stakeholders participate in the development of policies and regulations that affect them and the communities they live in (Jentoft 2006). Fisheries co-management has been identified as a realistic solution to many of the problems facing the world's small-scale fisheries (Gutiérrez et al. 2011). According to Evans et al. (2011), successful co-management involves the sharing of responsibility, authority, or possibly both, in varying degrees between resource users and another organization or entity (usually a government agency). This implies expansion of the typical fishers-government relationship to include other entities, such as non-governmental organizations (Nuñez Saravia 2000).

However, as Jentoft et al. (1998) indicated, "co-management is not so much about the rules per se as it is about the communicative and collaborative process through which these rules are formed". Stakeholder participation in small-scale fisheries co-management should be both quantitative (Béné and Neiland 2006; Béné et al. 2008; Neiland and Béné 2003; Sen and Raakjaer Nielsen 1996) and qualitative (Cohen and Uphoff 1980; Jentoft et al. 1998; Pinkerton 1989). Quantitative measures include the number of participants and degree of participation. Qualitative measures address questions such as: which part of the population participates (local population, local leaders, officials, external agents), when does participation occur (design and planning, implementation, monitoring phases), what kind of participation is allowed (instructive, consultative, cooperative, advisory, informative), and how does the process occur (its form, its extent and its local impacts)? From these elements, a typology of participation may be established, ranging from pseudo-participation to full-scale participation (Arnstein 1969; Pretty 1995).

Small-scale fisheries management in Central American countries has traditionally been imposed in "top-down" fashion by regulators who rely primarily on conventional management techniques, including species size limits, gear control, temporary fishing bans, zoning, fishing permits, and other management tools (Agüero 1992; Salas et al. 2007). All Central American states have laws that enable the participation of civil society in the management of protected areas, with varying degrees and modes of participation (CBM 2003; Estado de la Nación 2008; Luna 1999). However, Costa Rican law allows only marginal public participation in protected-area management (DFOE-AM 2005; Fonseca-Borrás 2009); the State may not delegate the administration, management of public funds, police power, definition of policies, and approval of management plans for protected areas. As a result, Costa Rica offers fewer opportunities for public participation in environmental management compared to other Central American nations (DFOE-AM 2005; Fonseca-Borrás 2009). The vertical and complex nature of its legal and institutional framework for natural resource management likely hinders active stakeholder participation.

In order to improve the situation in Costa Rica, Marine Areas of Responsible Fishing (Área Marina de Pesca Responsable—AMPR), created in 2008, constitute one of the most recent management tools developed by the Costa Rican government to involve fishers organizations in small-scale fisheries management (Fargier 2012). Based on the premise that successful small-scale fisheries management relies on participatory approaches, this chapter collates key lessons to be learned in regards to (1) the conditions favoring successful small-scale fisheries co-management processes in Costa Rica; and (2) the potential of AMPRs to achieve their purpose. Our approach involves a review of the recent history of participatory processes in small-scale fisheries management in Costa Rica. We compare an AMPR created in the Golfo Dulce (AMPR-GD) with AMPRs situated in the Golfo de Nicoya (AMPR of Tárcoles and Palito) and with participative management initiatives previously established in two no-take Marine Protected Areas (Cahuita and Marino Ballena National Parks). Finally, we present a matrix evaluating co-management potential from a synthesis of case studies reported in the scientific literature to describe fisheries participative processes. The matrix highlights key conditions necessary for the success of the co-management approach. To facilitate our analysis, we chose an analytical framework titled "*Principles for Sustainable Governance of Common-Pool Resources*", developed by Elinor Ostrom (Ostrom 1990; Ostrom and Ostrom 2003; Ostrom et al. 1994, 1999, 2002). The complete matrix and list of papers from which it was developed are available in Appendix 1.

10.2 Geography and Socio-Economic Characteristics of Artisanal Fishing

10.2.1 Geography and Biodiversity

The 1,254-km long Pacific coast of Costa Rica features diverse habitats such as headlands, cliffs, peninsulas, bays, islands, coral reefs, beaches, sand spits, and estuaries with sandy-muddy coasts and mangrove wetlands (Allen and Robertson 1998; Bergoeing 1998; Robertson and Allen 2008; Wehrtmann and Cortés 2009). These geographic features are the product of past tectonic activity and erosive action of the many rivers that drain the central mountains. Species diversity in this area is the highest in the Eastern Pacific region. With a marine area representing only 0.16 % of the world's ocean surface, Costa Rican waters harbor 3.5 % of the known marine bio-diversity (Wehrtmann and Cortés 2009). At least 4,745 species have been recorded in Costa Rican marine waters (Wehrtmann and Cortés 2009) with more than 800 marine fish species present from 0 to 200 m depth (Bussing and López 2009). A high level of endemism is also present (Mora and Robertson 2005a, b).

The Nicoya and Osa peninsulas encompass two major gulfs, the Golfo de Nicoya, which dominates the central-northern Pacific sector and the Golfo Dulce in the South Pacific zone near the Panamanian border. Annual precipitation increases from north to south, from ca. 2,000 mm with a distinct seasonality in the northern part of the Golfo de Nicoya to 4,500 mm in Golfito (Golfo Dulce), where the dry season is comparatively shorter (Quesada-Alpízar and Cortés 2006; Waylen et al. 1996). Both Gulfs drain wet tropical catchment basins and function as meta-estuaries. Golfo Dulce is relatively smaller with steep shorelines, restricted fjord-like circulation, few mangroves and relatively low biological productivity. It is one of the four known 'tropical fjords' (Richards et al. 1971; Vargas and Wolff 1996; Wolff et al. 1996). Golfo de Nicoya is a complex, partially mixed estuary with varied topography, extensive mangroves and high productivity (Voorhis et al. 1983; Wolff et al. 1998) (Table 10.1).

In 2006, the Costa Rican Interdisciplinary Marine Coastal Commission of the Exclusive Economic Zone (Nielsen Muñoz and Quesada Alpízar 2006) named Golfo Dulce and Golfo de Nicoya as two of eight biodiversity "hotspots" requiring urgent conservation. Three years later, both were still identified as important geographic areas that were needed to maintain the integrity of marine and coastal biodiversity in Costa Rica (SINAC 2009).

10.2.2 Historical and Socio-Economic Background of Costa Rican Small-Scale Fisheries

In Costa Rica, a country of inland traditions, commercial fishing is a relatively recent phenomenon and therefore is not culturally rooted (Elizondo Mora 2005; González Álvarez et al. 1993). The government offered incentives for development of industrial and semi-industrial fisheries, particularly that of shrimp fishing, but neglected the artisanal sector (Chacón et al. 2007; Elizondo Mora 2005; Jiménez 2013). The latter became established only during the last 30 years as an "occupation of last resort" for surplus rural labor, but has

Table 10.1 Basic environmental characteristics of Golfo Dulce and Golfo de Nicoya. (Bergoeing 1998; Cortés et al. 2010; Lei 2002; Blanco and Mata 1994; Nielsen Muñoz and Quesada Alpízar 2006; Quesada-Alpízar and Cortés 2006; Svendsen et al. 2006; Voorhis et al. 1983; Wolff et al. 1998)

Characteristic	Golfo Dulce	Golfo de Nicoya
Coordinates	8°32'N, 83°15'W	10°02'N, 85°00'W
Dimension (length × mean width)	50 × 14 km	80 × 25 km
Surface area	680 km^2	1,340 km^2
Catchment basin (main rivers, mean precipitation)	3,200 km^2 (4,147 mm a^{-1})	9,844 km^2 (6,2450 mm a^{-1})
Topography, depth (z)	Fjord-like deep inner basin (z_{max} = 215 m) outer gulf (sill: z = 65 m) minor islands and pinnacles	Inner gulf, z ≤ 25 m outer gulf (no sill, z to > 200 m) several inhabited islands
Circulation	Fjord-like, with a deep (> 150 m) seasonally anoxic zone	Partially to well-mixed estuarine, complex features
Net phytoplankton production (trophic status)	27 to 263 gC m^{-2} a^{-1} (oligo-mesotrophic)	610 gC m^{-2} a^{-1} (meso-eutrophic)
Mangroves, wetlands	< 1,000 ha, few mudflats	15,200 ha, extensive mudflats (inner gulf)
Hard bottom	Rocky shores, submerged basalt reefs, pinnacles, degrading coral communities (inner basin)	Rocky shores and reefs, fewer pinnacles, very little coral growth
Sandy beaches	Along the outer gulf	Along the outer gulf

developed considerable socio-economic importance (Elizondo Mora 2005; González Álvarez et al. 1993; Béné 2003). International projects facilitated the development of artisanal fishing, notably by promoting cooperatives (González Álvarez et al. 1993; López-Estrada and Breton 1991) and aiming to diversify local techniques, such as artisanal nets for jumbo-shrimp fishing that were previously reserved for the semi-industrial fleet (Elizondo Mora 2005; Marín Alpízar 2002).

Commercial fishing centers became established in a few coastal towns, notably Puntarenas on the Pacific coast and Limón on the Caribbean coast, where the necessary support services and infrastructure were concentrated (González Álvarez et al. 1993). The absence of market access in rural areas limited the development of commercial fisheries in these areas. However, some rural small-scale fishing communities with a sense of tradition and cultural identity arose, especially in Golfo de Nicoya, which was the cradle of commercial fishing in Costa Rica and the subject of numerous development projects for artisanal fisheries (Barguil Gallardo 2009).

Between 1996 and 2005, reported national annual fisheries production averaged 21,414 t. The Pacific coast accounted for ~97 % of the Costa Rican fisheries, and ca. 70 % was landed in Golfo de Nicoya essentially by offshore commercial fisheries (Araya et al. 2007). The latterfishery primarily catches mahi-mahi (*Coryphaena hippurus*), tuna (Scombridae, mostly yellowfin *Thunnus albacares*), billfish (Istiophoridae, Xiphiidae), oceanic sharks (mainly silky shark (*Carcharhinus falciformis*), blue shark (*Prionace glauca*), bigeye thresher (*Alopias superciliosus*), oceanic whitetip shark (*Carcharinus longimanus*), scalloped hammerhead (*Sphyrna lewini*), and also smooth-hounds (*Mustelus* spp.), tiger shark (*Galeocerdo cuvieri*), shortfin mako (*Isurus oxyrinchus*) and bonnethead (*Sphyrna tiburo*) and deep-water shrimp (white shrimps *Litopenaeus* spp., pinky shrimp *Farfantepenaeus brevirostris*, brown shrimp *Farfantepenaeus californiensis*, carabalí shrimp *Trachypeneneaus byrdii*, camellón shrimp *Heterocarpus* spp.) (Araya et al. 2007). The semi-industrial sector operates with trawlers and purse seines that target shrimp and sardine, respectively (Araya et al. 2007). Artisanal fishers on Costa Rica's west coast have historically targeted snappers (Lutjanidae), groupers (Serranidae), sharks, mackerels (Scombridae), cusk eels (Brotulidae) and shrimps in Golfo Dulce (Campos 1989; Fargier 2012; Guzmán-Mora 2013; Lagunas Vázques 2004; Poirout 2007) and weakfish and croakers (Sciaenidae), snappers, grunts (Haemulidae), snooks (Centropomidae), cusk eels, pike-congers (Muraenesocidae), sardines and white shrimp in Golfo de Nicoya (Araya et al. 2007; Wolff et al. 1998). Sport/tourist fishers target snappers, groupers and offshore game fish, especially billfishes (Magnin 2004).

Management measures in both gulfs have included declarations of marine and coastal/estuarine protected areas (Estado de la Nación 2011), species-specific fishing bans, gear restrictions especially concerning shrimp trawling (Álvarez and Ross Salazar 2010; González Álvarez et al. 1993), multiple use and fisheries management areas (Alvarado et al. 2012; Nielsen Muñoz and Quesada Alpízar 2006; Salas et al. 2012).

In Costa Rica, as in other Central American countries, fishing occurs mainly on the Pacific coast. The geographic distribution of people along the Pacific coast, especially fishers, is heterogeneous due to local variation in landscape features and biological productivity (ECLAC 2011; OSPESCA 2010; PNUD 2011). According to the most recent survey (OSPESCA 2010), 13,850 artisanal fishers live and work in 75 communities on the Pacific coast, compared to only 950 fishers in 11 communities on the Caribbean coast. Most artisanal fishers (about 94 %) are men (*ibid*). However, these estimates may be biased because neither the census methodology nor the types of fishing were specified (Fargier 2012).

The number of artisanal fishers in Golfo Dulce is relatively small due to the gulf's low productivity and geographic isolation. Less than 250 artisanal fishers live in a handful of isolated fishing communities on the Osa peninsula and in communties near the county capital of Golfito, a natural deep port formerly used for banana export (González Álvarez et al. 1993; Fargier 2012; Guzmán-Mora 2013). These fishers' communities include La Purruja, Puntarenitas de Golfito, Zancudo, Puerto Pilón, Cocal Amarillo, Río Claro de Pavones, as well as the Osa peninsula communities of Puerto Jiménez, La Palma-Playa Blanca and Rincón-Puerto Escondido. By contrast, roughly half of Costa Rica's artisanal fishers live in about thirty communities scattered around the Golfo de Nicoya (González Álvarez et al. 1993; OSPESCA 2010) that include the city of Puntarenas (Barrio el Carmen), as well as Tárcoles, Chomes, Costa de Pájaros, Puerto Níspero, Pochote, Palito and Montero (Chira Island), Isla Caballo, Isla Venado, Puerto Thiel and Paquera.

The term "community" can be defined geographically, politically or socially (Agrawal and Gibson 1999; Berkes et al. 2001a). Here, we use community as a village-type political unit where a group of fishers live. González Álvarez et al. (1993) identified three types of artisanal fishing communities in Costa Rica, each characterized by different development processes:

- Type A, a small community founded by squatters ("*precaristas*", small farmers or farm workers dismissed or evicted from elsewhere), for whom fishing is one of several sources of income;
- Type B, a bigger, older and more homogeneous community were artisanal fishing is the main commercial activity;
- Type C, a community where artisanal fishing is gradually being replaced by tourism activities.

Due to loss of employment caused by the departure of the banana industry in Golfo Dulce in the 1980s (United Fruit Company operated plantations on the coastal and inland plains of Costa Rica's South Pacific region between 1938 and 1984), a large number of former banana workers have joined the commercial fishing industry (Type A). Type C artisanal fishing communities have emerged in Puerto Jiménez, Río Claro de Pavones, Zancudo, and Golfito as bases for international sport fishers, surfers, National Park visitors, beach vacationers and, in the case of Golfito, duty-free shoppers. Fishing communities of Golfo de Nicoya are mostly Type A and Type B.

Small-scale fishing in Golfo Dulce ranks behind tourism in socio-economic importance, whereas in Golfo de Nicoya, it ranks well ahead of tourism (Marín Cabrera 2012). Indeed, tourism in Golfo Dulce exists in most communities, and supports about 60 % of the population, while small-scale fishing is practiced in about 60 % of the communities, and supports only 25 % of the population (*ibid.*). Conversely, in Golfo de Nicoya, tourism is present in only 10 % of the communities and supports less than 10 % of the population. Fishing, on the other hand, is practiced in most communities and supports about 60 % of the population (*ibid.*).

10.3 Coastal Activities Management: Integrating a Decentralized, Multi-Stakeholder Management System

Costa Rica, a centralized, land-oriented state, has historically focused on the management and conservation of its continental resources (SINAC 2009). The marine domain and its resources were only recently incorporated into the Costa Rican political agenda. The coastal theme has been integrated into national development plans of the last two consecutive governmental administrations (2006–2010 and 2010–2014) and into the initiative "Paz con la Naturaleza" (Peace with Nature) in 2007. Adoption of the "National Strategy for Integrated Management of Coastal Marine Resources" in 2007 marked the first major initiative, although hitherto inefficient, to regulate the use of coastal marine resources of the country. In 2012, a new Vice-Ministry of Waterand Seas was created under the Ministry of Environment and Energy (La Gaceta N° 162, 23 Aug 2012). Despite only recent initiatives to preserve them, marine-coastal resources are considered a national heritage and public property, and their exploitation is managed for public utility and social interest.

Three government agencies have jurisdiction over marine resources in coastal zones of Costa Rica: the Ministry of Environment, Energy, Waters, and Seas (MINAEM, as of 2013); the Costa Rican Institute of Fisheries and Aquaculture (Instituto Costarricense de Pesca y Acuicultura, INCOPESCA); and the Costa Rican Institute of Tourism (Instituto Costarricense de Turismo, ICT). In addition, the Coast Guard (Servicio Nacional de Guardacostas) and the Port Authority (Capitanía de Puertos) play supporting roles in enforcement.

Government policies and laws regulating marine fisheries in Costa Rica have been slowly evolving during the last decade. Despite the vertical structure, currently the management process involves multiple stakeholders including municipalities, non-governmental organizations (NGOs), universities, and community groups. The relative weight and role of each sector varies with the region, the issue at hand, and the target species under consideration (Marín Cabrera 2012).

10.3.1 Costa Rican Tourism Institute, ICT

The Institute of Tourism, created in 1955, monitors the Maritime-Terrestrial Zone (MTZ; Law N° 6043 of 1977), which is defined as "[...] the strip 200 meters wide along the entire

Atlantic and Pacific coasts of the Republic of Costa Rica, regardless of their nature, measured horizontally (landward) from the mean high tide level including areas and rocks uncovered by the sea at low tide". The MTZ is under the usufruct and administration of coastal municipalities. Nearly 45 % of Costa Rica's 1,466-km coastline (about 550 km) are subject to this law (Quesada-Alpízar 2006). Since many artisanal fishing communities are established in the MTZ, conflicts involving municipality-supported development are common, usually with the mass-tourism industry that has developed on the Pacific coast since the 1970s (e.g., Honey et al. 2010).

10.3.2 Ministry of Environment, Energy, Waters, and Seas, MINAEM

Since its inception in 1988 (as Ministry of Natural Resources, Energy and Mines—MIRENEM, successively Ministry of Environment and Energy—MINAE, Ministry of Environment, Energy and Telecommunications—MINAET, and currently Ministry of Environment, Energy, Waters and Seas—MINAEM), one of this ministry's major functions is to ensure the promotion and enforcement of environmental legislation in the country. It is responsible for creating and administering the Costa Rican continental and marine protected areas, which have been governed since 1994 by the National System of Conservation Areas (SINAC). Out of 166 protected areas, SINAC encompasses 62 areas bordering or containing a marine section, covering 50 % of the country's coastline (updated from Alvarado et al. 2012; Mora et al. 2006). While more than 25 % of the land area of Costa Rica is protected, only about 1 % of the marine area in the Exclusive Economic Zone (EEZ) is under some form of protection, corresponding to 17.5 % (5,296.5 km^2) of the territorial waters (Alvarado et al. 2012; Estado de la Nación 2010; SINAC 2009). Although a relatively small proportion of Costa Rica's EEZ is protected, the level of protection is ten times greater than the average for other countries in Latin America (Alvarado et al. 2012; Estado de la Nación 2010; SINAC 2009).

Costa Rica currently has 20 protected areas with a marine component (MPAs). Three of those MPAs –the National Parks Marino Las Baulas and, Marino Ballena, and the Playa Blanca National Wetland—are exclusively marine. Only about a third of the areas have a management plan (Alvarado et al. 2012; Estado de la Nación 2010). Due to the appropriation and global redistribution of park fees by the government (Alvarado et al. 2012), most MPAs lack enough human, technical and financial resources required for long-term conservation. Nine of the 23 existing MPAs are no-take zones that preclude artisanal fishing from approximately 16 % (3,000 km^2) of Costa Rica's coastal marine territory, which does not include Isla del Coco National Park (SINAC 2009). In addition, special restrictions on fishing are imposed by other MPA management plans in 15 % (550 km^2) of the coastal marine territory.

The creation of MPAs in Costa Rica used to be synonymous with expropriation and prohibition of extraction of natural resources, including fish, which caused distrust towards MINAEM and opposition of the public to creating new MPAs (TNC 2011). Under a new approach of sustainable use and absolute resource protection developed by MINAEM in 2008, two new types of protected areas allow fishing: Marine Reserves (MR) and Marine Management Areas (MMA; amendment to Art. 70 Decrees N° 34433-MINAET, April 8, 2008 [Regulation of the Law of Biodiversity N° 7788] and 35369-MINAET, May 5, 2009). Marine Reserves are intended for near-exclusive use by artisanal and tourist fisheries (i.e., assumed to be selective and of low impact), as well as other eco-touristic activities of low environmental impact. A broader spectrum of fishing activities is tolerated in MMAs. Prohibited activities include semi-industrial and industrial fishing, and high-impact touristic activities (e.g., marinas). However, while small-scale fishing is permitted under these two new management categories, artisanal fishers are yet to be invited to participate in their management or creation of management plans. As of 2013, only one MMA encompassing a large uninhabited area (9,640 km^2) adjacent to Cocos Island National Park had been implemented: the Seamounts Marine Management Area (MarViva 2011).

In its *Costa Rica por Siempre* (Costa Rica Forever) initiative, the Interdisciplinary Marine Coastal Commission of the Exclusive Economic Zone called for tripling the total size of marine protected areas in Costa Rica by 2016. Due to this inititiative, at least 12 new MPAs would be created, resulting in an expanded network of marine protected areas that would conserve 25 % of Costa Rica's EEZ (as stated in Executive decree 31832–MINAE). This level of protection would exceed the standard of 10 % protection of each of the world's eco-regions established inthe *Programme of Work on Protected Areas* (POWPA) of the 7th Conference of Parties of the Biodiversity Convention (Alvarado et al. 2012; TNC 2011).

10.3.3 Costa Rican Institute of Fisheries and Aquaculture, INCOPESCA

INCOPESCA is responsible for administering the 2005 Fishing and Aquaculture Law (which replaced the Maritime Fishing and Hunting Law of 1948) and the National Development Plan for Fisheries and Aquaculture. Previously, under the name of General Directorate of Fisheries Resources and Aquaculture, INCOPESCA was a department of the Ministry of Agriculture (MAG). With the exception of MPAs, which fall under MINAEM's authority, the entire Costa Rican EEZ falls under the jurisdiction of INCOPESCA. It corresponds to approximately 1,000 km of coastline. Mangrove forest areas are a special case in which INCOPESCA and MINAEM work together to create and implement management plans. Additionally, INCOPESCA is governed by a nine-member board of directors, five of whom are representatives of the fisheries sector. Decisions of the board only require simple majority, and quorum is achieved with at least five people. Directors are elected for 4 year terms, with no limit on the number of terms served. INCOPESCA's Executive Chairperson is appointed by the government.

In theory, this relatively balanced bipartisan composition of INCOPESCA could be interpreted as a special case of co-management of fisheries resources in Costa Rica. In reality, however, the fisheries interests on the board are represented by industrial, semi-industrial and large-to-medium scale fishing interests who may wield undue influence over board decisions. Small-scale artisanal fishers, despite making up more than 80% of Costa Rica's fishers, are absent from the board of directors. Their absence, and the potential for self interest and favoritism among standing board members, raises questions concerning the board's legitimacy as a fisheries management entity. Its critics consider INCOPESCA to be a de facto oligarchy that does little to protect the environment or public interest (Fargier 2012; Quesada-Alpízar 2006). While receiving government aid, semi-industrial fishing interests employ relatively few people, produce little wealth, and engage in fishing practices that degrade the environment (Álvarez and Ross Salazar 2010).

Although INCOPESCA administers a much larger marine area than MINAEM, it has significantly fewer resources, which limits its effectiveness (Barquero 2007; Caviedes 2013). Moreover, in 1995, just 1 year after INCOPESCA's creation, the Costa Rican Constitutional Court declared the article on sanctions in the Maritime Fishing and Hunting Law of 1948 unconstitutional. Thus, until the Fisheries and Aquaculture Law was enacted in 2005, INCOPESCA could not prosecute fishing violations. Making matters worse, the new law was not implemented until 2011.

Unfortunately, the prevalence, or at least the presumption of corruption, conflict of interest, and ineffectual decision-making combined with a lack of resources, capacity, and a clear legal mandate has undermined the authority of INCOPESCA since its inception (Guevara 1996; Defensoría de los Habitantes 2001, 2012). Its governance and operations have been repeatedly questioned by many, including the Comptrollership of the Republic (Contraloría General de la República[CGR]; DFOE-EC-IF 2012; DFOE-PGA 2006; DFOE-PGAA 2008), the Citizen's Ombudsman Office (Defensoría de los Habitantes 2001, 2012), and the Presidency of the Republic (Oviedo and Murillo 2013).

In response to an initiative from the Front For Our Seas (Frente Por Nuestros Mares, FNPM), which is a group of eight civil society organizations (Pretoma, Fundación Keto, Fundación Promar, International Student Volunteers Inc., Sea Save Foundation, The Leatherback Trust, UESPRA, Widecast) as well as interested citizens working to improve the administration and management of marine resources through a series of legal, scientific, political, and advocacy approaches, a Presidential Commission on Marine Governance was created in December 2011. This Commission is integrated by representatives of MINAEM, Public Security, MAG and the NGO Conservation International. The Commission has since publicly called for total reform of INCOPESCA (CPGM 2012), but at the time this chapter was written, had not completed its analysis of restructuring alternatives.

10.3.4 Institutional Coordination

Fisheries management in Costa Rica falls under the purview of multiple institutions with different and often conflicting goals, missions, and approaches, as well as some overlapping jurisdictions. For example, promotion of coastal tourism, including foreign investment by ICT, may clash with municipal coastal planning. Also, ICT and municipal development intiatives may conflict with efforts by MINAEM to conserve coastal marine resources. INCOPESCA's mandate to promote the fishing sector that uses these same resources, brings the two institutions into conflict. At other times, MINAEM and INCOPESCA join forces to create mangrove management plans and strengthen the existing MPA network.

Each agency pursues different interests in a common area of prime importance to coastal communities, in particular artisanal fishers. However, despite a few worthy exceptions, coordination and inter-agency cooperation is weak to non-existent, and small scale fishers are poorly considered (Caviedes 2013). To improve institutional coordination for the conservation of coastal marine resources, a nation-wide legal tool was created in 1995, the Multiple Use Marine Area (AMUM). Thirteen years later, in 2008, two Master plans were finally developed, AMUM *Golfo de Nicoya* and AMUM *Pacífico Sur* (which includes Golfo Dulce). As of today, neither has been implemented.

Meanwhile, the first formal national initiative for integrated coastal zone management took place in 2004 through the Interdisciplinary Marine Coastal Commission of the Exclusive Economic Zone (CIMC-ZEE). Restructured in 2005 as the National Marine Coastal Commission (CNMC), it published the *National Strategy for the Integrated Management of Coastal Marine Resources* in 2007 (Caviedes 2013). Likewise, INCOPESCA and MINAEM co-signed a directive for coordinating the development of MPA management plans in 2009. Other collegial bodies were created in 2010 with the goal of harmonizing coastal zone policies such as: the National Sea Council (CNM) and the Inter-institutional Commission for Marinas and Landing Sites (CIMAT). Delimitation of jurisdictions among all these entities is pending resolution.

At a regional level, the MINAEM's *Osa Conservation Area* (ACOSA, which encompasses Golfo Dulce) manages several marine protected areas (three national parks, a national wetland heritage site and a wildlife reserve). In 2006, ACOSA instituted its own *Inter-institutional Marine Commission* (CIMC-ACOSA) following an initiative from NGOs and university researchers. It convenes diverse actors and stakeholders (e.g., natural resource users, municipalities, state institutions, universities, NGOs, community-based associations), on a monthly basis, to coordinate initiatives and debate projects related to the marine environment in the area. One of its major objectives is to establish a general management plan for the AMUM *Pacífico Sur*. According the AMUM decree (Decree N° 32801 MINAE, La Gaceta N° 241, 14 Dec 2005), the procedure for developing the plan must be participative.

A recent comparative study about stakeholder roles and interactions for AMUMs of Golfo de Nicoya and Pacífico Sur concluded that the networks of key actors in the Golfo Dulce region are dominated by NGOs, government institutions, and the National Federation of Artisanal Fishers Organizations and Affiliates (FENOPEA), the latter as a bridge actor controlling the information flow rather than being information emitters (Marín Cabrera 2012).

10.4 Evolution of National Policies for Participative Management

Costa Rica ratified the Convention on Biological Diversity (signed in Rio de Janeiro in 1992) in June 1994 (Law 7416), with the implicit commitment to encourage participation of the civil society in environmental management, particularly within protected areas. While this commitment is reflected in many Costa Rican legal texts from the second half of the 1990s (i.e. Article 50 of the Political Constitution, Article 6 of the Organic Law of the Environment, Articles 22, 29 and 101 of the Biodiversity Law and The Law of Citizen Participation), a national policy clearly defining the different forms of civic participation in environmental management was still lacking. Furthermore, the identification, development and implementation of strategies, plans and budgets concerning conservation areas are considered exclusive powers of the State by the Constitution (DFOE-AM 2005).

Attempting to amend this void, three separate initiatives took shape, but none was ever formalized or implemented. The first one was led by the MINAEM divisions *of Civil Society* and *Gender and Environment* between 1999 and 2002 (Fonseca-Borrás 2009). Secondly, the IUCN coordinated the *National Policy on Shared Management* between 2003 and 2006 (MINAE-SINAC 2006). The third attempt was the submission of a draft *Law on Protected Areas* at the Costa Rican Legislative Assembly in November 2008, but it was not adopted. Meanwhile, in 2005, the Comptrollership of the Republic issued a precedent-setting report concluding that the Costa Rican Constitution did not recognize the concept of co-management of protected areas (DFOE-AM 2005).

Concurrently, a decline in the participation of civil society in environmental management was observed mainly during the 2006–2010 Oscar Arias administration (Fonseca-Borrás 2009). Its National Development Plan omitted the definition of actions promoting the participation of civil society in environmental management. In late 2008, the President exercised the first veto of his mandate, precisely on articles of the *Law of Citizen Participation* concerning environmental management; he argued that they were unconstitutional (Fonseca-Borrás 2009). At the same time, regulations of the *Law of Biodiversity* were modified such that the "free, prior and informed consent" of communities was no longer required for the implementation of a project affecting their immediate environment. Thus, civil participation in environmental management is currently possible only at local councils and regional committees of the SINAC-governed Conservation Areas (Article 39 of the Biodiversity Law) but these have been characterized as politicized bodies and poor representatives of the communities (Solís-Rivera et al. 2012). According to Fonseca-Borrás (2009), such practices cannot be considered co-management. Some Costa Rican cooperative and political actors see this strategy reversal as one of the indicators of policy change in the country, that might be related to the ratification of the Free Trade Agreement (FTA) between the United States, Central America and the Dominican Republic in October 2007 (Fonseca-Borrás 2009).

10.5 Small-Scale Fishers' Participation in Coastal Fisheries Management

10.5.1 Experiences within MINAEM-Created MPAs

Although co-management is not legally recognized in Costa Rica, some forms of civic participation in environmental management have emerged that are associated with protected areas managed by SINAC. Due to the combination of increased pressure on natural resources, limited government capacity, devolution and decentralization of some central government functions, civil society has become pro-active in natural resource management. Most initiatives arose from informal local processes that were *de facto* institutionalized, some of which have benefitted from legal recognition that has allowed them to persist.

Examples of these experiences are seen in areas with less restrictive protection status, such as the National Wildlife Refuges (RNVS). Two well-known cases are located on the Pacific Coast of the Nicoya peninsula: the participation of the Artisanal Fishers Association of Puerto Coyote (ASPECOY)

Fig. 10.1 Marine Protected Areas (*MPAs*) studied in this chapter. MPAs and National Wildlife Refuges and National Parks (*RNVS*) are the ones created by the Ministry of Environment, Energy, Waters, and Seas (*MINAEM*). Marine Areas of Responsible Fishing (*AMPR*) are the ones created by are the Costa Rican Institute of Fisheries and Aquaculture (*INCOPESCA*)

in the creation of the Caletas-Ario National Wildlife Refuge; and, further north, the Integral Development Association of Ostional (ADIO) that sustainably harvests Olive Ridley turtle (*Lepidochelys olivacea*) eggs in the Ostional National Wildlife Refuge (Fig. 10.1). Two other cases concern National Parks: Cahuita National Park on the southern Caribbean coast (Weitzner and Fonseca-Borrás 2001) and Ballena Marine National Park on the South Pacific Coast (Fig. 10.1).

The government declared the coral reefs near Cahuita as a *National Monument* in 1970, without consulting the community. Local discontent arose following the associated use-restrictions (Girot et al. 2000), which in 1974 led to the creation of an *ad hoc* commission involving government officials and influential members of nearby communities. However, the commission's recommendations were disregarded during the designation of the Cahuita National Park in 1978: a combination of the banning of resource exploitation, expropriations and the increase of park entry fees for foreign visitors led to recurring serious conflicts with local communities between 1978 and 1994 (Girot et al. 2000; Weitzner 2000; Weitzner and Fonseca-Borrás 2001). Eventually, negotiations mediated by the Citizen's Ombudsman Office (*Defensoría de los Habitantes*) enabled legal recognition of a Management Committee in charge of administering the entire park made up by local leaders and other stakeholders (Fonseca-Borrás 2009).

Likewise, the Ballena Marine National Park had a long history of conflicts. Created in 1989 as one of the first marine national parks in Latin America, local communities were not consulted in the process, giving rise to conflicts between local fishers and MINAEM officials over resource use restrictions (Fonseca-Borrás 2009). The Association for the Development of Ballena Marine National Park (ASOPARQUE, consisting of 22 local organizations) was finally established in 1997 (*ibid.*). Negotiations between MINAEM and local authorities began in 1998, through a liaison committee, with the objective of initiating a co-management process. In 2002, the liaison committee requested a external technical expert, CoopeSoliDar (a social-environmental consultancy organized as a cooperative) to develop a co-management plan for the park. As a result, ASOPARQUE participated *de facto* in the park's management. However, unlike Cahuita, this initiative was not legally recognized (Fonseca-Borrás 2009). The 2005 CGR's negative ruling on the co-management process (DFOE-AM 2005) focused on the lack of representativeness of ASOPARQUE, the illegality of fundraising and control of the park access by ASOPARQUE. The CGR report urged MINAEM to regain control of the park and to regularize the situation. In the aftermath, 10 years of efforts to establish a co-management for the Ballena Marine National Park were ended.

10.5.2 Experiences Related to Artisanal Fishers

Outside of marine protected areas, the government has developed various initiatives for small-scale fisher participation in coastal resource management: during the 1980s, through the cooperative movement and Local Committees of Artisanal Fishers (COLOPES); since 2008, by the creation of Marine Responsible Fishing Areas (AMPRs).

10.5.2.1 Cooperative Movement

In the 1980s, Central American artisanal fishers, traditionally suspicious of organizational structures, became interested in the cooperative movement, which had gained "virtually indestructible" institutional faith in Costa Rica (López-Estrada and Breton 1991). Beltrán (2005) explained this attitude change with three reasons:(1) enhanced local capacity achieved by training, (2) the need for alliances to improve the bargaining power of small-scale fishers in marketing, and above all, (3) the need to comply with the requirements of national and international aid.

However, despite the influx of international financial aid for developing countries, the cooperative initiative was short-lived in Costa Rica, due to a combination of conditions prevailing then: (a) the top-down nature of the cooperative creation process, promoting collective ownership of production means (contrary to the expectations and interests of artisanal fishers); (b) prioritization of fish production at the expense of social welfare; (c) little room for training; (d) unorganized increase in artisanal fishing effort, despite the signs of resource decline; (e) organizational problems; (f) financial opportunism of local leaders; and (g) cessation of external funds (Breton 1991; Chauveau and Jul-Larsen 2000; Elizondo Mora 2005; Villalobos-Chacón 2011, School of Biological Sciences, National University of Costa Rica, pers. comm.). Once the aid programs ended, out of the 20 cooperatives established in the early 1980s, only three remained at the end of the decade, including CoopeTárcoles, the Tárcoles' fishermen cooperative (Villalobos-Chacón 2011, pers.comm.).

10.5.2.2 Local Committees of Artisanal Fisheries

Local Committees of Artisanal Fishers (COLOPES) were created in 1989 (Decree N° 19141-MAG, La Gaceta N° 162, 28 Aug 1989) by the first Oscar Arias administration (1986–1990) to remedy the lack of organization of the artisanal fisher's sector and the failure of the fishing cooperatives. With ambitious goals, COLOPES were to function as a liaison between artisanal fishers and the former General Directorate of Fisheries Resources and Aquaculture (today's INCOPESCA). One COLOPES could be created per fishing community, with a minimum membership of 40 fishers.

Lack of realistic decree specifications and institutional and legal support brought about conflicts among artisanal fishers. Moreover, the following government (the 1990–1994 Rafael A. Calderón administration) did not support the initiative. By 1995, only five out of 40 COLOPES remained (González 2011, INCOPESCA, Golfito Regional Office, pers. comm.; Villalobos-Chacón 2011, pers.comm.).

In summary, Costa Rican government efforts to organize artisanal fisheries through "top-down" initiatives such as cooperatives and COLOPES may be understood as an attempt by the State to anchor its presence in these communities and/or to delegate its responsibilities. After two failed attempts in the 1980s, the disillusioned Costa Rican artisanal fishers would not try new forms of organization until the late 1990s, when they began creating Associations on their own, without state support. Nowadays, the most common types of artisanal fisher organizations are associations and cooperatives (OSPESCA 2010).

10.5.2.3 Marine Areas for Responsible Fishing (AMPRs)

In 2007, fishers of the CoopeTárcoles R.L. cooperative, with support of CoopeSoliDar, submitted to INCOPESCA a custom-made proposal for the creation of a fisheries management area, a "Community Marine Area for Responsible Fishing", that would recognize the efforts of the Tárcoles fishers for the sustainable exploitation of their resources. This prompted the appointment of a mixed government-cooperative-NGO commission (INCOPESCA, MINAE, CoopeTárcoles, CoopeSoliDar, and marine resource conser-

Table 10.2 Data about the first seven Marine Areas of Responsible Fishing (AMPRs) officially established in Costa Rica. Acronyms and their definitions include: ASOPECUPACHI, the Association of Line-fishers of Palito de Chira; FENOPEA, the National Federation of Artisanal Fishers Organizations and Affiliates; CoopeTárcoles, Tárcoles' fishermen cooperative; ASOMM, the Mixed Association of Montero; and AJDIP, the Agreement of the Costa Rican Institute of Fisheries and Aquaculture (INCOPESCA) Governing Board

AMPR	Communities/fisher organizations involved	Size (km^2)	Date of creation	AJDIP
Palito de Chira	Palito/ASOPECUPACHI	0.01	10/22/2009	315-2009
Golfo Dulce	La Palma, Puerto Jiménez, Puerto Pilón, Río Claro de Pavones, Zancudo, Golfito, Puntarenitas de Golfito, La Purruja/FENOPEA	≈700	06/11/2010	191-2010
Tárcoles	Tárcoles/CoopeTárcoles	273.2	05/27/2011	193-2011
Palito-Montero	Montero/ASOMM	6.31[a]	03/29/2012	154-2012
Puerto Níspero	Puerto Níspero/Aso. pesc. local	2.6	03/29/2012	160-2012
Isla Caballo	Isla Caballo/Aso. pesc. local + tourism development	1.48	04/13/2012	169-2012
San Juanillo	San Juanillo, Lagarto, Punta Guiones, Playa Pelada, Nosara Asociación de Pescadores de San Juanillo	56.2	02/15/2013	068-2013

[a] Size of the combined Palito and Montero AMPR, datum provided by E Ross Salazar and M Castro 2013, pers. comm., Fundación MarViva

vation NGOs) to develop a national proposal. However, during the negotiations, INCOPESCA dismissed the community role in management, characterizing it as an exclusive use of fisheries resources, which is a public good (CoopeSoliDar 2010a).

By April 2008, the INCOPESCA Board of Directors approved the regulations for the establishment of Marine Areas for Responsible Fishing (AMPRs) at national level. This decision meant a positive step for participatory governance of natural resources in Costa Rica in contrast with the trends seen for national protected areas (see Sect. 10.4), even though it came from an institution concerned with resource exploitation rather from one focused on conservation. According to the Decree N° 35502-MAG, an AMPR has "significant biological, fisheries or socio-cultural characteristics" and well-defined geographic boundaries. Under this management regime, "fishing activity is regulated to ensure the use of fisheries resources for the long term and in which, for its conservation, use and management, INCOPESCA can count on the support of coastal communities and/or other institutions". Under this decree, INCOPESCA shall give priority to AMPR proposals from fishers' organizations. Requirements for submission include: (1) documents proving valid legal standing of the organizations requesting the AMPR; (2) biological, fisheries and socio-cultural characteristics justifying the AMPR creation; (3) socio-economic diagnoses of the organizations' members; and (4) a spatially-explicit management plan.

According to the Decree, the AMPR proposals are to be drafted in a participatory manner with the support of INCOPESCA, or any other institution or organization. Based upon the information provided, an *ad hoc* 6-member Working Group, made up of four representatives of INCOPESCA, one from MINAEM and two from local fishers' organizations, has 2 months to develop a fisheries management plan (Plan de Ordenamiento Pesquero, POP). The management plan implementation is overseen by a local Monitoring Commission (Comisión de Seguimiento), consisting of two INCOPESCA representatives and one member appointed by the fishing community, all with their corresponding substitutes. Access, fishing and other maritime and coastal activities including tourism are allowed for any stakeholder in good standing, as long as it is authorized by the POP. Seven AMPRs have been approved between 2009 and 2013, all of them located on the Costa Rican Pacific coast, in or surrounding Golfo de Nicoya and Golfo Dulce waters (Table 10.2). The first three AMPRs implemented are discussed here, with their histories analyzed and compared.

AMPR Tárcoles

Tárcoles, at the southern end of Golfo de Nicoya, is a Type B fishing community where artisanal activities form an important part of the community's business (*sensu* González Álvarez et al. 1993; cf. Sect. 10.2.2). The petitioning organization, CoopeTárcoles R.L. founded in 1985, is one of the three surviving from that decade's cooperative boom. Its members' commitment to the cooperative movement, along with some operative changes, preserved it from dissolution during the 1990s (Herrera-Ulloa et al 2011; Villalobos-Chacón 2011, pers. comm.).

Currently, Tárcoles is the only Costa Rican fishing group integrating the entire food chain, from production to marketing to selling. The cooperative strengthened since 2000 through its association with CoopeSoliDar. Focusing initially on the socio-economic benefits for its members from fish production and sales, it currently seeks to advance sustainable fishing, improve the quality of life of Tárcoles fishers and residents, promote the artisanal fisheries culture, and develop economic alternatives (Bowman 2011; CoopeSoliDar 2006, 2010b; López et al. 2007; Rodríguez Chaves 2008). Cooperation between both groups generated many projects, e.g., Code of Responsible Fishing, community monitoring of landings, socio-economic diagnostic, cultural celebrations, product certification (CoopeSoliDar 2010b). Collaboration was consolidated in 2007 through the foundation of Por La Mar Consortium R.L., a small marine ecotourism business.

Following the publication of the AMPR decree, CoopeTárcoles, joined by CoopeSoliDar, was the first fishers' organization to submit an AMPR proposal in May 2008. Yet, instead of the 2 months stipulated in the decree, three more years passed until its official creation (May 2011) due to laborious negotiations to exclude the semi-industrial trawling fleet from the coast to the 15-m isobath. Meanwhile, the first AMPR was officially declared in *Palito*, followed thereafter by the AMPR *Golfo Dulce*.

AMPR Palito, Isla Chira

In the mid-1990s, hand-line fishers of Palito of Isla Chira (Golfo de Nicoya, Fig. 10.1) declared a zone of rocky reefs in front of the village as an "Area of Responsible Fishing". In this area, only hand-line fishing was allowed, and community members enforced it by patrolling. The community is classified as Type B, similar to Tárcoles (*sensu* González Álvarez et al. 1993). The initiative was not granted official recognition nor legal or technical support. In 1995, at the request of the fishers, the then newly created INCOPESCA declared the zone exclusive for hand line fishing.

Given insufficient government support, together with the Chira Ladies Ecotourism Association (AEDC, *Asociación Ecoturística Damas de Chira*), the Association of Handline Fishers of Palito, Chira Island was created (ASOPECUPACHI, *Asociación de Pescadores Cuerderos de Palito Isla Chira*). In 2003, the Global Environmental Fund Small Grants Programme (SGP-UNDP-GEF) funded the purchase of buoys to improve monitoring and protection of the zone (Hernández 2011). The establishment of the association and the visible boundaries of the area brought obvious benefits: (i) fishing from outsiders declined, (ii) Coast Guard surveillance, albeit sporadic, gradually increased, and (iii) artisanal fishing tours with the AEDC generated additional income.

Nevertheless, institutional support remained weak and the community surveillance of the zone left the fishers feeling unrewarded and disenfranchised. Eventually, Palito fishers' efforts received recognition nearly 15 years after declaring their protective zone, by becoming Costa Rica's first AMPR in October 2009, "only" 9 months after submitting their application to INCOPESCA. It was then visited by government dignitaries and received international attention (Babeu et al. 2012; Solís-Rivera et al. 2012).

Monitoring by INCOPESCA showed that conservation efforts of the Palito fishers since the mid 1990s resulted in (i) sustainable fish landings, (ii) increased abundances of spawners of large-size prime commercial species (*primera grande* category) and (iii) creation of secondary income (from tourism and associated benefits) (Marín Alpízar et al. 2010). In addition, artisanal fishers of Palito noticed anecdotal changes in 2011: "Before, there were only 8 days of fishing each month, four before and four after full moon. Now the situation has changed, fishing is good almost every day. You can almost hear the fish and shrimp breed" (María Eugenia Fernández, Esteban García and Gabriel Cruz, artisanal fishers of Palito, 2011, pers.comm.). This example inspired nearby fishing communities to propose their own AMPRs to the government. The AMPR Palito was extended to Montero on the north side of Chira Island in March 2012, while others were created elsewhere in the gulf (Puerto Níspero, Isla Caballo, Table 10.2).

AMPR Golfo Dulce

Artisanal fishers of Golfo Dulce have been organized in seven local associations since early 1990s. Notwithstanding, artisanal catches have been declining while sport, tourism, and above all, semi-industrial fishing were expanding. In 2007, following a decade of cooperation with NGOs and universities (Feutry et al. 2010; Germain 2004; Guzmán-Mora 2013; Guzmán-Mora and Molina Ureña 2008; Hartmann et al. 2002; Lagunas Vazques 2004; Magnin 2004; Poirout 2007; Silva and Carillo 2004; Stern-Pirlot and Wolff 2006), the collaboration between small-scale fishers and the Osa Socio-environmental Center (CSAO, a local grassroots group) gave rise to the Commission of Artisanal Fishers of Golfo Dulce (COMPESCA-Golfo Dulce, Gómez Quijano and Tavares Almeida 2007), with advisory support from consultants and researchers. COMPESCA consolidated the fishers opposition to a MINAEM-proposed extension of the marine portion of Piedras Blancas National Park, on the northeastern shores of Golfo Dulce (Fig. 10.1). It also voiced the fishers socio-economic concerns about other government conservation projects in the area, in particular, a management plan for the South Pacific Multiple-Use Marine Area (AMUM-Pacífico Sur) (Gómez Quijano and Tavares Almeida 2007). COMPESCA then multiplied initiatives (e.g., inter-agency meetings and workshops, integration into a national federation of small-scale fishers), submitted counter-proposals for resource conservation and addressed the decline in Gulf fisheries and conflicts of use with semi-industrial and sport-fishing fleets (Glénard 2008; OSPESCA 2008).

The Gulf's artisanal fishers regrouped in March 2009 as the National Federation of Artisanal Fishers Organizations and Affiliates (FENOPEA), with the aid of the National Workers' Union (CMTC) (Fargier 2009, pers. obs.). The same year, the Tourist Fishing Association of Costa Rica (APTC) proposed to fund sustainable fishing in Golfo Dulce through foreign donations. APTC lobbied the Gulf's artisanal fishers to seek the creation of the AMPR Golfo Dulce under several management approaches and actions towards this sector. The proposed management approaches included exclusion of shrimp-trawler fleets from the Gulf, gill netting ban, grants and training for alternative selective gear, fishing licenses, exclusive access to fishery resources, and long-term financial assistance to each small scale fisher's association. FENOPEA accepted the proposal, while negotiating and expecting further terms, such as the development of new conservation projects, business options for small-scale fishing tourism, and a United States Dollar (USD) $ 40,000 compensation per fisher for the buy-back of gill nets and unrealized income (Fargier 2009, pers. obs., (Rocha 2009, FENOPEA President, pers. comm.)

The AMPR *Golfo Dulce* was officialized by INCOPESCA in June 2010, 10 months after the proposal submission. Comprising the entire gulf, it is the largest of its kind in Central America to date (Table 10.2). INCOPESCA issued new licenses to most local artisanal fishers (Cordero 2010) and the Costa Rican Touristic Fishing Federation (FECOPT, created in 2009) awarded a USD $ 20,000 grant to each of the participating Fishers Associations.

In 2012, FECOPT, including APTC among other countrywide sports fishing association, updated its name to FECOP (Costa Rican Federation of Fishing), accordingly with the expansion of its goals scope and interests.

Unlike the previous two areas (Palito and Tárcoles), this AMPR declaration process fulfilled none of the AMPR decree's application requirements, nor did it meet FENOPEA's expectations. The FECOPT-funded proposed management plan did not reflect the products of a collaborative process. The semi-industrial shrimp fleet was excluded from the Gulf before the formal declaration of the AMPR, by means of an undisclosed financial compensation process. As a result, the artisanal fishers took their disappointment to the regional INCOPESCA office in Golfito and street-protested to stop the AMPR declaration procedure (the "voice" response, as called by Hirschman 1970; PRETOMA 2009).

After the initial implementation of the AMPR regulations, the artisanal sector has retired twice from the Monitoring Commission (2012 and 2013). At the time of writing this chapter, artisanal fishers have no representation in the Commission. Furthermore, by the end of 2013, FENOPEA was showing evidence of disintegration, with its original associations regrouping under a different structure (Molina-Ureña, pers. obs.).

10.6 Comparative Analysis

We carried out a five-way comparative study of co-management processes for sustainable coastal activities among AMPRs Golfo Dulce, Palito and Tárcoles (Gulf of Nicoya) and the National Parks of Cahuita and Marino Ballena. The comparison was based on 32 conditions considered necessary for a successful co-management process, as detailed in Fargier (2012). A score of 100 % indicates that all 32 conditions were fulfilled. It must be noted that the list of conditions contributing to the decision matrix is not exhaustive, and all conditions do not need to be fulfilled to accomplish a successful co-management process. Nevertheless, most authors agree that the probability for a successful co-management process increases with the number of fulfilled key conditions (Gutiérrez et al. 2011; Pomeroy and Andrew 2011). Based on this comparison, the three top-scores, Tárcoles (85 %), Cahuita (76 %) and Palito (62 %) were clearly distinguished from those of Golfo Dulce (34 %) and Marino Ballena (29 %). We then identified ten key conditions to validate the development of a small-scale co-management fishery process in Costa Rica (Table 10.3).

Remarkably, the ranking of the five case studies remained unchanged when restricting the co-management analysis to these ten key conditions (Table 10.3). The analysis matrix also revealed that few conditions were favorable to the success of a co-management process in Golfo Dulce. Below we discuss and analyze the factors leading to this apparent lack of success.

10.6.1 Size, Complexity and Support

The size and geographic diversity of Golfo Dulce, the number of communities involved, and fluctuating support throughout the process are important factors that could explain the poor potential for small-scale fisheries co-management in the area. Indeed, defining clear and appropriate spatial boundaries of the area to be managed is considered a key factor in the success of the co-management process, that should also take into account ecological factors (e.g., common pool resources), management considerations, and the local fishing communities (Carlsson and Berkes 2005; Govan 2008; Jentoft et al. 1998; Noble 2000; Ostrom 1990; Pomeroy et al. 2003, 2011). However, the AMPR Golfo Dulce is vast (about 70,000 ha), almost three times larger than Tárcoles (27,320 ha), and more than 60,000 times larger than AMPR Palito (1.14 ha). Currently, AMPR Golfo Dulce is the largest marine area of its kind in Central America (Cordero 2010). Thus, while the people living around Golfo Dulce, especially fishers, have recognized its ecological and heritage value (van den Hombergh 1999), the majority of them do not know it entirely. Additionally, although the boundaries of the AMPR-GD are precisely and formally defined, fishers that did not participate in their creation do not know them well.

The hasty creation of the very large AMPR Golfo Dulce differs from the slower and more gradually established, smaller AMPRs in the Gulf of Nicoya (Fig. 10.2). A slower process facilitates better adaptation to the local socio-economic and environmental conditions and fosters ownership by the local fishers' communities (Marín Cabrera 2012). The complexity, size and fast creation of the AMPR-GD arguably

Table 10.3 Key conditions required for the realization of a small-scale fisheries co-management process in Costa Rica. *Green*: Condition fulfilled. *Yellow*: Condition partially fulfilled. *Red*: Condition not fulfilled. Case studies definitions are: *GD* Marine Area of Responsible Fishing (AMPR)—Golfo Dulce, *T* AMPR—Tárcoles, *P* AMPR—Palito, *C* Cahuita National Park, *MB* Marino-Ballena National Park

Key conditions	GD	T	P	C	MB
1. Clearly defined boundaries	Yellow	Green	Green	Green	Red
2. Pre-existing local organisations (legitimate & representative)	Yellow	Green	Green	Yellow	Yellow
3. Good local & political will	Yellow	Green	Green	Green	Yellow
4. Creation of an adequate management commission with legal backing	Yellow	Green	Yellow	Green	Red
5. Participation to decision making (collective choice arrangements)	Red	Green	Yellow	Green	Red
6. Technical support (external agent: non-governmental organization (NGO), barefoot ecologist, etc.)	Yellow	Green	Yellow	Yellow	Yellow
7. Congruence in the distribution of costs & benefits rules	Red	Green	Green	Green	Yellow
8. Transparency and trust	Red	Green	Yellow	Yellow	Red
9. Leadership	Yellow	Green	Yellow	Green	Red
10. Territorial identity	Yellow	Green	Green	Green	Red

let to difficulty in its management, monitoring, and a sense of "topophilia" (i.e., people's identification with the geographical area including their sense of belonging and cultural roots; Fonseca-Borrás 2009; Pinkerton 2005). A contributing factor is that most fishing communities in Golfo Dulce are Type A (i.e., recently established after the arrival of squatters; *sensu* González Álvarez et al. 1993). The latter aspect may also hinder "ownership" (i.e., the engagement and participation of fishers in the management process (Govan et al. 2008; Pomeroy et al. 2001, 2011).

Due to Golfo Dulce's geography, the dispersal of fishing communities and the deficient public transportation infrastructure, a fisher requires 1–3 days to attend a meeting anywhere along the Gulf. This travel time represents a loss in fishing days, which complicates the ability to assemble members of the co-management group. Thus, the Gulf's artisanal fishers meet less frequently than those of Tárcoles, Palito, Cahuita and MarinoBallena. Moreover, most meetings are attended only by the association presidents. At the community level, in Palito or Tárcoles, where only one local organization is involved, fishers may speak for themselves and legitimacy is not an issue. However, for Golfo Dulce, where FENOPEA represents nine fisher's organizations, representatives speak for their peers and legitimacy might be a problem (Jentoft 2000).

Legitimacy issues at the grass-roots level come about regularly within fisheries co-management of Costa Rica (Quintero et al. 2009). For Golfo Dulce, it primarily concerns trust toward FENOPEA's delegates and the legitimacy of their decisions, especially in the face of existing conflicts within and among the member organizations (Fargier 2012; Fonseca-Borrás 2009; Jentoft 2000; Jentoft et al. 1998). In Marino-Ballena National Park, the lack of representativeness of ASOPARQUE, composed of 22 organizations, was one of the reasons given by the Comptrollership of the Republic to put an end to its co-management process (DFOE-AM 2005).

Unlike CoopeTárcoles, Golfo Dulce artisanal fishers' associations have not benefitted from long-term and steady external assistance. External support is considered another key factor in the success of a co-management process (Carlsson and Berkes 2005; CBM 2003; Chuenpagdee and Jentoft 2007; Govan 2008; Luna 1999; McConney and Baldeo 2007; Nuñez Saravia 2000; Pomeroy and Berkes 1997; Pomeroy et al. 2001, 2004, 2011). The type of support varies and it can include organization of meetings and workshops, initial site assessments, appraisals, environmental education and training, facilitating and mediating dialogue between resource users and government to aid the search for partners and financial aid. In Central America, external organizations commonly provide such assistance (i.e., an NGO, development agency, sponsor, foundation, university or other institution; Nuñez Saravia 2000).

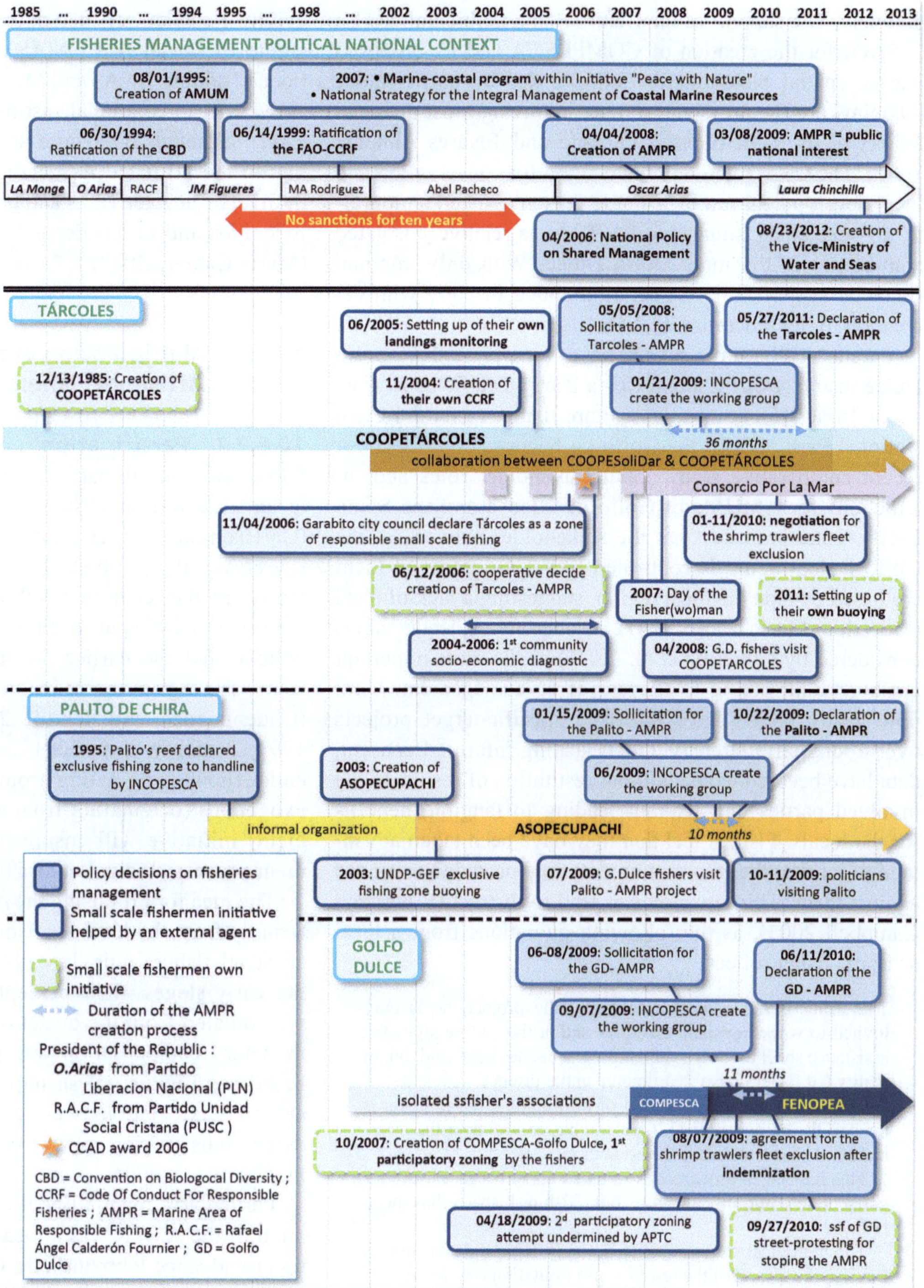

Fig. 10.2 Timeline of the declarations of the Marine Areas of Responsible Fishing (AMPRs) of Palito, Tárcoles and Golfo Dulce

The social service cooperative CoopeSoliDar has been working with the Tárcoles community for more than a decade, showcasing particularly its cooperation with CoopeTárcoles (CoopeSoliDar 2010a; Rodríguez Chaves 2008). What began as a pilot project received recognition and diversified to become a full-fledged and continued engagement. The cooperation between the two organizations obtained in 2006 the Prize of Technological Innovation from the Central American Commission for Environment and Development. In 2007, the Por La Mar Consortium was founded, a micro-enterprise facilitating local eco-tourism. Through this process, CoopeSoliDar seemingly passed from a simple external agent to a leading party in the co-management process of the AMPR Tárcoles.

In Golfo Dulce, despite regular initiatives by NGOs and universities (Feutry et al. 2010; Germain 2004; Guzmán-Mora 2013; Guzmán-Mora and Molina Ureña 2008; Hartmann et al. 2002; Lagunas Vazques 2004; Magnin 2004; Poirout 2007; Silva and Carillo 2004), long-term and continued assistance to small-scale fishing communities has been deficient. The support of the Centro Socio-ambiental de Osa (CSAO) for the creation of COMPESCA and its proposed marine spatial planning for Golfo Dulce by participatory mapping, are the only ones to date involving most artisanal fishers of the Gulf (Gómez Quijano and Tavares Almeida 2007). The CSAO was an organization for conservation and local development run by a single person assisted temporarily by specialists, situated at La Palma, a relatively isolated community of the inner Golfo Dulce. With only minimal financial support, this cooperation ended in 2009 with the creation of the federation FENOPEA.

Paradoxically, more NGOs have been present in Golfo Dulce than elsewhere along Costa Rica's Pacific coast (i.e., about 15 active organizations, more than for Isla del Coco National Park and for the Golfo de Nicoya; TNC 2011). A recent comparative study about stakeholder roles and interactions for AMUMs of Golfo de Nicoya and the South Pacific zone indicates that the stakeholder network in the Golfo Dulce region is controlled by NGOs. Most of them are based outside the local region, suggesting a lack of local leadership (Marín Cabrera 2012). Nevertheless, leadership is considered by Gutiérrez et al. (2011) as the most important attribute contributing to success of co-management. Until now, many NGOs have preferred specific-target projects over a long-term strategy. Often lacking informed consent, data have been collected without restitution of results to the involved parties and without leading to tangible benefits for the locals. Fishers feel that they have been taken advantage of, or that the information that they provided was used against them. This situation can lead to distrust (Silver and Campbell 2005), as the following quotations from a local artisanal fisher reflect:

> The fisher is never incorporated into the project, he is only invited to workshops and meetings and at the end he signs the assistance sheet used as a validation to what has been said, but in reality the fisher doesn't participate in the project.
>
> They make us to give all sort of information, but afterwards none of the organizations [of fishers] has control over the information it has provided, it has no authority over it.
>
> This has led to a distrust of organizations, because we have collaborated but in exchange our [fishing] space has been reduced.
>
> The biological studies should not only look at the resources but also the effects of the resource use limitations on the fishers. It is not possible that fishers collaborate and that then the information becomes a string around their own neck! .
>
> -Juan, artisanal fisher, Río Claro, 2011 pers. comm.

Because of distrust by local fishers, FENOPEA became an important actor for Golfo Dulce and has been trying to position itself since 2009 as a "development broker" (Chauveau and Jul-Larsen 2000; Olivier de Sardan 1995). FENOPEA appears as a meeting and coordination entity but it doesn't carry out concrete actions or strong alliances. In order to obtain external resources and redistribute them to local fisher's organizations, the federation continuously tries to expand collaborations with NGOs (e.g., MarViva, PRONATURE, Tsikita Foundation), national institutions (e.g., Mixed Institute of Social Aid [IMAS], National Institute of Vocational Training [INA] and Ministry of Agriculture–[MAG]), as well as international institutions (e.g., UNDP-GEF and FAO), although with little success. Instead of concentrating on the affiliated Associations' necessities, this strategy of diversifying activities and pursuing alliances weakens the federation and fosters dependence on NGOs and institutions (Marín Cabrera 2012).

10.6.2 Manipulation of the Golfo Dulce Artisanal Fishers

10.6.2.1 Participation

Participation of all parties, and particularly resource users, is an essential ingredient of a co-management process (Carlsson and Berkes 2005; Fonseca-Borrás 2009; Pomeroy et al. 2001, 2003, 2011; Pomeroy and Carlos 1997). However, the recommended period of intervention during the process varies according to the authors. Many authors believe that the earlier the involvement of the resource users, the higher the probability of a successful outcome (Chuenpagdee and Jentoft 2007; Pomeroy and Carlos 1997; Pomeroy et al. 2001, 2011). In addition, neither an endogenous (originating from the resource users) nor an exogenous (originating from an external agent or government) initiative will ensure the success of a co-management process (Weigel et al. 2007).

The idea for proposing the AMPR Golfo Dulce arose externally from the local tourism fishing sector. Even though artisanal fishers quickly embraced it and got involved in the early stages, their perception is that their sector was manipulated from the outset of the AMPR process. Larger monetary compensation was provided to the Puntarenas-based semi-industrial shrimp trawler owners, while local artisanal associations were compensated to a much lesser extent, thus making apparent that the tourism fishing sector had its own agenda.

Particularly, many FENOPEA associates feel that the creation of the AMPR was a means for tourism fishing operators to advance lobbying efforts and assert their claims at a national scale. Case in point, previously agreed gear restrictions aiming at the conservation of game fish and their prey within the Gulf were unilaterally changed at the last minute before being submitted to the INCOPESCA board, at the potential expense of the livelihood of artisanal fishers.

> The AMPR Golfo Dulce was not an initiative of the fishers, but of foreign institutions. We fishers backed it up to be able to fish more, eliminate gillnets, regulate hooks and longlines, fish with handlines and fish for subsistence, but now they pressed and squeezed us until they hardly still let us fish.
>
> Chichi, artisanal fisher, Puerto Pilón, 2011, pers. comm.

Moreover, collective choice rights, i.e., the possibility to participate in modifying operational rules that structure day-to-day management activities—also called "right of management" (Ostrom 1990) or "first order governance" (Kooiman 2003), are particularly important for achieving good governance of a common resource. For example, fishers in Tárcoles and Palito proposed zoning and associated operational rules to the corresponding AMPR working groups. Tárcoles incorporated balanced numbers of CoopeTárcoles fishers, government officials, unassociated local fishers, municipality, and Coast Guard representatives, with CoopeSoliDar as observers. This relatively broad representation favored transparency and improved the chance of success for the co-management process. Besides, the CGR's Cahuita study welcomed the opening of the Management Committee to other groups in the community, fostering transparency in the park's management, and thus increasing the confidence and sense of belonging of the community to it (Fonseca Borrás 2009).

Similarly, the composition of the AMPR Tárcoles Monitoring Commission appears balanced, with an equal number of fishers and INCOPESCA representatives. Thus, fishers participated fully in the development of "collective-choice rules" and could be classified as "proprietors" (*sensu* Schlager and Ostrom 1999). Furthermore, since CoopeTárcoles' fishers participated in the development of their community fishing proposal leading to the national AMPR decree, we can consider that they also participated in the "supreme" level of Schlager and Ostrom's (1999) property rights classification, namely the development of constitutional rules determining how the collective choice rules are designed (or "second order governance", Kooiman 2003). In Palito, the AMPR Monitoring Commission is less balanced, as small-scale fishers are in the minority. Here, the latter may be classified as *authorized users,* holding only operational rights of access and withdrawal (Schlager and Ostrom 1999).

By contrast, in Golfo Dulce only one artisanal fisher participated in the 8-member AMPR *ad hoc* Working group. Fishers did not participate in the zoning process despite various participative zoning workshops backed up by university researchers (Fargier 2009, 2012; Glénard 2008; Poirout 2007) nor did they in the elaboration of the associated operational rules. Moreover, the Monitoring Commission consisted of 11 members (instead of the decree-specified three members), including two sport-fishing representatives and only one artisanal fisher, who was not systematically convoked to all meetings. Thus, the artisanal sector may be classified as *authorized users* who did not participate in the development of zoning and operational rules.

> The fishers ought to be part of the management actions of an area of responsible fishing. But for the AMPR Golfo Dulce, fishers don't participate in management decisions. As I see it, it's the MINAEM, MarViva and the tourism sector who manage it.
> -Elvis, artisanal fisher, Puerto Pilón, 2011, pers. comm.

In summary, the role of small-scale fishers in the AMPR Golfo Dulce process could be characterized as manipulative, according to Pretty's (1995) typology of participation. Participation is then considered as a pretense and users do not have any power, whereas fishers should take part in joint analysis (interactive participation, *ibid.*). As CoopeSoliDar advocated since its inception of collaboration with CoopeTárcoles, participation should be seen as a right, and not just as a means (Solís and Madrigal 2004).

10.6.2.2 Economic Incentives

In Golfo Dulce, promises concerning the development of economic alternatives and improving the quality of life of artisanal fish workers were not honored. Yet, it is essential that the resident populations involved in the process of declaring a protected area are not prejudiced, given their dependence on access to natural resources for livelihood, especially in remote rural areas, where peoples' welfare are intimately linked to nature (Weigel et al. 2007).

> The fishers have supported the creation of the AMPR to get the shrimp trawlers out [of the Gulf] and to improve the fishing. But this was done with commitments that never were met. These included, among others, the payment for our fishing nets, but since we did not have licenses, they said they could not pay us.
> -Santos, artisanal fisher, Puerto Pilón, 2011, pers. comm.

One of Ostrom's eight principles (Ostrom 1990; Pomeroy et al. 2011) stresses the balance between benefits and costs of management measures for the fishing households. If management measures are to be sustainable, profits must be higher than the costs they generate. While in the long term, resource conservation will be beneficial to all fishers, the short-term goal of any resource user involved in a process of co-management is to maintain or improve his/her quality of life (Nuñez Saravia 2000). Analyzing Southeast Asian fisheries co-management, Evans et al. (2011) revealed that the best indicators for evaluating success of a co-management process were revenue and well-being, rather than resource status. The authors attribute such improvements to the development of economic alternatives and accompanying measures (e.g., micro-credits).

Thus, in Palito and Tárcoles, fishing household benefits outweighed costs due to the creation of coastal ecotourism projects. Conversely in Golfo Dulce, the new AMPR management measures were not accompanied by alternative economic projects. As a result, Fargier (2012) observed that the socio-economic situation of most fishers had degraded, less selective and environmentally more damaging bottom gear was introduced, disobeying rules (notably, introduction of illegal small artisanal bottom trawls, which represent a form of "exit response" as defined by Hirschman 1970).

10.6.3 Transparency of Procedure

From its beginning, the procedure for the creation of the AMPR in Golfo Dulce lacked transparency. For example, the preparatory working group was formed with only three of the seven legally required conditions of the decree for establishing an AMPR. INCOPESCA accepted the application as soon as two conditions had been met: an *a priori* acceptance for establishing the AMPR from the artisanal fishers, and a commitment to cease fishing in Golfo Dulce from the semi-industrial shrimp-trawling sector. Both were negotiated with financial compensation.

As written, the Golfo Dulce fisheries management plan (POP), ostensibly funded through the sport and tourism fishery sector (APTC-FECOPT), appears largely copied from previous reports (Fargier 2012; Gómez Quijano and Tavares Almeida 2007; Morera Quesada and Vargas Bonilla 2009). Artisanal fishers did not participate in its elaboration and were not informed about its progress. Moreover, an INCOPESCA high ranking officer was given a leave of absence for a consulting job with FECOPT to elaborate the POP, raising the question of conflict of interests.

Finally, the issue of legitimacy of the monetary compensation process to shrimpers has been raised by several sectors of stakeholders, leading to distrust by civil society and artisanal fishers in particular (Constitutional Court Resolution No. 2010-1315, July 20, 2010). Given the tenure of the semi-industrial fishing sector in the INCOPESCA Board of Directors, the apparent fast-track completion of declaration of the AMPR-GD skipped other alternatives, such as specific legislative or executive decisions, as was the case in the Golfo de Nicoya and the National Parks.

Indeed, if we compare the three AMPRs examined in this study, Golfo Dulce is undoubtedly the most complex (the largest, with more professional groups, and one of the least studied), though the declaration procedure lasted only 10 months, not longer than for the AMPR Palito, 60,000 times smaller and involving only one fisher's association (Fig. 10.1). The declaration of the AMPR Golfo Dulce in June 2010 certainly recognized the 10-year effort of artisanal fishers for a sustainable use of its coastal resources. However, their participation in management during conception and implementation of the AMPR appears fictitious, revealing a diversion from the original purpose of this protected-area category.

Being major users of the in-gulf fishery resources, the artisanal sector's *de facto* absence from the AMPR-GD Monitoring Commission defeats the purpose of the participatory management approach designed by INCOPESCA for AMPRs. It also endangers the basis of consented enforcement: if the fishers feel strongly about the perceived lack of legitimacy to the regulations, it is nearly impossible to ensure they will respect them. Up to now, no concerted effort has been explicitly expressed by government agencies regarding the reincorporation of the small-scale fishing associations to the Commission. This neglect exerts a negative influence on the mindset of a sector that already feels outcast and marginalized by the state and the civil society.

10.7 The AMPR Golfo Dulce: What Lessons can be Learned?

10.7.1 The AMPR, a Tool for Small-Scale Fisheries Co-Management in Costa Rica?

As noted by the Estado de la Nación (2008) report, Costa Rica is the country with the lowest proportion of co-managed protected areas in Central America. This statistic concerns mainly terrestrial areas, since conservation of the marine environment historically was neglected in Costa Rica and in the rest of the region. Given this precedent, we may ask: Would the AMPR current figure allow the participation of small-scale fishers' organizations in coastal fisheries co-management?

The original intention of the "AMPR—Decree" (N° 35502 MAG) was to recognize the efforts of artisanal fishing communities and to formulate a path towards sustainable use of coastal marine resources. The decree prompted fishing organizations to initiate the creation of an AMPR (Art. 2, *ibid.*) in a national political context where the idea of civil society participation in environmental management was losing ground. In other words, a bottom-up approach was established in a country where top-down policy has dominated environmental management.

According to the AMPR decree, through participative methods, fishers define their "Maritory" (Parrain 2012) and propose appropriate zoning and management measures justified on biological, fisheries and socio-cultural grounds (Art. 2, *ibid.*). The decree invites them to participate in the surveillance of the AMPR (Art. 7, *ibid.)*. Finally, the artisanal fishers are supposed to participate in the Monitoring Commission, albeit as a minority (1/3 artisanal fishers against 2/3 INCOPESCA members (Art. 11, *ibid.*).

Under this procedure, the current AMPR would allow co-management as defined by Evans et al. (2011). In addition, our analysis shows that this system is best classified as *advisory co-management*, where users advise government on decisions to be made and government endorses or adapts these decisions (Sen and Raakjaer Nielsen 1996; Berkes et al. 2001a). Indeed, while fishers' organizations propose a management plan, INCOPESCA alone validates, modifies, or rejects it while maintaining a clear majority in the Montoring Commisssion. The three AMPR cases analyzed here reveal that the type of co-management implemented locally varied *de facto* according to the application of the decree texts, as evidenced by the difference in the process for establishing the management plan (participative or not) and by the different composition of their respective *ad hoc* Working Groups and Monitoring Commissions.

The feeling of empowerment for local communities usually improves their sense of stewardship, thus facilitating the achievement of a sustainable activity. Conversely, it would be expected that a perception of being powerless, manipulated, or under an illegitimate and repressive regulatory process, would bring about unsuccessful outcomes for sustainable fishing (Jentoft 2006). Thus, on the continuum of power sharing of authority and responsibility between government and community as described by Berkes et al. (2001a), we considered the co-management process in Tárcoles as *advisory* while it could be classified as *cooperative* in Palito, where users have some input into management and *informative* in the Golfo Dulce, where users are informed about decisions that the government had already made.

10.7.2 What is the Future for the AMPR in Costa Rica?

The AMPR declarations of Palito, Golfo Dulce and Tárcoles caused great interest among Costa Rican artisanal fishing communities and are now acclaimed by a number of them. Four new AMPRs have been declared, including three more AMPRs in Golfo de Nicoya: *Puerto Níspero, Montero-Chira* and *Isla Caballo*; the fourth, *San Juanillo,* is along the outer Nicoya Peninsula (Table 10.2). Additional AMPRs are being planned.

For interested communities, the AMPR would be a means to appropriate management of resources they exploit, and empower themselves in the stewardship structure; assure conservation of these resources, in particular through exclusion of the semi-industrial fleet from their maritory; develop non-extractive economic alternatives such as ecotourism; and validate their community efforts to organize themselves and apply good practices for sustainable fishing. Moreover, many see the AMPR accreditation process more streamlined than those of MINAEM's Marine Reserves (MR) and Marine Management Areas (MMA). In fact, the MINAEM categories require more constraints and steps along the way and need to be declared through legislative or executive processes.

Notwithstanding, the apparent AMPR success exacerbates two institutional problems, already mentioned previously. First, INCOPESCA lacks the institutional capacity and means to properly process all of the new AMPR applications. Second, the apparent success of AMPR overshadows MINAEM's efforts, including its *Costa Rica por Siempre* initiative, and accentuates the long-standing interagency conflicts for jurisdiction between both institutions (Cicin-Sain and Knecht 1998). Despite the two new marine management categories allowing artisanal fishing (MR and MMA), a majority of the artisanal fishers continue to mistrust MINAEM and oppose new protected areas it proposes. To complicate matters, AMPRs would not qualify as a management category contributing to the objectives of POWPA (TNC 2011).

A new conflict has arisen between INCOPESCA and MINAEM for administering MINAEM's new management categories. Currently it is unclear whether MRs and MMAs could superpose AMPRs and how they would be administered in detail. MINAEM would be responsible for overall management of the MPAs while INCOPESCA would define the respective fisheries management plans (e.g., zoning, gear, boats, access, etc.) in collaboration with the artisanal fishers organizations.

The implementation of MRs and more MMAs would help test these mechanisms of inter-institutional coordination. Two artisanal fisher's associations of communities on the outer coast of the Nicoya Peninsula (Coyote and Bejuco, north of the Caletas Ario Wildlife Refuge, Fig. 10.1), assisted by a national NGO, participate in a proposal for creating the *Nanyadure* MMA (Arauz 2013, PRETOMA, pers. comm.). If created, it would be the first MMA involving fishing communities. Furthermore, a pilot project extending south from Golfo Dulce to Punta Burica near the Panamanian border is being considered (TNC 2011). If successful, it could set the stage for a transboundary MPA, originally proposed in 2011 at an international regional participative workshop involving artisanal fishers, NGOs and institutions from both countries (Hartmann et al. 2012; Documentary "Si el mar me da, yo le doy", https://vimeo.com/33205185 last accessed 23 Apr 2014).

One last aspect could put the declaration of some AMPRs at risk, despite their current success: financial compensation to shrimp-trawlers for the creation of the AMPR-Golfo Dulce. As other artisanal fisher groups also want to exclude trawlers from their areas, the Golfo Dulce case could set a precedent that may inhibit the creation of an AMPR that is not able to rely on an external donator, as was the case for Golfo Dulce.

10.8 Conclusions

The AMPR management category represents a legal framework that formalizes, for the first time, small-scale fishing activities and co-management in Costa Rica. Based on the experience with Costa Rica's first AMPRs, this management figure appears beneficial to small-scale fishers (see 10.5.2) and constitutes a step to recognize over 10 years of efforts for the conservation of the marine coastal resources they exploit. The type of co-management developed locally, from advisory to cooperative or even informative, according to the 'quality' of participation, appears to depend essentially on the actors and interested parties involved.

In the case of Golfo Dulce, fishers were deprived of proper initiatives including their right to participation in the process, and felt manipulated by other actors. Unlike Tárcoles and Palito, Golfo Dulces' fishers short-term economic situation declined, partly due to a lack of economic alternatives.

However, over the long term, exclusion of shrimp trawling from gulf waters, *a priori* could assure better resource conservation by artisanal fishers, by reducing fishery conflicts and better applying good fishing practices. The AMPR-GD creation also legalized for the first time the activities and lives of many artisanal fishers (licenses and relocation of illegal precarious housing).

Concerning environmental effects, it is still too soon to evaluate them precisely. Resource assessment studies are undergoing at all three sites, in cooperation with universities, NGOs and the government that will allow comparisons with historical data (Campos 1989; CoopeSoliDar 2006, 2010a, b; Fargier 2012; Guzmán-Mora 2013; Lagunas Vazques 2004; Marín Alpízar et al. 2010; Molina-Ureña, unpubl. data; Poirout 2007). Currently, MPAs on the Pacific side are separated, on average, by 22.4 km and have a median size close to 54 km^2. When compared to those presented by Halpern (2003) and proposed by Shanks et al. (2003), Costa Rica is above average with good conditions for an incipient network of MPAs that would allow exchange between marine organisms populations (Alvarado et al. 2011). Notwithstanding, lack of buffer zones and unknown by catch of target fish species by shrimp trawlers operating around current AMPRs' boundaries make some observers skeptical about demonstrating scientifically sound ecosystem benefits (Arauz 2013, PRETOMA, pers. comm.).

Based on our comparative analysis along with recent publications about stakeholders in Costa Rica's Pacific AMUMs (Marín Cabrera 2012) and social dimensions of MPA management in four countries of Central America (Solís-Rivera et al. 2012), we propose five key recommendations for small-scale fishing co-management improvement in the region:

1. Increase the participation of artisanal fishers in the collective choice rules development process. In particular, the AMPR-Monitoring Commissions should balance the weight of the members. Tárcoles Commission could be used as a model.
2. Improve distribution of costs and benefits of management measures for artisanal fishers, in particular by introducing legal alternative fishing gear or economic options such as ecotourism.
3. Provide conditions fostering positive leadership skills, emphizising Golfo Dulce artisanal fishing groups. Appeal to recognized leaders to set up a group of stakeholders with a vision of community welfare.
4. Ensure transparency of actions (e.g., official invitations, committee minutes, prompt and open communication channels, decision-making process, etc.), in order to create a climate of trust and respect, which is currently lacking in the process.
5. In the Golfo Dulce case study: formalize the elaboration of an agreed strategic plan (mission, vision, objectives) that would provide guidance to establish strong alliances and prevent manipulation or dependence of FENOPEA on NGOs or other institutions.

As better biological data become available, additional recommendations include reformulating AMPR boundaries and creation of buffer zones set by biological criteria, to minimize the deleterious effects of external fishing and resource extraction (e.g. by catch, low water quality) on target species.

The future of the AMPRs, together with MINAEM's MR and MMA, will depend on the efficacy of inter-institutional collaboration among INCOPESCA, MINAEM, ICT, Coast Guard, Port Authority, Municipalities, Universities, NGOs, grass-roots groups, as well as the ability to coordinate their respective coastal resources conservation projects and visions (Alvarado et al. 2011, 2012). In August 2013, a precedent-setting ruling from Costa Rica's Constitutional Chamber of the Supreme Court banned shrimp bottom trawling in Costa Rican waters, until a significant by-catch reduction method could be demonstrated by scientifically sound studies (Sala Constitucional 2013). This judgment constitutes a keystone incentive for improving collaboration towards responsible marine coastal management.

Acknowledgments This research was partially supported by GEF (Global Environmental Fund) project N° 3340 *Good practices and portfolio learning in GEF transboundary freshwater and marine legal and institutional frameworks*, University of Costa Rica research projects by CIMAR N° 808-B0-038 *Fishing communities from the South Pacific of Costa Rica: participative processes towards fishery resource co-management*, and 808-A7-135 *Spatio-temporal variability of fish communities in Golfo Dulce and surrounding waters, Costa Rica*, UCR Golfito campus facilities, University of La Rochelle and LIENSs research grants to H. J. Hartmann and V. Duvat-Magnan, the French Ministry of Higher Education, Research and Technology (Ph. D. national research fellowship to L. Fargier) and the Acroporis Foundation (Paris, France). We are indebted to Bo Bricklemyer Jr., Richard Paisley, Glen Hearns and Susan Bazilli for facilitating much of the project's impetus and administration, to Fabien Monteils for stimulating discussions and to Virginie Duvat-Magnan for critical and fruitful participation and advice throughout the process. We deeply appreciate the feedback and participation of fishers and fishworkers from Golfo Dulce, members of FENOPEA, represented by their corresponding fishers association officers (Victor Julio Rocha, Manuel Ramón Loaiza, Benancio Martínez Sanabria, William Mata, Juan Jiménez, Emiliano González, Cirilo Quintero, Juan Eduardo Barroso, Maximiliano Guerrero) and other non-associated fishers (the Sanabria family). From Tárcoles and Palito, we would like to thank David Chacón and María Eugenia, Gabriel Cruz, Esteban García, Osvaldo Sequeira and Liliana Martínez. We especially acknowledge the Loaiza family for sharing their home to Luc Fargier. Other people that helped in the data gathering or participatory processes are Iván Pérez Rubio, Annabelle Espinoza Ovares, Georgina Moreira Quesada, Sonia Vargas, Ana Lucía Gutiérrez Espeleta, Thomas Poirout, Zoé Glénard, Sarah Fery, Jean Baptiste Livenais, Mickael Mora Arias, Erick Vindas. The authors would like to sincerely thank the thorough anonymous reviewers for their comments that helped improve the manuscript. Thanks also to our families, friends & BGB. Welcome Natty! We would finally like to salute Macho Pulpo, Tchuss and Nicolas.

Appendix

Co-management potential evaluation matrix from a literature synthesis of case studies analyzing fisheries participative processes. Numbers in the references column correspond to the following publications: **1**, Chuenpagdee and Jentoft (2007) ; **2**, Carlsson and Berkes (2005) ; **3**, Govan (2008); **4**, Nielsen et al. (2004); **5**, McConney and Baldeo (2007); **6**, Geoghegan and Renard (2002); **7**, Geoghegan et al. (1999); **8**, Renard (2001); **9**, Cumberbatch (2001); **10**, Mahon and Mascia (2003); **11**, Ravndal (2002): **12**, Renard (1991); **13**, Govan (2003); **14**, Brown and Pomeroy (1999); **15**, CAR-

				References
Ex-ante	Cause	Resource decrease, conflicts		35, 36, 37
Ex-ante	Initiative	Local		1, 3, 26
Ex-ante	Initiative	Mixed (local + external agent)		1
Ex-ante	Initiative	Government		1
Implementation	Supra-communautary	Backing	Legal	6, 11, 12, 13, 18, 19, 20, 21, 22, 27, 28, 29, 33, 37
Implementation	Supra-communautary	Backing	Financial	11, 13, 15, 18, 19, 21, 22, 27, 28, 29, 31, 33
Implementation	Supra-communautary	Backing	Technical	11, 13, 15, 18, 22, 27, 28, 29, 31, 33
Implementation	Supra-communautary	Rights to organize		30, 37
Implementation	Supra-communautary	External agents		1, 3, 5, 22, 33
Implementation	Communautary	Organized groups		9, 11, 15, 17, 22, 33
Implementation	Communautary	Leadership		21, 22, 32, 33, 34, 38
Implementation	Communautary	Clearly defined boundaries	User group	8, 21, 22, 24, 25, 30, 33
Implementation	Communautary	Clearly defined boundaries	Management area	3, 5, 21, 22, 24, 25, 30, 33, 34
Implementation	Communautary	Clearly defined boundaries	Resource	2, 21, 22, 24, 30, 32, 33, 34
Implementation	Communautary	Clearly defined roles		2, 3, 5, 6, 7, 8, 9, 10, 13, 14, 17, 21, 26, 28, 31, 33
Implementation	Communautary	Participation		6, 7, 8, 9, 12, 14, 17, 20, 21, 22, 23, 26, 33
Implementation	Communautary	Local knowledge		3, 4, 5, 14, 21
Implementation	Communautary	Collective-choice arrangements		30, 32
Implementation	Communautary	Conflict-resolution mechanisms		21, 22, 26, 30, 33
Implementation	Communautary	Local will		27, 28, 29, 31, 32, 33
Implementation	Communautary	Graduated sanctions		27, 30, 33
Implementation	Communautary	Monitoring		30, 33
Implementation	Communautary	Nested enterprises		30, 33
Implementation	Ind.	Benefits > Costs		22,30, 33
Values	Supra-communautary	Equity		8, 19, 20, 21, 26
Values	Supra-communautary	Transparence		5, 6
Values	Supra-communautary / Communautary	Trust		5, 21, 26, 27
Values	Supra-communautary / Communautary	Respect		5, 18, 21, 26
Values	Supra-communautary / Communautary	Cooperation		28, 29, 31
Values	Communautary	Ownership		3, 21, 26
Values	Communautary	Social cohesion		34
Values	Communautary	Empowerment		3, 4, 26
Values	Communautary	Legitimity		8, 24, 26
Values	Ind.	Topophily		26, 32
Values	Ind.	Right to participate		26

ICOM-CFRAMP (1995); **16**, Almerigi et al. (1999); **17**, White et al. (1994); **18**, McConney (1999); **19**, Begossi and Brown (2003); **20**, (Renard, 1991); **21**, Pomeroy et al (2003); **22**, Pomeroy et al. (2001); **23**, Pomeroy and Carlos (1997); **24**, Jentoft et al. (1998); **25**, Noble (2000); **26**, Fonseca-Borrás (2009); **27**, Luna (1999); **28**, Nuñez Saravia (2000); **29**, (2003); **30**, Ostrom (1990); **31**, Nuñez Saravia (2005); **32**, Pinkerton (2005); **33**, Pomeroy et al. (2011); **34**, Gutiérrez et al. (2011); **35**, Kikuchi and Hayami (1980); **36**, Hayami and Kikuchi (1981); **37**, Pomeroy and Berkes (1997); **38**, Weitzner (2000). Ind.: Individual.

Almerigi, S, Mahon, Alleyne, Y, Atherley, K, Cumberbatch, J, Mahon, S (1999) Barbados Coastal Conservation Programme (Phase 1), Demonstration Projects. Participatory coastal zone management in Barbados. . (ed C. C. Z. M. Unit), pp. 30

Begossi, A, Brown, D (2003) Fisheries co-management experiences in Latin America and the Caribbean. *The Fisheries Co-management Experience. Accomplishments, Challenges and Prospects* (eds D C Wilson, J R Nielsen, P Degnbol), pp. 348. Kluwer, Dordrecht

Brown, D N, Pomeroy, R S (1999) Co-management of Caribbean Community (CARICOM) fisheries. *Marine Policy*, 23, 549–570

CARICOM-CFRAMP (1995) Report of the strategic planning mission on community involvementand education sub-project. (ed C F M Unit) Belize

Carlsson L, Berkes F (2005) Co-management: concepts and methodological implications. Journal of Environmental Management 75 (1):65–76

CBM (2003) Estado Actual del Comanejo de Áreas Protegidas en Mesoamerica. Proyecto para la consolidación del Corredor Biológico Mesoamericano. CCAD/PNUD/GEF, Managua, Nicaragua

Chuenpagdee R, Jentoft S (2007) Step zero for fisheries co-management: What precedes implementation. Marine Policy 31 (6):657–668

Cumberbatch, J (2001) Case study of the Folkstone Marine Park and Reserve, Barbados. (ed CANARI). CANARI Technical Report, St Lucia

Fonseca-Borrás M (2009) Comanejo de areas silvestres protegidas en Costa Rica: Analisis comparativo de las experencias del parque nacional Cahuita y el parque nacional Marino-Ballena. Universidad de Costa Rica, San José, Costa Rica

Geoghegan, T, Renard, Y (2002) Beyond community involvement: lessons from the insular Caribbean. *Parks*, 12, 16–25

Geoghegan, T, Renard, Y, Brown, N, Krishnarayan, V (1999) Evaluation of Caribbean experiences in participatory planning and management of marine and coastal resources. *CANARI Technical Report* (ed CANARI). CANARI, St. Lucia

Govan, H (2003) Co-management of natural resources in Central America: The road to "equitable distribution of the benefits of biodiversity" or another empty discourse of the technical elite? *Ecosystems, Protected Areas and Peoples Initiative: Lessons learned in community-based management and co-manegement of natural resources and protected areas: focus on coastal and marine resources and Indigenous People's communities in Central American and the Caribbean.* (ed CMWG-IUCN), pp 36 IUCN, San José, Costa Rica

Govan H (2008) Overview: Reclaiming "Protected Areas" as a livelihood tool for Pacific Island people. In: Cohen P, Valemei AD, Govan H (eds) Annotated bibliography on socioeconomic and ecological impacts of Marine Protected Areas in Pacific Island countries. WorldFish Bibliography, vol No. 1870. The WorldFish Center, Penang, pp 1–5

Gutiérrez NL, Hilborn R, Defeo O (2011) Leadership, social capital and incentives promote successful fisheries. Nature 470 (7334):386–389

Hayami, Y, Kikuchi, M (1981) *Asian Village Economy at the Crossroads: An Economic Approach to Institutional Change*, University of Tokyo Press, Tokyo

Jentoft S, Mc Cay BJ, Wilson DC (1998) Social theory and fisheries co-management. Marine Policy 22 (4-5):423–436

Kikuchi, M, Hayami, Y (1980) Inducements to Institutional Innovation in an Agrarian Economy. *Economic Development and Cultural Change*, 29, 21–36

Luna R (1999) El comanejo de áreas protegidas en Centroamérica. In: PROARCA/CAPAS/AID (ed) Memoria del Taller Centroamericano sobre Comanejo de Areas Protegidas, realizado en Panamá, entre el 22 y el 24 de noviembre de 1999. Panamá

Mahon, R, Mascia, M (2003) The Barbados Marine Reserve: a late bloomer? *Gulf and Caribbean Research*, 14, 171–180.

McConney P, Baldeo R (2007) Lessons in co-management from beach seine and lobster fisheries in Grenada. Fisheries Research Evaluation and Management of Coastal Fisheries in Latin America and the Caribbean, 1st CoastFish Conference 87 (1):77–85

McConney, P A (1999) Organising fisherfolk in Barbados without completing a clean round. *Proceedings of the Gulf and Caribbean Fisheries Institute* 52, 290–299

Nielsen, J R, Degnbol, P, Viswanathan, K K, Ahmed, M, Hara, M, Abdullah, N M R (2004) Fisheries co-management--an institutional innovation? Lessons from South East Asia and Southern Africa. *Marine Policy*, 28, 151–160

Noble BF (2000) Institutional criteria for co-management. Marine Policy 24 (1):69–77

Nuñez Saravia, O M (2000) El comanejo y la participación de la sociedad civil en las áreas protegidas de Centroamérica. (ed F D N-T N C-. PROARCA/CAPAS/AID), pp. 46 Guatemala

Nuñez Saravia OM (2000) El comanejo y la participación de la sociedad civil en las áreas protegidas de Centroamérica. Guatemala

Ostrom E (1990) Governing the Commons, The evolution of Institutions for collective action. The Political economy of institutions and decisions, 22d ed. Cambridge University Press, New York

Pinkerton E (2005) Les partenariats de gestion. In: Cochrane KL (ed) Guide du gestionnaire des pêcheries: Les mesures d'aménagement et leur application, vol 424. Document technique sur les pêches. FAO, Rome, p 235

Pomeroy RS, Berkes F (1997) Two to tango: The role of government in fisheries co-management. Marine Policy 21 (5):465–480

Pomeroy RS, Carlos MB (1997) Community-based coastal resource management in the Philippines: A review and evaluation of programs and projects, 1984-1994. Marine Policy 21 (5):445–464

Pomeroy RS, Cinner JE, Raakjær J (2011) Conditions for Successful Co-management : Lessons Learned in Asia, Africa, the Pacific, and the wider Caribbean. In: Pomeroy R, Andrew NL (eds) Small-scale Fisheries Management Frameworks : Frameworks and approaches for the Developing World. CABI Publishing

Pomeroy RS, Katon BM, Harkes I (2001) Conditions affecting the success of fisheries co-management: lessons from Asia. Marine Policy 25 (3):197–208

Pomeroy RS, McConney PA, Mahon R (2003b) Comparative analysis of coastal resource co-management in the carribean. CCA guidelines project.

Ravndal, V (2002) Community co-managed park system for Belize. *Final Project Evaluation* UNDP/GEF, Belize

Renard, Y (1991) Institutional challenges for community-based management in the Caribbean. *Nature and Resources*, 27, 4–9

Renard, Y (2001) Case of the Soufriere Marine Management Area (SMMA). *Technical Report No. CANARI, St Lucia* CANARI, St Lucia

Weitzner V (2000) Probing the Powers at Play in Co-Management from teh Bottom Up: The case of Cahuita National Park, Costa Rica. In: Vlllth Biennial Conference of the International Association for the Study of Common Property "Constituting the Commons: Craning Sustainable Commons in the New Millennium", Indiana University in Bloomington, Indiana, May 31-June 4, 2000.

White, A, Hale, L Z, Renard, Y, Cortesi, L (1994) *Collaborative and community-based management of coral reefs*, West Hartford, Connecticut, USA

References

Agrawal A, Gibson CC (1999) Enchantment and disenchantment: the role of community in natural resource conservation. World Dev 27:629–649

Agüero M (1992) La pesca artesanal en América Latina: una visión panorámica. In: Agüero M (ed) Contribuciones para el estudio de la pesca artesanal en América Latina. Proceedings of the Mini-Symposium on Small Scale Fisheries of the 46th International Congress of Americanists 4–8 July 1988, Amsterdam, The Netherlands, vol 35. ICLARM Conference Proceedings. Manila, Philippines, pp 1–27

Allen GR, Robertson DR (1998) Peces del Pacifico Oriental Tropical. Conabio, Agrupación Sierra Madre y Cemex, México

Alvarado JJ, Herrera B, Corrales L et al (2011) Identificación de las prioridades de conservación de la biodiversidad marina y costera en Costa Rica. Rev Biol Trop 59:829–842

Alvarado JJ, Cortés J, Esquivel MF et al (2012) Costa Rica s Marine protected areas: status and perspectives. Rev Biol Trop 60:129–142

Álvarez J, Ross Salazar E (2010) La Pesca de Arrastre en Costa Rica. Fundación MarViva, San José

Andrew NL, Béné C, Hall SJ et al (2007) Diagnosis and management of small-scale fisheries in developing countries. Fish Fish 8:227–240

Araya UH, Vásquez AR, Marín BE et al (2007) Reporte del manejo de los recursos pesqueros en el Golfo de Nicoya. In: INCOPESCA—UNA—JICA (eds) Proyecto Manejo sostenible de la pesqueria para el Golfo de Nicoya, Costa Rica. Presentacion de conclusiones y recomandaciones. Comité de Evaluacion de Recursos Pesqueros. Puntarenas, Costa Rica, pp 6–67

Arnstein SR (1969) A ladder of citizen participation. JAIP 35:216–224

Babeu A, Cabral K, Hartmann J, Poti K (2012) Establishing socioeconomic baselines of sustainable fishing communities. A study of the fishermen of Palito and Montero. Dissertation, Worcester Polytechnic Institute, Worcester, Massachusetts

Barguil Gallardo D (2009) Coastal artisanal fisheries and community conservation in Costa Rica. Dissertation, Institute of Social Studies, The Hague, Netherlands

Barquero SM (2007) Incopesca está saturado de lanchas decomisadas. La Nación (Costa Rica), 24 July 2007. Grupo Nacion GN, San José, Costa Rica

Beltrán C (2005) Evaluación de la pesca de pequeña escala y aspectos de ordenación en cinco países seleccionados de América Latina: El Salvador, Costa Rica, Panamá, Colombia y Ecuador. Penodos 1997–2005. FAO Circular de Pesca 957/2:71

Béné C (2003) When fishery rhymes with poverty: a first step beyond the old paradigm on poverty in small-scale fisheries. World Dev 31:949–975

Béné C (2004) Contribution of small scale fisheries to rural livelihoods in a water multi-use context (with a particular emphasis on the role of fishing as "last resort activity" for the poor). In: Satia BP, Staples D (eds) Advisory Committee on Fisheries Research. Papers presented at the second session of the Working Party on Small-scale Fisheries. Bangkok, Thailand, 18–21 November 2003, vol 735. FAO Fisheries Report. FAO, Rome, p 198

Béné C, Neiland AE (2006) From participation to governance: a critical review of the concepts of governance, co-management and participation, and their implementation in smallscale inland fisheries in developing countries. World Fish Center Studies and Reviews. Penang

Béné C, Macfadyen G, Allison EH (2007) Increasing the contribution of small-scale fisheries to poverty alleviation and food security. FAO Fisheries Technical Paper. FAO, Rome

Béné C, Russell A, Emma B et al (2008) Governance reforms: a review of small-scale inland fisheries experiences in Lake Chad and Zambezi Basins. Food security and poverty alleviation through improved valuation and governance of river fisheries in Africa. The World Fish Center—GTZ, Cairo

Bergoeing JP (1998) Geomorfología de Costa Rica. Instituto Geográfico Nacional, San José, Costa Rica

Berkes F, Mahon R, McConney P et al (eds) (2001a) Managing small scale fisheries: alternatives directions and methods. International Development Research Center, Ottawa

Berkes F, Mathias J, Kislalioglu M et al (2001b) The Canadian arctic and the oceans act: the development of participatory environmental research and management. Ocean Coast Manage 44:451–469

Blanco O, Mata A (1994) La Cuenca del Golfo de Nicoya. Editorial de la Universidad de Costa Rica, San José, Costa Rica

Bowman L (2011) Possible advances towards fair trade in COOPETÁRCOLES R. L., an artisan fishing cooperative in Central Pacific, Costa Rica. Dissertation, Ripon College, Wisconsin, United States of America

Breton Y (1991) Economic anthropology and interdisciplinatrity in Costa Rica Fishing. The pifalls of problem formulation. MAST 4:1–12

Bussing W, López M (2009) Marine fish. In: Wehrtmann IS, Cortés J (eds) Marine biodiversity of Costa Rica, Central America. Springer, Dordrecht, pp 453–458

Campos JA (1989) Evaluación de la pesca artesanal del Golfo Dulce. Primera parte: Golfo interno. Informe final de proyecto. CIMAR, San José

Carlsson L, Berkes F (2005) Co-management: concepts and methodological implications. J Environ Manage 75:65–76

Carvalho N, Edwards-Jones G, Isidro E (2011) Defining scale in fisheries: small versus large-scale fishing operations in the Azores. Fish Res 109:360–369

Caviedes V (2013) Diagnóstico de la gestión integrada en Costa Rica. In: Chica Ruiz JA, García Onetti J, García Sanabria J, Pérez Cayeiro ML (eds) I Congreso Iberoamericano de Gestión Integrada de Áreas Litorales, Libro de comunicaciones, University of Cadiz, pp 1471–1483

CBM (2003) Estado Actual del Comanejo de Áreas Protegidas en Mesoamerica. Proyecto para la consolidación del Corredor Biológico Mesoamericano. CCAD/PNUD/GEF, Managua

Chacón CA, Araya UH, Vásquez AR et al (2007) Estadisticas pesqueras del Golfo de Nicoya, Costa Rica, 1994–2005. Proyecto Manejo Sostenible de la Pesquería para el Golfo de Nicoya. INCOPESCA—UNA—JICA, Puntarenas

Chauveau JP, Jul-Larsen E (2000) Du paradigme halieutique à l'anthropologie des dynamiques institutionnelles. In: Chauveau JP, Jul-Larsen E, Chaboud C (eds) Les pêches piroguières en Afrique de L'Ouest. Editions Karthala, IRD, CMI, Paris, pp 9–85

Chuenpagdee R, Jentoft S (2007) Step zero for fisheries co-management: what precedes implementation. Mar Policy 31:657–668

Chuenpagdee R, Liguori L, Palomares MLD et al (2006) Bottom-up, global estimates of small-scale fisheries catches. Fisheries Center Research Report, vol 14. Fisheries Center, University of British Columbia, Vancouver

Cicin-Sain B, Knecht RW (1998) Integrated coastal zone and ocean managementconcepts and practicesIsland Press, Washington DC. Island Press, Washington, DC

Cohen JM, Uphoff NT (1980) Participation's place in rural development: seeking clarity through specificity. World Dev 8:213–235

CoopeSoliDar RL (2006) Tárcoles: una comunidad de pescadores artesanales en Costa Rica, aportes a la conservación de los recursos marino costeros del golfo de Nicoya. CoopeSoliDar R. L., San José

CoopeSoliDar RL (2010a) Plan de ordenamiento de la pequeña pesquería. Area marina de pesca responsable de Tarcoles. CoopeSoliDar R. L., San José

CoopeSoliDar RL (2010b) Tarcoles +5: Linea de base de la comunidad de Tarcoles. CoopeSoliDar R. L., San José

Cordero R (2010) Incopesca entregó 105 licencias de pesca sostenible a vecinos del Golfo Dulce. La Nación. Online news 5 July 2010; http://www.nacion.com/archivo/Incopesca-licencias-sostenible-Golfo-Dulce_0_1132686734.html. Accessed 30 Jan 2014

Cortés J, Jiménez CE, Fonseca AC, Alvarado JJ (2010) Status and conservation of coral reefs in Costa Rica. Rev Biol Trop 58(1):33–50

CPGM (2012) Informe de la Comisión Presidencial para la Gobernanza Marina. Presidencia, República de Costa Rica, San José

Defensoría de los Habitantes (2001) Informe Anual de Labores 2000–2001, Addendum. La Defensoría de Los Habitantes de La República, San José, Costa Rica

Defensoría de los Habitantes (2012) Informe Anual de Labores 2011–2012. La Defensoría de Los Habitantes de La República, San José, Costa Rica, pp 152–154

DFOE-AM (2005) Informe sobre los resultados del estudio especial efectuado en el Ministerio del Ambiente y Energía sobre el "comanejo" del Parque Nacional Marino Ballena. Report DFOE-AM-38-2005, División de Fiscalización Operativa y Evaluativa, Contraloría General de la República de Costa Rica, San José, Costa Rica

DFOE-EC-IF (2012) Informe sobre auditoría de carácter especial efectuada en el Instituto Costarricense de Pesca y Acuicultura (INCOPESCA) relacionado con el cumplimiento de sus funciones en materia de conservación del recurso marino. Report DFOE-EC-IF-14-2012, División de Fiscalización Operativa y Evaluativa, Contraloría General de la República de Costa Rica, San José, Costa Rica

DFOE-PGA (2006) Informe sobre la evaluación de la gestión del Instituto Costarricense de Pesca y Acuicultura en relación con la vigilancia y control de la pesca del tiburón. Report DFOE-PGA-86-2006, División de Fiscalización Operativa y Evaluativa, Contraloría General de la República de Costa Rica, San José, Costa Rica

DFOE-PGAA (2008) Informe acerca de los resultados del análisis de procedimiento para el trámite de combustible a precio preferencial otorgado a la flota pesquera nacional no deportiva. Report DFOE-PGAA-6-2008, División de Fiscalización Operativa y Evaluativa, Contraloría General de la República de Costa Rica, San José, Costa Rica

Durand JR, Lemoalle J, Weber J (eds) (1991) La recherche face à la pêche artisanale—Symposium international ORSTOM/IFREMER Montpellier (France) 3 au 7 juillet 1989. Collection colloques et séminaires. ORSTOM, Paris

ECLAC (2011) Statistical yearbook for Latin America and the Caribbean 2010. United Nations, Santiago

Elizondo Mora S (2005) Pesca y procesos de trabajo. El caso de a los pescadores de isla Caballo, Golfo de Nicoya. Costa Rica. Dissertation, Universidad de Costa Rica, San José, Costa Rica

Estado de la Nación (2008) El desafío regional de proteger el patrimonio natural. In: Programa Estado de la Nación (ed) Estado de la Región en Desarrollo Humano Sostenible un informe desde Centroamérica y para Centroamérica. Programa Estado de la Nación, San José, Costa Rica, p 407–435

Estado de la Nación (2010) Armonía con la naturaleza. In: Programa Estado de la Nación (ed) Decimosexto Informe Estado de la Nación en Desarrollo Humano Sostenible. Programa Estado de la Nación, San José, Costa Rica, p 169–224

Estado de la Nación (2011) Panorama Ambiental. In: Programa Estado de la Nación (ed) Cuarto Informe Estado de la Región en Desarrollo Humano Sostenible. Programa Estado de la Nación, San José, Costa Rica, p 361–402

Evans L, Cherrett N, Pemsl D (2011) Assessing the impact of fisheries co-management interventions in developing countries: a meta-analysis. J Environ Manage 92:1938–1949

FAO (1999) Numbers of fishers 1970–1996. FAO Fisheries Circular, vol 929 Rev. 1. FAO, Rome

FAO (2005) Increasing the contribution of small-scale fisheries to poverty alleviation and food security. FAO Technical Guidelines for Responsible Fisheries, No. 10. FAO, Rome

FAO (2007) Social issues in small-scale fisheries. In: FAO (ed) Report of the twenty-seventh session of the Committee on Fisheries. Rome, 5–9 March 2007. FAO Fisheries Report No. 830. FAO, Rome, pp 53–62

FAO (2009) Securing sustainable small-scale fisheries: bringing together responsible fisheries and social development. In: FAO (ed) Report of the twenty-eighth session of the Committee on Fisheries. Rome, 2–6 March 2009. FAO Fisheries and Aquaculture Report No. 902. FAO, Rome, pp 74–83

FAO (2011a) Discussion document: towards voluntary guidelines on securing sustainable small-scale fisheries. FAO, Rome

FAO (2011b) Good practices in the governance of small-scale fisheries: sharing of experiences and lessons learned in responsible fisheries for social and economic development. In: FAO (ed) Report of the twenty-ninth session of the Committee on Fisheries. Rome, 31 Jan–4 Feb 2011. FAO Fisheries and Aquaculture Report. No. 973. FAO, Rome, pp 53–60

FAO (2011c) Report of the Latin America and Caribbean regional consultative meeting on securing sustainable small-scale fisheries: bringing together responsible fisheries and social development, San José, Costa Rica, 20–22 Oct 2010. FAO Fisheries and Aquaculture Report. No. 964. San José

FAO (2011d) Report of the Africa regional consultative meeting on securing sustainable small-scale fisheries: bringing together responsible fisheries and social development, Maputo, Mozambique, 12–14 Oct 2010. FAO Fisheries and Aquaculture Report. No. 963. FAO, Maputo

FAO (2012) State of world fisheries and aquaculture 2012. FAO, Rome

FAO, World Fish Center (2008) Small scale capture fisheries—a global overview with emphasis on develoing countries. A prelimanry report of the big number projects. World Bank, Washington, DC

Fargier L (2009) Hacia el manejo sostenible en el AMUM Pacifico Sur—Informe del taller participativo para la zonificación de la parte marina del Golfo Dulce por parte del gremio de pesca artesanal, Golfito, abril 2009. Universidad de La Rochelle, Francia/Universidad de Costa Rica, Golfito

Fargier L (2012) La participation des pêcheurs artisanaux à la gestion des activités halieutiques artisanales tropicales. Etude de cas dans le Golfo Dulce, Costa Rica. PhD Dissertation, Université de La Rochelle, La Rochelle

Feutry P, Hartmann HJ, Casabonnet H, Umaña G (2010) Preliminary analysis of the fish species of the Pacific Central American Mangrove of Zancudo, Golfo Dulce, Costa Rica. Wetl Ecol Manage 18:637–650

Fonseca-Borrás M (2009) Comanejo de areas silvestres protegidas en Costa Rica: analisis comparativo de las experencias del parque nacional Cahuita y el parque nacional Marino-Ballena. Dissertation, Universidad de Costa Rica, San José, Costa Rica

Germain N (2004) Importancia de la calidad de agua del río Conte y recomendaciones para un plan de acción de gestión sostenible, Playa Blanca, Golfo Dulce, Costa Rica. Dissertation, Universidad de Costa Rica, San José, Costa Rica

Girot OP, Weitzner V, Fonseca-Borrás M (2000) From conflict to collaboration: the case of Cahuita National Park. Limon

Glénard Z (2008) Evaluación del potencial de co-manejo comunitario de los recursos marino-pesqueros, Golfo Dulce, Costa Rica. Dissertation, Université de La Rochelle, La Rochelle

Gómez Quijano E, Tavares Almeida R (2007) Planificación para la conservación y manejo de los recursos pesqueros del Golfo Dulce–Percepción de la comunidad de pescadores artesanales locales—Informe del Primer Taller Participativo COMPESCA-GD. Golfito, Septiembre 2007. ASOPEZ—Centro Socioambiental Osa, La Palma

González Álvarez L, Harrera-Ulloa A, Villalobos Chacón L et al (eds) (1993) Comunidades pesquero-artesanales en Costa Rica. Colleción Cubujuquí. Universidad Nacional, Heredia

Govan H (2008) Overview: reclaiming "Protected Areas" as a livelihood tool for Pacific Island people. In: Cohen P, Valemei AD, Govan H (eds) Annotated bibliography on socioeconomic and ecological impacts of Marine Protected Areas in Pacific Island countries. WorldFish Bibliography, vol 1870. The World Fish Center, Penang, pp 1–5

Govan H, Aalbersberg W, Tawake A, Parks J (eds) (2008) Locally-managed marine areas: A guide to supporting Community-Based Adaptive Management. The locally-managed marine area (LMMA) network. http://www.lmmanetwork.org/files/lmmaguide.pdf. Accessed 24 Apr 2014

Gréboval DF (2007) Ordenación de la capacidad pesquera: panorama general. In: Agüero M (ed) Capacidad de pesca y manejo pesquero en América Latina y el Caribe, vol 461. Doc. Téc. Pesca. FAO, Roma, pp 3–17

Guevara M. HD (1996) Incopesca: Una red cargada de denuncias. In: La Nación (Costa Rica), 10 Nov 1996. Grupo Nacion GN, San José, Costa Rica

Gutiérrez NL, Hilborn R, Defeo O (2011) Leadership, social capital and incentives promote successful fisheries. Nature 470:386–389

Guzmán-Mora AG (2013) Diagnostico de la composicion de capturas de la pesca artesanal de Golfo Dulce, Pacifico Sur, Costa Rica. Dissertation, Universidad de Costa Rica, San José, Costa Rica

Guzmán-Mora AG, Molina Ureña H (2008) La Pesquería Artesanal de Peces en Golfo Dulce, Costa Rica. Paper presented at the The 60th Gulf and Caribbean Fisheries Institute Punta Cana, Dominican Republic, November, 5–9, 2007

Halpern BS (2003) The impact of marine reserves: do reserves work and does reserve sizes matter? Ecol Appl 13:S117–S137

Hartmann HJ, Govan H, Vazquez D et al (2002) Hacia el manejo sostenible del Golfo Dulce. Informe Primer Taller Participativo, Golfito, Septiembre 2001. Pto. Jiménez, Costa Rica

Hartmann HJ, Rivera S, Molina Ureña H et al (2012) Taller sobre las mejores prácticas para el desarrollo de iniciativas locales en el manejo de los recursos pesqueros costeros. Del 28 de Febrero al 5 de Marzo de 2011. Golfito, Costa Rica—David, Panamá. Université de La Rochelle, La Rochelle

Hernández MG (2011) Swimming against the currents: entrepreneur women of Chira Island, Costa Rica. Univ. Florida Digital Collection, http://ufdc. ufl. edu/AA00001648. Accessed 8 Aug 2013

Herrera-Ulloa A, Villalobos-Chacón L, Palacios-Villegas J et al (2011) Coastal fisheries of Costa Rica. In: Salas S, Chuenpagdee R, Charles A et al (eds). Coastal fisheries of Latin America and the Caribbean. FAO Fisheries and Aquaculture Technical Paper No. 544. FAO, Rome, p 137–153

Hirschman AO (1970) Exit, voice, and loyalty responses to decline in firms, organizations and states. Harvard University Press, Cambridge

Honey E, Vargas M, Durham WH (2010) Impacto del Turismo Relacionado con el Desarrollo en la Costa Pacífica de Costa Rica. Informe Ejecutivo. Center for Responsible Travel, San José

Jacquet J, Pauly D (2008) Funding priorities: big barriers to small-scale fisheries. Conserv Biol 22:832–835

Jacquet J, Pauly D, Ainley D et al (2010) Seafood stewardship in crisis. Nature 467:28–29

Jentoft S (2000) Legitimacy and disappointment in fisheries management. Mar Policy 24:141–148

Jentoft S (2006) Beyond fisheries management: the phronetic dimension. Mar Policy 30:671–680

Jentoft S, Mc Cay BJ, Wilson DC (1998) Social theory and fisheries co-management. Mar Policy 22:423–436

Jiménez J (2013) Crisis de ecosistemas marinos y costeros en Costa Rica. Ambientico 230–231:4–8

Johnson DS (2006) Category, narrative, and value in the governance of small-scale fisheries. Mar Policy 30:747–756

Kooiman J (2003) Governing as governance. Sage, London

Kurien J (2003) People and the sea: a "Tropical-Majority" world perspective. MAST 12:9–26

Lagunas Vazques M (2004) Análisis socio-ambiental de la pesca ribereña como actividad de subsistencia en las comunidades costeras de Rincón de Osa y Puerto Escondido-La Palma, Península de Osa, Costa Rica. Dissertation, Universidad de Costa Rica, San José, Costa Rica

Lei Z (2002) Marine coastal dynamics and primary production response in Golfo Dulce, Costa Rica. International Institute for Geoinformation Science and Earth Observation, Enschede

López-Estrada E, Breton E (1991) Saber técnico, organización social e identidad de los Chomeños. Dinámica social y comunidades pesqueras en el pacífico costarricense. Faculté des sciences sociales, Université de Laval, Laval

López A, LeFoulgoc L, Arauz R (2007) Analysis of the annual landings of shark and rays in the coast fisheries of the Pacific Costa Rica (Tárcoles). Paper presented at the 11th annual scientific meeting of the European Elasmobranch Association, Brest, France, 23–26 Nov 2007

Luna R (1999) El comanejo de áreas protegidas en Centroamérica. In: PROARCA/CAPAS/AID (ed) Memoria del Taller Centroamericano sobre Comanejo de Areas Protegidas, 22–24 Nov 1999, Panamá

Magnin V (2004) Evaluation initiale de l'impact de la pêche sportive dans les eaux du Golfo Dulce, Costa Rica. Dissertation, Université Catholique d'Angers, France, Angers, France

Marín Alpízar BE (2002) Diagnostico de los recursos pesqueros en el Golfo de Nicoya, Costa Rica. Dissertation, Universidad Nacional, Heredia—Puntarenas, Costa Rica

Marín Alpízar BE, Araya Umaña H, Vásquez Arias AR et al (2010) Informe del estado actual de las pesquerias en el Área Marina de Pesca Responsable de Palito, Chira. INCOPESCA, San José

Marín Cabrera M (2012) Identificación y caracterización de actores institucionales y de la sociedad civil claves en la gestión de las AMUM Golfo de Nicoya y Pacífico Sur. Manejo integrado de los recursos marino costeros en la Provincia de Puntarenas. Puntarenas

Martin KS (2001) Making space for community resource management in fisheries. Ann Assoc Am Geogr 91:122–142

MarViva (2011) Costa Rica seamounts marine management area. Outreach brochure. http://www.marviva.net/Publicaciones/Folleto.pdf. San José, Costa Rica. Accessed 6 July 2013

McConney P, Baldeo R (2007) Lessons in co-management from beach seine and lobster fisheries in Grenada. In: Charles A et al (eds) Evaluation and management of coastal fisheries in Latin America and the Caribbean. Fish Res 87:77–85

MINAE-SINAC (2006) Política de manejo compartido de las áreas silvestres protegidas de Costa Rica. UICN Oficina Regional para Mesoamérica, San José

Mora A, Fernández C, Guzmán AG (2006) Áreas Marinas Protegidas y Áreas Marinas de Uso Múltiple de Costa Rica -Notas para una discusión. Fundación MarViva, San José

Mora C, Robertson DR (2005a) Causes of latitudinal gradients in species richness: a test with the endemic shorefishes of the Tropical Eastern Pacific. Ecology 86:1771–1782

Mora C, Robertson DR (2005b) Factors shaping the range-size frequency distribution of the endemic fish fauna of the Tropical Eastern Pacific. J Biogeogr 32:277–286

Morera Quesada G, Vargas Bonilla S (2009) Informe de línea base de condición socioeconómica de miembros de organizaciones interesadas en el establecimiento del Área Marina de Pesca Responsable Golfo Dulce. Asociacion de Pescadores Turisticos de Costa Rica, Golfito

Neiland AE, Béné C (2003) A review of fisheries management performance in developing countries, with particular reference to issues of policy and governance. A report prepared for the ACP Fish II Programme Feasibility Study, Support Unit for International Fisheries and Aquatic Research (SIFAR)

Nielsen Muñoz V, Quesada Alpízar MA (eds) (2006) Ambientes Marino Costeros de Costa Rica. Informe técnico. CIMAR, CI, TNC, San José

Noble BF (2000) Institutional criteria for co-management. Mar Policy 24:69–77

Nuñez Saravia OM (2000) El comanejo y la participación de la sociedad civil en las áreas protegidas de Centroamérica. Guatemala

Olivier de Sardan J-P (ed) (1995) La politique du terrain—Sur la production des données en anthropologie. Enquête, Les terrains de l'enquête. http://www.enquete.revues. org/263, doi:10. 4000/enquete, p 263. Accessed 21 June 2013

OSPESCA (2008) Conformación de la Federación de Pescadores de Pequeña Escala de Costa Rica. El Cardumen

OSPESCA (2010) Encuesta estructural de la pesca artesanal y la acuicultura en Centroamérica—2009—Informe final preliminar. La Libertad, El Salvador

Ostrom E (1990) Governing the commons, the evolution of Institutions for collective action. The Political economy of institutions and decisions, 22nd edn. Cambridge University Press, New York

Ostrom E, Ostrom V (2003) Rethinking institutional analysis: interviews with Vincent and Elinor Ostrom. Mercatus Center at George Mason University, Arlington

Ostrom E, Gardner R, Walker J (eds) (1994) Rules, games, and common-pool resources. University of Michigan Press, Bloomington

Ostrom E, Burger J, Field CB, Norgaard RB, Policansky D (1999) Revisiting the commons: local lessons, global challenges. Science 284:278–282

Ostrom E, Dietz T, Dolšak N et al (eds) (2002) The drama of the commons—committee on the human dimensions of global change. Division of behavioral and social sciences and education—national research council edn. National Academy Press, Washington, DC

Oviedo E, Murillo Á (2013) A Chinchilla le simpatiza la idea de eliminar directivas. In: La Nación (Costa Rica), 19 Jan 2013. http://www.nacion.com/archivo/Chinchilla-simpatiza-idea-eliminar-directivas_0_1318468175.html. Last Accessed 25 Apr 2014

Parrain C (2012) La haute mer: un espace aux frontières de la recherche géographique. EchoGéo 19, http://echogeo. revues. org/12929, doi:10. 400/echogeo, p 12929. Accessed 8 June 2013

Pauly D (1997) Small-scale fisheries in the tropics: marginality, marginalization and some implication for fisheries management. In: Pikitch EK, Huppert DD, Sissenwine MP (eds) Global trends: fisheries Management. American Fisheries Society Symposium 20. American Fisheries Society, Bethesda, pp 40–49

Pauly D (2006) Major trends in small-scale marine fisheries, with emphasis on developing countries, and some implications for the social sciences. MAST 4:7–22

Pauly D, Agüero M (1992) Small-scale fisheries in the neotropics: research and management issues. In: Agüero M (ed) Contribuciones para el estudio de la pesca artesanal en América Latina, Proceedings of the Mini-Symposium on Small Scale Fisheries of the 46th International Congress of Americanists 4–8 July 1988, Amsterdam, The Netherlands, 35, ICLARM Conference Proceedings, pp 28–36

Pinkerton E (1989) Introduction: attaining a better fisheries management through co-management-prospects, problems, and propositions. In: Pinkerton E (ed) Co-operative management of local fisheries: new directions for improved management and community development. University of British Columbia Press, Vancouver, pp 3–33

Pinkerton E (2005) Partnerships in management. In: Cochrane KL (ed) A fishery manager's guidebook—management measures and their application. Fisheries technical paper No. 424. FAO, Rome, pp 159–174

PNUD (2011) Human Development Report 2011. PNUD, New York

Poirout T (2007) Diagnostic socio-environnemental de la pêche artisanale pour une gestion participative de la zone côtière de Zancudo (Golfo Dulce, Costa Rica). Dissertation, Université de La Rochelle, La Rochelle, France

Pomeroy RS, Andrew NL (eds) (2011) Small-scale fisheries management—frameworks and approaches for the developing World. CABI, Wallingford

Pomeroy RS, Berkes F (1997) Two to tango: the role of government in fisheries co-management. Mar Policy 21:465–480

Pomeroy RS, Carlos MB (1997) Community-based coastal resource management in the Philippines: a review and evaluation of programs and projects, 1984–1994. Mar Policy 21:445–464

Pomeroy RS, Cinner JE, Raakjær J (2011) Conditions for successful co-management: lessons learned in Asia, Africa, the Pacific, and the wider Caribbean. In: Pomeroy R, Andrew NL (eds) Small-scale fisheries management frameworks: frameworks and approaches for the developing world. CABI Publishing, Wallingford

Pomeroy RS, Katon BM, Harkes I (2001) Conditions affecting the success of fisheries co-management: lessons from Asia. Mar Policy 25:197–208

Pomeroy RS, McConney PA, Mahon R (2003) Comparative analysis of coastal resource co-management in the Caribbean Caribbean. Coastal Co-management Guidelines Project. Caribbean Conservation Association, Barbados

Pomeroy RS, McConney PA, Mahon R (2004) Comparative analysis of coastal resource co-management in the Caribbean. Ocean Coast Manage 47:429–447

PRETOMA (2009) Golfo Dulce Artisanal fishermen demand incopesca uphold their rights. Press Releases 11/26/2009; http://www.pretoma.org/golfo-dulce-artisanal-fishermen-demand-incopesca-uphold-their-rights/. Accessed 30 June 2013

PRETOMA (2012) Tide of Change (PRETOMA's Newsletter) N°1, 12/2012. http://www.pretoma.org/wp-content/uploads/2013/03/Newsletter-2012-Pretoma1.pdf, San José. Accessed 23 May 2013

Pretty JN (1995) Participatory learning for sustainable agriculture. World Dev 23:1247–1263

Quesada-Alpízar MA (2006) Participation and fisheries management in Costa Rica: from theory to practice. Mar Policy 30:641–650

Quesada-Alpízar MA, Cortés J (2006) Los ecosistemas marinos del Pacífico sur de Costa Rica: estado del conocimiento y perspectivas del manejo. Rev Biol Tropi 54(1):101–145

Quintero QC, García G, Barroso Chavarría JE et al (2009) Pescadores del golfo Dulce aclaran posición sobre críticas a Incopesca. Boletin El Cardumen

Richards FA, Anderson JJ, Cline JD (1971) Chemical and physical observations in Golfo Dulce, an anoxic basin on the Pacific coast of Costa Rica. Limnol Oceanogr 16:43–50

Robertson DR, Allen GR (2008) Shorefishes of the Tropical Eastern Pacific online information system. Smithsonian Tropical Research Institute, Balboa

Rodríguez Chaves M (2008) Pescadores artesanales en Tárcoles. Diagnóstico legal. San José

Ruddle K (2011) Coastal marine resources in human ecological context: the scale and modes of integration. In: Christensen V, Maclean J (eds) Ecosystem approaches to fisheries: a global perspective. Cambridge University Press, Cambridge, pp 241–271

Ruddle K, Hickey F (2008) Accounting for the mismanagement of tropical nearshore fisheries. Environ Dev Sustain 10:565–589

Ruddle K, Satria A (2010) Conclusion: errors and Insights. In: Ruddle K, Satria A (eds) Managing coastal and inland waters: pre-existing aquatic management systems in Southeast Asia. Springer Netherlands, Heidelberg, pp 161–173

Sala Constucional (7 Aug 2013) Sala Constitucional prohibe el uso de redes de arrastre para la pesca de camaron. Sala Constitucional Press Release SC-CP-30-13, San José

Salas E, Ross SE, Arias A (eds) (2012) Diagnóstico de áreas marinas protegidas y áreas marinas para la pesca responsable en el Pacífico costarricense. Edición MarViva, San José

Salas S, Chuenpagdee R, Seijo JC et al (2007) Challenges in the assessment and management of small-scale fisheries in Latin America and the Caribbean. Fish Res 87:5–16

Salas S, Chuenpagdee R, Charles A et al (eds) (2011) Coastal fisheries of Latin America and the Caribbean. FAO Fisheries and Aquaculture Technical Paper No. 544. FAO, Rome

Schlager E, Ostrom E (1999) Property rights regimes and coastal fisheries: an empirical analysis. In: McGinnis MD (ed) Polycentric governance and development: readings from the Workshop in political theory and policy analysis. Institutional Analysis. University of Michigan Press, Bloomington, p 433

Sen S, Raakjaer Nielsen J (1996) Fisheries co-management: a comparative analysis. Mar Policy 20:405–418

Shanks AL, Grantham BA, Carr MH (2003) Propagule dispersal distance and the size and spacing of marine reserves. Ecol Appl 13:S159–S169

Silva AM, Carillo NN (2004) El manglar de Purruja, Golfito, Costa Rica: un modelo para su manejo. Rev Biol Trop 52(2):195–201

Silver JJ, Campbell LM (2005) Fisher participation in research: dilemmas with the use of fisher knowledge. Ocean Coast Manage 48:721–741

SINAC (2009) Análisis de vacíos de conservación en Costa Rica vacíos en la representatividad e integridad de la biodiversidad marina y costera. Grúas II. Propuesta de ordenamiento territorial para la conservación de la biodiversidad de Costa Rica, vol III. SINAC, San José

Smith IR (1979) A research framework for traditional fisheries. ICLARM, Manila

Solís V, Madrigal P (2004) Comanejo: Una reflexión conceptual desde Coope SoliDar R. L. San José

Solís-Rivera V, Fonseca-Borras M, Barguil Gallardo D et al (2012) Estudio Regional sobre las Dimensiones Sociales en el Manejo de Áreas Marinas Protegidas: Casos en Costa Rica, Nicaragua, Honduras y Panamá. ICSF, San José

Staples D, Satia B, Gardiner PR (eds) (2004) A research agenda for small-scale fisheries. FAO, Regional Office for Asia and Pacific, Bangkok

Stern-Pirlot A, Wolff M (2006) Population dynamics and fisheries potential of Anadara tuberculosa (Bivalvia: Arcidae) along the Pacific coast of Costa Rica. Rev Biol Trop 54(1):87–99

Svendsen H, Rosseland R, Myking S et al (2006) A physical-oceanographic study of Golfo Dulce, Costa Rica. Rev Biol Trop 54(1):147–170

Teh LSL, Teh LCL, Sumaila UR (2011) Quantifying the overlooked socio-economic contribution of small-scale fisheries in Sabah, Malaysia. Fish Res 110:450–458

TNC (2011) Análisis de factibilidad de los Acuerdos de Conservación Marina para el Pacífico Tropical Oriental. Fundación de la Familia Walton, San José

van den Hombergh H (1999) Guerreros del Golfo Dulce: industria forestal y conflicto en la Península de Osa, Costa Rica. San José

Vargas JA, Wolff M (1996) Pacific Coastal Ecosystems with emphasis on the Golfo Dulce and adjacent areas: a synoptic view based on the R/V/ Victor Hensen—expedition 1993/4 and previous studies. Rev Biol Trop 44(3):187–214

Voorhis A, Epifanio CE, Maurer D et al (1983) The estuarine character of the Gulf of Nicoya, an embayment on the Pacific coast of Central America. Hydrobiologia 99:225–237

Waylen PR, Quesada ME, Caviedes CN (1996) Temporal and spatial variability of annual precipitation in Costa Rica and the Southern Oscillation. Intern J Climatol 16:173–193

Wehrtmann IS, Cortés J (eds) (2009) Marine biodiversity of Costa Rica, Central America. Monographiae Biologicae 86. Springer and Business Media B. V., Berlin

Weigel JY, Féral F, Cazalet B (eds) (2007) Les aires marines protégées d'Afrique de l'Ouest: gouvernance et politiques publiques. Perpignan, France

Weitzner V (2000) Probing the powers at play in co-management from the bottom up: the case of Cahuita National Park, Costa Rica. In: Vlllth Biennial Conference of the International Association for the Study of Common Property "Constituting the Commons: craning Sustainable Commons in the New Millennium", Indiana University in Bloomington, Indiana, May 31–June 4, 2000

Weitzner V, Fonseca-Borrás M (2001) Cahuita (Limón), Costa Rica: D'une situation de conflit à un climat de collaboration. In: Buckles D (ed) Cultiver la paix: conflits et collaboration dans la gestion des ressources naturelles. IDRC, Ottawa, pp 143–166

Wolff M, Hartmann HJ, Koch V (1996) A pilot trophic model for Golfo Dulce, a fjord-like tropical embayment, Costa Rica. Rev Biol Trop 44(3):215–231

Wolff M, Koch V, Chavarria JB et al (1998) A trophic flow model of the Golfo de Nicoya, Costa Rica. Rev Biol Trop 46(6):63–79

World Bank, FAO, Center WF (2010) The hidden harvests. The global contribution of capture fisheries. World Bank, Washington, DC

Wetland Conservation in Northern Sonora, Mexico: Legal Tools and Active Communities

11

Peggy J. Turk-Boyer, Hiram Peña-Bonilla, Hem Nalini Morzaria-Luna, Paloma A. Valdivia-Jiménez, Hitandehui Tovar-Vázquez and Alejandro Castillo-López

Abstract

During the global economic boom of the last decade (2000s), the shoreline of northern Sonora became one of the most rapidly growing coastal areas in Mexico, second only to the Caribbean. Many high impact development projects were proposed for the estuaries near Puerto Peñasco between Bahía Adair and Bahía San Jorge. Because of the recognized value of these wetlands, a multi-layered suite of legal and management instruments were applied to guarantee long-term protection of these essential habitats. Natural Protected Areas, Federal Zone Concessions and Ramsar Site Designations have been implemented to protect different wetland sites in northern Sonora, Mexico. Development projects are required to present Environmental Impact Studies for approval, and need vigilance to assure maintenance of coastal integrity. These different instruments are being brought to life in a biological corridor near Puerto Peñasco, Sonora by active participation of local communities and civil society organizations. The coastal inhabitants of these wetlands have united in processes to zone their use, develop low impact economic activities (resource monitoring, ecotourism and handcrafts), and participate in training and education programs, which also involves the youth. A Ramsar wetland network was formed to link wetland users with each other. Environmental contests and campaigns have engaged the region's youth in working with their communities to solve environmental problems. Fishers are participating in management initiatives for individual species and a vision for ecosystem-based management is growing. The impact of these programs points to an emerging wetland conservation ethic that may prove as important for long–term protection as the other instruments used. We present this process as a model for social participation in management of natural resources that can be used with a variety of user groups and legal tools.

P. J. Turk-Boyer (✉) · H. Peña-Bonilla · H. Morzaria-Luna · P. A. Valdivia-Jiménez · H. Tovar-Vázquez
Centro Intercultural de Estudios de Desiertos y Océanos (CEDO), Apartado Postal #53, 83550 Puerto Peñasco, SON, Mexico
e-mail: peggy@cedointercultural.org

H. Peña-Bonilla
e-mail: hiram@cedointercultural.org

H. Morzaria-Luna
e-mail: hmorzarialuna@gmail.com

P. A. Valdivia-Jiménez
e-mail: paloma@cedointercultural.org

H. Tovar-Vázquez
e-mail: hita@cedointercultural.org

A. Castillo-López
Pronatura Noroeste A.C., Programa de Conservación Marina y Pesca Sustentable, Calle Décima 60, 22800 Ensenada, B.C., Mexico
e-mail: acastillo@pronatura-noroeste.org

11.1 Introduction

Wetlands are recognized as important ecosystems for their high productivity and biodiversity, and for the ecosystem services they offer. They provide habitat for breeding, feeding and development for many marine species, commercial and non-commercial, as well as nesting and feeding areas for birds (Glenn et al. 2006). They are known as key sites for marine fisheries production (Houde and Rutherford 1993; Aburto-Oropeza et al. 2008). Wetlands help mitigate floods and prevent coastal erosion (Brusca et al. 2006). Local communities make use of them in many ways as well.

Because wetlands are readily accessible from the coast they are highly vulnerable to multiple types of land-based

F. Amezcua, B. Bellgraph (eds.), *Fisheries Management of Mexican and Central American Estuaries*, Estuaries of the World, DOI 10.1007/978-94-017-8917-2_11, © Springer Science+Business Media Dordrecht 2014

Keywords

Social participation · Upper Gulf of California · Wetland conservation · Ramsar · Environmental impact assessment · Estuaries

pressures. Threats include upstream diversion of riparian habitats, watershed pollution, local water extraction that exceeds aquifer recharge, construction of marinas and sea level changes (Glenn et al. 2006; Brusca et al. 2006). Management of upstream freshwater resources is considered a key component to sustaining a productive marine system, since freshwater discharge impacts physical, chemical and biological processes in ocean ecosystems, as well as circulation patterns and primary productivity (Drinkwater and Frank 1994). Estuarine and coastal fisheries are especially vulnerable to overfishing because of their proximity to the coast (Houde and Rutherford 1993; Aburto-Oropeza et al. 2008; Barbier et al. 2011). Overexploitation, incidental catch, and lack of adequate management for all life stages of fisheries species threaten the long-term viability of coastal fisheries. To adequately address such diverse issues, management of coastal fisheries requires an integrated approach that considers watershed-level management and uses coastal, marine and fisheries instruments and legal tools as well.

Along the northeastern coast of the Northern Gulf of California near Puerto Peñasco, Sonora, Mexico a distinct fisheries biological corridor has been defined which is dominated by extensive wetland systems, known hereafter as the Peñasco Corridor (Fig. 11.1). These wetlands include two bays (Bahía Adair and Bahía San Jorge) and the ephemeral Sonoyta and Asunción riparian systems, with a total of eight estuaries along a 280 km stretch of coast (Fig. 11.2). Here the predominant wetlands are marshes without mangroves, known as esteros (negative hyper-saline lagoons or estuaries). The marshes are located within coastal lagoons that have a high tidal variation and no regular fresh water input. They have been identified as key sites for spawning of commercially important fisheries species such as shrimp, blue crab and several species of sharks and rays (Cudney-Bueno and Turk-Boyer 1998). They serve for breeding and/or feeding of various species of benthic mollusks (Martínez et al. 2011) and support life stages of other commercially important species within the region. These wetlands and other habitats of the Corridor support diverse fisheries harvested by six coastal communities. These complexes are considered priority wetlands for conservation by the National Commission for Understanding and Use of Biodiversity (Comisión Nacional de Conocimiento y Uso de Biodiversidad, CONABIO: Arriaga et al. 2000) and other authors (Glenn et al. 2006; Brusca et al. 2006).

Connectivity between estuarine dependent species and habitats and the adjacent marine ecosystem is evident through food web interactions and other mechanisms (Spackeen 2009; Glenn et al. 2006), while in the marine system oceanographic currents distribute larvae, nutrients and food, connecting species between different habitats and ecosystems (Marinone et al. 2008). Ecosystem-based fisheries management seeks to protect and manage the ecological function of a suite of habitats, targeted species, and their predators and prey and strives to assure that ecological processes and ecosystem services such as connectivity and trophic relations are maintained (Crowder et al. 2008; Pikitch et al. 2004).

Fisheries related pressures have increased gradually over the last 100 years since the first fishing camps were established in the Upper Gulf in the early 1920s. Three fishing sectors operate out of the Puerto Peñasco community including recreational, industrial and small-scale fishers (Cudney-Bueno and Turk-Boyer 1998). The small-scale fishers, also known as artisanal or riparian fishers, are the most important in terms of the social and economic impact that their activities have on local populations and the environment. The five other communities in the Corridor, to the south of Puerto Peñasco, all depend on small-scale fishing as the primary economic activity. In 2011 a total of 1,158 skiffs were working in the entire Peñasco Corridor harvesting a variety of species, including blue crab, gulf drum, guitarfish and flounder (Downton-Hoffmann et al. 2013; Cudney-Bueno and Turk-Boyer 1998; CONANP 2007; Ruiz-López 2009; De la Cruz-González et al. 2011).

There is ample evidence of decline in fisheries production in the region (Cudney-Bueno and Turk-Boyer 1998; PANGAS 2012; Sáenz-Arroyo et al. 2005; Erisman et al. 2012; Rupnow 2008). Overfishing and open access conditions are prevalent; in fact, it is estimated that 75% of the fishing activity within the Peñasco Corridor occurs without the appropriate fishing permits (Downton-Hoffmann et al. 2013). Environmental impacts are also of concern as some fishing techniques cause damage or destroy habitats and impact the abundance and distribution of non-target species (De la Cruz González et al. 2011) which could represent prey of other species. Ultimately these conditions threaten the integrity of marine and coastal ecosystems, and damage the economic base of these communities, impacting the lives of thousands of families in the Peñasco Corridor (Ruiz-Lopez 2009). Better fisheries management is called for. Information has been lacking to advance management for many species, and where management rules do exist, a commitment to follow them and surveillance and enforcement of laws by government authorities is lacking (Cinti et al. 2010). These components

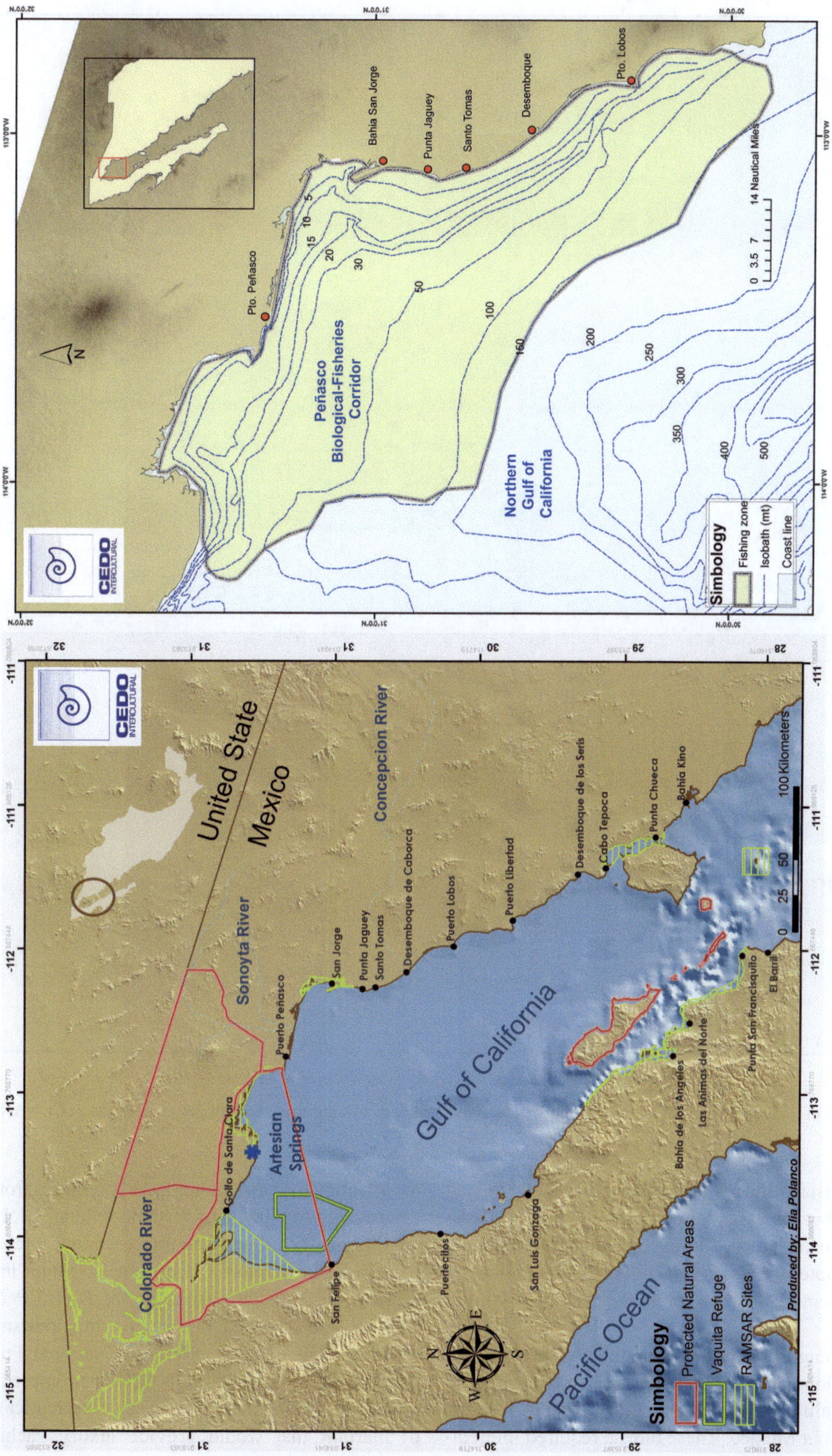

Fig. 11.1 Location of the Peñasco Fisheries Biological Corridor in the Northern Gulf of California, Sonora, Mexico

Fig. 11.2 Wetlands of the Peñasco Fisheries Biological Corridor, Northern Gulf of California, Sonora, Mexico. (Source: CONANP)

of managing coastal fisheries are not specifically addressed in this chapter, but are essential components of ecosystem-based fisheries management and integrated coastal management. In this chapter we address the direct threats impacting wetlands as essential habitats for maintenance of coastal fisheries.

At the beginning of the millennium, the pressures related to tourism development accelerated at a phenomenal speed in the Gulf of California, threatening destruction of wetland habitats throughout Mexico. These threats required immediate attention to prevent land-based pressures from potentially destroying large tracts of wetland habitat. Global economic conditions, including availability of low interest loans and the associated economic boom, fueled major investments in tourism infrastructure especially at Puerto Peñasco, which was Mexico's second fastest growing tourism destination after Cancun. In 2005, Mexico's National Fund Trust for Tourism Development (FONATUR) promoted a project to create the Nautical Ladder (Escalera Nautica), a stairway of marinas that would provide luxury yachting services throughout the Gulf of California and the Pacific coast of Baja California. A total of 14 new marinas were projected to be built, mostly in wetland habitats, and 13 sites were to be enhanced throughout the Gulf and the Pacific Coast (FONATUR 2003). This resulted in land speculation and promotion of large mega developments that marketed their projects using marinas and golf courses as the lure. From September 2006 to September 2007 tourism investment at Puerto Peñasco was $ 481 million U.S. dollars distributed in 22 projects (SECTUR 2007) including hotels, condominiums, golf courses and marinas. At this time there was a marina projected for construction in four of the eight estuaries in the Peñasco Corridor, each positioning itself as one of FONATUR's Nautical Ladders (though they were unrelated projects). Threats for each wetland in the region were distinct as were local stakeholder interests, calling for specific strategies and actions for each wetland system. Generally the proposed projects were characterized by their potential to physically alter or eliminate wetland habitats.

11.1.1 Strategic Wetland Conservation Initiatives

To respond to these rapidly growing pressures, in 2007 community and institutional involvement for conservation of coastal wetlands was promoted following three strategic initiatives:

- Promotion of adequate review and decision-making for approval of development projects threatening critical habitats. All new development projects are required to submit an Environmental Impact Assessment (EIA) to Federal authorities and obtain approval. This strategy involved promoting use of legal actions associated with EIA such as participation in public hearings, consultation processes, and appeals to assure adequate evaluation of a project's impact, mitigation if the project is approved, and follow up if the project is developed.
- Creation of legal tools, policy or management instruments to offer long-term protection for critical habitats. Two types of instruments were applied in wetland habitats:
  - Promotion of wetland designation as Ramsar sites by the Federal Government under the Convention of Wetlands of International Importance and development of management programs for these sites with strong community participation.
  - Request of Federal Maritime Zone rights for land adjacent to wetlands, either as concessions or destination agreements designated for conservation use.
- Active support and engagement of local communities in wetland management and sustainable economic alternatives that have low impact on wetland habitats.
  - Leadership training for community groups and partners that offers tools for active participation in restoration, conservation and management projects in wetlands, such as wildlife monitoring and surveillance.
  - Awareness strategies directed to potentially affected stakeholders to identify threats and involve them in EIA consultation processes.
  - Development of innovative economic projects that promote economic diversification and environmental responsibility.

To compliment these strategies research was conducted to generate needed information to understand wetland function. We developed and strengthened partnerships with municipal, state and federal governments, other social organizations, NGO's and stakeholders to support these research efforts. Training and education programs were offered to local stakeholders to foster a deep appreciation for the environment and to empower them to become agents of change and transformation in their own communities.

11.2 Peñasco Corridor Wetlands: Values and Threats

The wetlands of the Peñasco Corridor in northern Sonora include: (1) the Bahía Adair complex and its estuary systems which are San Judas-Las Lisas esteros in the north, and Cerro Prieto and La Cholla at the southern end of the bay; (2) the estuarine complex Morúa—La Pinta; and (3) Bahía San Jorge, which includes La Cinita, Almejas-La Salina esteros, and San Francisquito (Fig. 11.2).

11.2.1 Bahía Adair

Bahía Adair is an extensive wetland that covers an area of 42,430 ha with approximately 100 km of coastline (about 62 miles) (Ramsar 2009). It is located within the Upper Gulf of California and Colorado River Delta Biosphere Reserve (UGBR) and consists of three estuaries: Las Lisas-San Judas (treated as one system), Cerro Prieto and La Cholla (Fig. 11.2). This bay is characterized by high primary productivity and the presence of nesting areas of Least Tern (*Sterna antillarum browni*), a species under special protection (NOM-059-SEMARNAT-2010 in DOF 2010). The area is an important resting place for migratory birds, and provides nesting and feeding conditions as well (Glenn et al. 2006).

Bahía Adair is a breeding and nursery area for fisheries species of commercial importance such as crab, shrimp and various fish species (Cudney-Bueno and Turk-Boyer 1998; Calderón-Aguilera et al. 2003). A tidal amplitude of 10 m in the bay facilitates the export of food and energy into the surrounding desert and marine ecosystems, providing organic materials and nutrients for the food chain (Okin et al. 2005). The bay's extensive marshes host the Gulf of California's second largest stand of wild wheat (*Distichlis palmerii*) (Brusca et al. 2006) as well as other halophyte species. Freshwater artesian wells in the western bay area support flora and fauna distinct from the surrounding desert and provide essential freshwater for birds and mammals (Felger 2000).

The bay has a history of traditional use by the Tohono O'odham or Papago indigenous community, who made extensive journeys across the Sonora Desert to the bay to extract salt and conduct rite of passage ceremonies (Addison-Sorey 1989). Today two permanent human settlements have less than 10 people living at the Carlos Salinas and López Aceves ejidos. Most of the land around the bay is owned communally by six ejidos that are slowly disincorporating into small private properties. In Esteros La Cholla and Cerro Prieto, in the southern part of the bay, a large colony of tourists own residential vacation homes.

Only a few economic activities take place here, including fishing from nearby Puerto Peñasco and La Cholla communities. Two blue crab cooperatives use a temporary camp near Cerro Prieto estero during open crab season. Artisanal clam collectors work in La Cholla estero. A mining company in the community of Carlos Salinas, approximately 38 km northwest of Puerto Peñasco, has a concession for trona (soda salt) exploration. Infrastructure was built at the site, though the mine did not have commercial permits, but there has been no activity there since 2010.

The threat of primary concern in the Bahía Adair wetland complex, despite the fact that it lies within the Upper Gulf of California / Colorado River Delta Biosphere Reserve, is the construction of tourism infrastructure, mainly condominiums, golf courses and marinas. This is especially true at La Cholla and Cerro Prieto esteros located nearest Puerto Peñasco. The northern-most section of La Cholla estuary was impacted in 2003 by the construction of a golf course. The southern-most section was targeted for the construction of a marina by a large development group. The marina project as initially conceived was not approved, but it has re-appeared in different forms. Actions taken to prevent destruction of this wetland will be detailed in later sections. These and other development projects built at Puerto Peñasco from 2000 to 2006 caused a rapid appreciation of the land, stimulated land speculation, and detonated large-scale investment in real estate. Such investments, however, didn't always translate into increased tourism revenue for the local tourism economy (Guido 2006).

The government also has invested in infrastructure to support development along the Bahía Adair coast, which further fueled land speculation. In 2007–2008 a coastal highway was constructed from Puerto Peñasco to El Golfo de Santa Clara, along the perimeter of Bahía Adair. This stimulated the sale of large tracts of former ejido lands, some within the Bahía Adair wetland systems. In 2011 the Federal Commission of Electricity (Comisión Federal de Electricidad, CFE) proposed the construction of a high-voltage line through the Upper Gulf Biosphere Reserve along the coast, though the project did not materialize. With a look towards development opportunities, both projects prompted widespread land speculation.

11.2.2 Esteros Morúa and La Pinta

The Morúa and La Pinta estuarine complex, east of Puerto Peñasco, Sonora, is characterized by two elongated lagoons of approximately 30 km^2 separated by the deltaic deposits of the Sonoyta River, which used to flow between them (Fig. 11.2; Turk-Boyer 1989). Today only occasional surface flow enters the Gulf from this river system, mostly in El Niño years, as was seen in 1983 (Felger 2000). A stabilized dune system on the outer coast separates each lagoon from the open sea (Turk-Boyer 1989).

The Morúa estuary hosts 12 species of birds with protected status in Mexico, including the savannah sparrow (*Passerculus sandwichensis*) (Castillo-López et al. 2007) and two colonies of nesting Least Terns (Alcock 1992; Palacios and Mellink 1996). It also is a breeding and rearing area for commercial species such as crab, shrimp and rays (Brusca et al. 2006), and 46 species of plants have been identified, including many halophytes (Castillo-López et al. 2007).

Existing economic activities include three oyster farms each with their own small restaurant, ecotourism and educational activities. This estuary represents an important weekend recreation area for the Puerto Peñasco community. Land ownership around Estero Morúa consists of small private properties, two large residential areas for American tourist vacation homes, and one ejido with communal properties. Development companies own some property, which represents the greatest threat to the estuary. From 2005 to 2007 several proposals for construction of a marina, a golf course, and high-density condos at the mouth of the estuary became serious threats. Solid waste drainage from septic systems of residential communities and solid waste accumulation at oyster farms represent potential problems for contaminating the estuary. All-terrain vehicle (ATVs) traffic in the estero impacts soils and destroys vegetation. These vehicles and tourists walking in the area also threaten the nesting activities of the protected Least Tern.

La Pinta is an estuary with essential habitats including open ponds, flood channels, mudflats and marshes. It has areas of high productivity and nursery areas. It produces organic matter and is key for nutrient recycling. The mullet, anchovy and sardines which are found here are important food for commercial species and form the basis for productivity in the Northern Gulf ecosystem. Benthic invertebrates include shrimp, crabs, snails, clams and worms that offer food for resident and migratory birds. Protected bird species in this wetland include the endemic Heermann's Gull (*Larus heermanni*), Yellow-Footed Gull (*Larus livens*), and the Reddish Egret (*Egretta rufescens*), all subject to special protection, and the Savannah Sparrow (*Passerculus sandwichensis*) and Prairie Falcon (*Falco mexicanus*), which are threatened.

Most of the land around La Pinta is owned by the Mayan Palace resort; the western portion is private property and in the northeast there are two ejidos: El Tiburón and La Pinta. Some vacation residences for tourists are found along the outer sand bar. The main threats to this estuary are the construction of a golf course and a marina and the damming and dredging of the estuary.

11.2.3 Bahía San Jorge Wetland Complex

The Bahía San Jorge wetland system covers 12,198.74 ha with a coastline of approximately 38 km (23 miles) and includes eight types of habitats: estuaries, tidal mudflats, salt marshes, coastal dunes, sandy beaches, permanent shallow marine waters and wetland-terrestrial ecotones (Ramsar 2010). It is separated from the open sea by a dune system approximately 10 km long (Zuria and Mellink 2005).

This wetland complex integrates three estuaries (La Cinita, Almejas-La Salina and San Francisquito (Fig. 11.2) that are important for their biodiversity and their role as sites for breeding, feeding and life cycle development for marine and terrestrial species (Glenn et al. 2006). The sand bar that separates the bay from the open ocean is an important nesting site for one of northwestern Mexico's largest colonies of the protected Least Tern (*Sternula antillarum browni*) (Zuria and Mellink 2005). The bay's estuaries are used as feeding grounds, reproduction or other life cycle needs for many species of resident and visiting fish species (Zuria and Mellink 2005; Iris-Maldonado 2011).

The primary estuary of Bahía San Jorge, La Salina, is located near the Ejido Roberto Campodónico. It has large shallow subtidal areas and marshes. Approximately 151 species of migratory and resident birds have been identified in this wetland (Mellink and Palacios 1993; Zuria-Jordan 1996); 12 of these species are protected. Seven endemic species from the Gulf of California can be found here: the short-jawed mudsucker, (*Gillichthys seta*) (Barlow 1961), Gulf anchovy, (*Anchoa mundeoloides*) (Iris-Maldonado 2011), false grunion (*Colpichthys regis*) and Gulf grunion (*Leuresthes sardine*) (Zuria and Mellink 2005); the plants wild wheat (*Distichlis palmeri*) and saltwort (*Suaeda puertopeñascoa*) (Felger 2000); and the fishing bat, (*Myotis vivesi*) (Arriaga et al. 2000).

La Salina estuary has deeper channels than other estuaries in the region; water remains in the main channels even during low tide, making it an important area for reproduction and nursery of commercially important species such as crab, shrimp, clams and flounder (Brusca et al. 2006). Fifty species of plants are found in the surrounding terrestrial area, notably the endemic and protected senita cactus (*Lophocereus shotii*) and the saguaro (*Carnegiea gigantea*).

The Ejido Rodolfo Campodónico, also known as the Bahía San Jorge community, is located near the bay. Its members are very unified and organized into one fishing cooperative which integrates 30 small-scale boats or pangas. Fishing is their primary economic activity and crab, flounder, and rays are the most relevant fisheries (Downton-Hoffmann et al. 2013). During closed fishing seasons ejido members work in agricultural fields and a gold mine that was reactivated in 2010.

The La Pinta gold mine, 18 km east of Bahía San Jorge in the nearby Sierra Pinta mountain range, is a growing concern to nearby communities, as it uses cyanide to extract the gold. Cyanide is a toxic material that can potentially leak into the soil and groundwater, and might contaminate the wetlands and damage oyster farming and fishing. Before operations were begun the mine presented an Environmental Impact Assessment for construction as required, but no public consultation process took place to analyze if the potential impacts were adequately addressed and mitigated. Many clandestine mines have popped up in the surrounding desert plains, even closer to the wetland systems than the primary operation. These are likely not following any environmental protection protocols.

La Cinita and Almejas estuaries each have an oyster farm managed by local cooperatives, who also collect clams. Both La Cinita and Almejas cooperatives sell their products in Puerto Peñasco and Caborca, the largest communities nearby. La Cinita is a settlement of 20 ejido members. Between 2005 and 2007 CONANP offered them support to construct a laboratory for production of oyster larvae. The Cinita cooperative is in the process of certifying their oysters as an organic product and recently they have formalized a social organization called Unión de Productores Ostrícolas, AC. Two restaurants on the main highway connecting Puerto Peñasco to Caborca sell oysters supplied by the farms, but the owners are independent.

In summary, we found that wetlands in the Peñasco Corridor are under various kinds and degrees of threats. Land and development speculation have increased pressure to change land use, spurring the sale of ejido lands to developers, and in some cases resulting in significant wetland habitat loss. Nonetheless various opportunities for conservation of wetlands were presented at different sites, and the approach taken in each wetland depended on the type and status of land ownership, land use regulations and economic interests of stakeholders.

11.3 Participatory Processes for Evaluating Development Projects in Critical Habitats

In 2007, 22 tourism infrastructure projects ranging from hotels, condominiums, golf courses and marinas, were proposed for development at Puerto Peñasco, representing a threat to beaches, dunes, wetlands and rocky reefs. Such projects are required to submit an Environmental Impact Assessment or EIA (Manifestación de Impacto Ambiental, MIA in Spanish), and by law the public has an opportunity to review and comment on the impact of a project. For many of the projects presented in the Peñasco Corridor public consultations were requested and organized to provide diverse opinions on the impact of a project to authorities.

Mexico started to use Environmental Impact Assessments in 1977, but it was not until 1988 that all development projects were required by law to submit an EIA to the Secretariat of the Environment and Natural Resources (Secretaría del Medio Ambiente y Recursos Naturales, SEMARNAT) before starting construction. When a developer submits an EIA, SEMARNAT then evaluates potential impacts and determines whether those impacts represent significant threats to the environment and are in some way in violation of existing regulations. The evaluation process can include a public consultation (if requested) and a public hearing (if requested and authorized), which provide opportunities for interested parties to review and comment on the EIA and the impact of a project. At the end of the evaluation process, the environmental authorities can approve the project as presented, approve it with specific conditions or reject it. In cases where a project is authorized but there is reason to believe that the approval was outside the boundaries of the law, there are other opportunities for stakeholders to challenge the authorization.

There are still many challenges for using the EIA as an effective conservation and sustainable development tool. For example, developers often do not adequately analyze the project's real ecological impacts, nor do they present mitigation proposals consistent with realistic environmental objectives. Also, EIAs rarely take into account cumulative impacts, so they do not address long-term ecological damage. In the context of EIA there is no generally accepted practice for quantitative assessment of goods and services to help evaluate the degree of an impact, except in cases where the action in question causes a damage that is a measurable quantity, such as the concentration of a contaminant (DOF 1988). Finally, environmental authorities do not always have all the necessary capacities to properly evaluate EIA and decisions are often discretionary. Coordination with other agencies and the capacity to monitor a project are also lacking. Even with these shortcomings, there are many benefits to EIA when the public actively and professionally participates in the process.

In response to the development and potential destruction of critical habitats in the Peñasco Corridor, a strategy was developed and initiated that included on the ground surveillance of construction activities, monitoring electronic presentation of EIAs, and reviewing their content. Where high impact projects were presented public consultations and hearings were requested that involved local fishers, tourists, scientists and legal and environmental groups in the evaluations. Technical comments on environmental and legal issues were integrated and submitted to support a thorough evaluation of the project by the authorities. In some cases, these comments requested that the project authorization be rejected, while in others mitigation actions to minimize impacts were suggested. Here we present a case study of successful wetland protection using this tool.

In 2006 Mexico's Federal Prosecutor for Environmental Protection, PROFEPA (Procuraduría Federal de Protección al Ambiente) was advised about a channel being proposed for construction as part of a residential marina project near the La Cholla estuary that did not have an EIA. PROFEPA detained the project requiring the developer to present an EIA to environmental authorities. Once the EIA was submitted a public consultation and a hearing were requested to review the project. The project proposed dredging 235 ha to build a residential marina in the La Cholla estuary, located within the Upper Gulf Biosphere Reserve. Public participation processes successfully integrated comments from researchers, commercial divers, whose fishing grounds were being threatened, members of the adjacent La Cholla community, and environmental and legal (Northwest Environmental Defense-DAN and Mexican Environmental Law Center-CEMDA) civic organizations. The result of this public evaluation was submitted to the authorities stating that the EIA under estimated the severe environmental impacts that marina construction would have, that the project would violate several laws, and requested that the project be rejected. This broad public participation was a key factor to the rejection of the project by environmental authorities.

Shortly after the resolution was given, the developer submitted a new EIA for the project with some modifications. This time the environmental authorities approved the project conditionally, but did not authorize construction of a residential marina on the estuary. This meant that critical wetland habitat within the UGBR had been successfully protected and La Cholla estuary could continue to provide important environmental services.

Dredging on the rocky reef fishing grounds which were outside the Reserve, however, was authorized. The resolution was appealed to force authorities to conduct a deeper analysis of the potential environmental damage of the rocky reef habitat and fishing activities. In 2009, the EIA was eventually rejected. In 2012 a new version of the project was presented for approval by the State of Sonora Tourism Bureau. The project proposed construction of a Home Port for the arrival of a projected 48 cruise ships per year, carrying 48,000 tourists (Villacaña-Yepéz 2012). Though this new Home Port does not propose dredging of the rocky habitats, it includes construction of a seawall that could alter sedimentation patterns and impact important fishing grounds. A new public consultation was organized with input from various NGOs and the local community, but this time the project was approved with some modifications and conditions.

In general, EIAs are meant to detect and prevent possible impacts to the environment including the loss of habitats such as wetlands and reefs, critical for reproduction and recruitment of species of commercial importance for the community. As seen in this section, public scrutiny is a vital element to assure that the tool serves its purpose. Maintaining the needed level of surveillance, participation and the technical capacity to conduct thorough external evaluations and legal follow-up is extremely costly, however, and this tool should be considered only as a short-term solution for protection of key habitats. Training local stakeholders to be involved in such processes can offer needed vigilance and support. These actions need to be combined with long-term legal conservation tools and social participation strategies.

11.4 Legal Tools for Wetland Conservation

Although there are numerous laws, regulations and standards that contribute to the protection, conservation and sustainable use of wetlands, Mexico does not have specific legislation for non-mangrove wetlands. The establishment of natural protected areas and fisheries reserves under the environmental or fisheries laws and the designation of Ramsar sites under the Convention of Wetlands of International Importance are some of the ways to increase long-term legal protection for wetlands. These instruments operate under the framework of a management program, a guiding instrument for planning and regulation that establishes the activities, actions and basic guidelines for management and administration. These can be used in combination with other instruments, such as Federal maritime zone concessions and destination agreements, which can restrict use in the coastal zone for conservation purposes. Ecological zoning is an additional planning tool that can be used for coastal conservation of wetlands by defining use restrictions for critical sites. All of these instruments, except fisheries refuges as defined by SAGARPA, are currently being used in the Upper Gulf of California and Peñasco Corridor as a way to offer protection for wetlands.

11.4.1 Reserves: A Tool for Wetland and Fisheries Management

The legal framework to establish marine reserves, refuges or more generally any spatial zoning for the promotion of sustainable fisheries integrates a wide range of regulations including laws, norms, rules, management programs and procedures. The use of these instruments requires development of complex, multi-stakeholder and multi-sector strategies that include significant socioeconomic, scientific and legal components to ensure the preservation of species and the livelihoods of fishers, and their communities (Uribe et al. 2010; Ruiz-Lopez 2009). In addition to official reserves, there are "indirect reserves", that are not officially registered, nor do they have a management plan. By the nature of their remoteness, difficulty of access or travel restrictions such areas have been restored, preserved, and repopulated indirectly, independent of any direct conservation actions.

In the Upper Gulf of California spatial management instruments have been used since 1955, when the Mexican government established a refuge zone at the mouth of the Colorado River, extending from Bahía Ometepec, Baja California to the mouth of the Colorado River near El Golfo de Santa Clara, Sonora (DOF 1955a, b). This designation highlighted the importance of this vast wetland as a spawning habitat for commercial fisheries such as totoaba (*Totoaba macdonaldi*) and shrimp. Additional refuge areas were established in 1974, but there was very little enforcement and no recognition of these instruments by local communities.

In 1993 the Upper Gulf of California and Colorado River Delta Biosphere Reserve (UGBR) was established by Presidential decree to protect this important spawning habitat and other key components of biodiversity (DOF 1993). This is by far the most important legal instrument at work in the region today and the one with the greatest impact on the region's communities and their fishing activities (Cudney-Bueno and Turk-Boyer 1998; Rodríguez and Bracamontes 2008). Established as a Natural Protected Area the UGBR follows the model of the United Nations Man and the Biosphere Program. The area is managed by the National Commission of Protected Areas (CONANP), which operates under the Secretary of the Environment and Natural Resources (SEMARNAT). The model sets aside a core area for conservation of the most important resources with very restricted use, while the outlying buffer zone allows for controlled use, research and other activities in order to achieve a balance between protection and use (DOF 1988, 2000). This category of protected area is designated as a place to learn how to use resources in sustainable ways for the benefit of local communities. In 1995 the first Management Program for the Reserve was published, providing a guide for integrating policies and actions for the conservation, protection and controlled use of the Reserve's resources. The program was revised and updated in 2007 (SEMARNAP 1995; CONANP 2007).

At the time the Upper Gulf Reserve was established it was thought that the core zone at the mouth of the Colorado River was the primary habitat of the critically endangered vaquita porpoise (*Phocoena sinus*), as well as the primary spawning habitat for the endangered totoaba population, and other commercially important species. Fishing was legally restricted from this area, but allowed in outlying areas. In the subsequent decade however, the scientific community learned more about the vaquita's population distribution (Jaramillo-Legorreta et al. 1999) and status (567 individual in 1997) and the species was listed as critically endangered by the International Union for the Conservation of Nature (IUCN 2004), the Convention on International Trade in Endangered Species of Wild Flora and Fauna (CITES 2011), and the Mexican Endangered Species Act (NOM-059-SEMARNAT-1994) under the category of endangered. The Vaquita Refuge was created in 2005 to address the growing concern for conservation of the vaquita porpoise; setting aside as a no-fishing zone the area of greatest concentration of the species, as it was known at the time (DOF 2005a, b). The Management Program for the Refuge establishes the rules and general guidelines for use of this area and promotes a series of measures and mechanisms to regulate fishing activities and develop productive activities compatible with the conservation and recovery of the vaquita.

The activities to protect the vaquita are coordinated through an integrated "Program for Conservation of the Species: Vaquita", known as PACE Vaquita (SEMARNAT 2008). The program operates as a regulatory mechanism and limits fishing effort through a variety of programs. The PACE Vaquita has offered voluntary buyout programs for fishers in the communities of the Upper Gulf Reserve with three options, technological and productive conversion projects, technological innovation projects, and a fourth option which gives payments to fishers for conservation of biodiversity (SEMARNAT 2008). The goal of all programs is to reduce and eventually eliminate the primary cause of mortality of vaquita, incidental capture in gillnets (Avila-Forcada et al. 2012).

When the General Law for Ecological Equilibrium (LGEEPA) was revised in 2000, new regulations affected the implementation of fisheries policies in protected areas in Mexico, specifically outlawing fisheries with bycatch greater than 1:1, and fisheries that caused destruction of sea floor habitat. Environmental Impact Assessments would be required to address the impacts of fishing on many components of biodiversity. With these laws in place a number of actions have been taken to change the dynamic of fishing within the context of the Upper Gulf of California. Fisheries management in the Upper Gulf Reserve communities has also evolved in the context of conservation of vaquita and the restrictions and opportunities that were offered through the Vaquita Refuge, PACE, and the UGBR. It has also impacted fisheries management in the Peñasco Corridor, as fishing grounds of the community of Puerto Peñasco lie both inside and outside of the Reserve.

The Secretary of the Environment and Natural Resources and the Secretary of Agriculture, Livestock, Rural Development, Fisheries and Food (Secretaría de Agricultura, Ganadería, Desarrollo Rural y Alimentación, SAGARPA), signed an agreement to work together to implement the PACE Vaquita. With the publication of this Agreement, the National Commission of Aquaculture and Fisheries (Comisión Nacional de Acuacultura y Pesca, CONAPESCA), is using various tools in fisheries management to contribute to the conservation and responsible use of the fisheries resources within the UGBR. In addition to environmental laws, SAGARPA enacted the Ley General de Pesca y Acuacultura Sustentable (General Law for Sustainable Fisheries and Aquaculture), LGPAS in 2007, which offers general guidelines for the ordering of fisheries.

The LGPAS allows for establishment of Refuge Zones for fisheries and defines them as delineated areas in Mexico's federal waters to conserve and contribute to the development of fishery resources through their reproduction, growth or recruitment, and to preserve and protect the surrounding environment (DOF 2007). In 2012, the LGPAS was modified to permit the decree of fisheries refuges and in November 2012, the first fishing refuge network was established (DOF 2012). This is a management tool under consideration for management of critical habitats in the Peñasco Corridor and potentially for offshore extensions of wetland habitats.

11.4.2 Designation of Ramsar Sites

The Convention on Wetlands of International Importance (also known as the Ramsar Convention) sets a framework for national action and international cooperation for wetland conservation (Ramsar 1996). Mexico joined the Convention in 1986. As a party to this agreement, Mexico is committed to: (1) propose wetlands for inclusion in the Ramsar list according to current criteria, and guarantee their effective management; (2) implement policies, management and legislative actions that promote rational wetland use; and (3) cooperate in trans-border wetland issues (Travieso-Bello 2009). The Ramsar Secretariat provides support in policy and legislative measures, training, public participation, and funding for various projects including wetland restoration (Astrálaga 2006). In accordance with the Convention's strategic plan, Mexico has established the National Wetlands Council to share information with stakeholders and to promote wetland conservation and to implement a national strategy on communication, education, and participation to support application of the Ramsar Convention (CONANP 2012).

The Ramsar Convention considers a broad definition for wetlands, which encompasses coastal areas up to 6 m deep during low tide, continental wetlands, and man-made sites. Wetlands are designated as Ramsar sites according to criteria in five categories, whether wetlands are unique, rare or representative, criteria based on species and ecological communities, and criteria specific to water birds, fish, and other species (Ramsar 2008). Two criteria specifically address fish populations: (1) the wetland should support a significant proportion of indigenous fish, life-history stages, species interactions, or populations representative of wetland values and that contribute to global biodiversity; and (2) the wetland should be an important source of food for fishes, spawning ground, nursery, or a migration path. Additionally, the criteria for species and ecological communities can also include fish species, such as rare or threatened species and those that contribute to the biological diversity of a particular biogeographic region. Although not specifically addressed in the designation criteria, member states are encouraged by the guidelines to emphasize the importance of wetlands in sustaining coastal fisheries.

To date, 138 wetlands have been declared Ramsar sites in Mexico; including four coastal wetlands along the coast of Northern Sonora (Fig. 11.1): wetlands of the Colorado River Delta (250,000 ha extending into the state of Baja California + 127,614 ha of riparian wetlands added a decade later); Wetlands of Bahía Adair (42,429 ha); Canal del Infiernillo and esteros in the Comcaac region (Xepe Coosot) (29,700 ha) and Wetlands of Bahía San Jorge (12,197 ha) (CONANP 2012). The importance of coastal wetlands to fish populations has been used to justify the designation of the sites in northern Sonora as Wetlands of International Importance, with a particular emphasis on their importance as nurseries for species of commercial importance (Iris-Maldonado 2011) and their socioeconomic importance for local communities dependent on fisheries (Cudney-Bueno and Turk-Boyer 1998).

The Bahía Adair wetland complex, within the Upper Gulf Biosphere Reserve (UGBR), was designated a Ramsar site in 2009, while the designation of Bahía San Jorge was granted in 2010. This was the first formal step towards recognizing the environmental value of the San Jorge wetland area. Both cases serve as examples to describe the establishment and planning process for Ramsar sites in Mexico.

The first step for establishing a Ramsar site is to document the importance of the wetland following the Ramsar application process. For these two wetlands, vulnerable, endangered or critically endangered and threatened ecological communities were identified, as well as plant and animal species important for the maintenance of biological diversity, especially where critical life stages occur. Native fish species, their life stages and their interactions with other species and representative populations that contribute to the biodiversity of the region were also identified along with nutrient sources for fish, spawning areas, areas of fish maturation and growth, and migration routes that fish stocks depend upon whether within or outside the wetlands.

In addition to characterizing the flora and fauna, key social and cultural values associated with these wetlands and the status of land tenure and current land use were also identified, all important information for determining their suitability for long-term preservation. This process helped assess the threats to the wetlands and the conservation status. Information regarding ongoing communication and education activities and level of awareness about the importance of wetlands of the primary stakeholders was also collected. Once the application is completed and the site complies with the Ramsar criteria, the National Commission of Protected Areas (CONANP) proposes the designation of a given area to the Convention, who then evaluates the application. If the criteria are met, the site is incorporated into the List of Wetlands of International Importance.

Once registered under the Convention, Ramsar sites within already established natural protected areas (NPAs) are managed using existing tools, such as the reserve's establishment decree and, if it exists, the reserve's management program, providing great legal certainty for such wetlands. Sites outside the boundaries of existing NPAs are subject to guidelines and resolutions under the convention, in addition to national, state and municipal regulations and ecological ordinances (CONANP 2013).

All federally designated Natural Protected Areas in Mexico and federal conservation areas, such as Ramsar Sites, require management and conservation plans to guide management and use. These plans support ecosystem and biodiversity conservation and are an integral part of the process for setting conservation goals within protected areas (CONANP s. f.). Plans must define the technical and legal guidelines that regulate the creation and fulfillment of conservation objectives and include stakeholder needs and perspectives (García-Frapolli et al. 2009). They should specify which activities are allowed within the area (i.e. zoning), the management framework (in accordance with local, state, national and international policies), strategies needed for environmental protection (Travieso-Bello 2009), and the program to monitor its progress.

Plans are generally structured to cover three main areas: (a) a description of the biophysical and socioeconomic characteristics of the site; (b) the management and conservation program, which defines objectives, activities, and operations such as monitoring, research, education, social development, and addressing knowledge gaps, among others; and (c) the action plan that details how management and conservation actions will be implemented. The plan also includes an analysis of existing challenges and management options and establishes administrative rules for individual management units and proposed zoning (usually within the management and conservation program section). In the case of Ramsar sites, the development of management and conservation plans is part of the Ramsar Convention guidelines (Ramsar 2008).

The Ramsar designation can be a management driver at the national and local level, used to leverage, encourage, and promote citizen participation, economic benefits, and conservation, as occurs in other countries (Fletcher et al. 2011), as well as for future management planning (CONANP 2013). During the preparation of the management and conservation plan for the Bahía Adair Ramsar site in 2010 and 2011, a robust process to consolidate social participation was conducted. Funding for the planning process was provided by the Commission of Ecology and Sustainable Development of the State of Sonora (CEDES), following SEMARNAT guidelines. A group of five ejido members, that later integrated into a Network for the Wetlands of Bahía Adair (REHBA, Red de Humedales de Bahía Adair), were key in actively engaging local ejidos in management of this wetland and implementation of numerous projects, such as a monitoring of water birds and creation of an educational center; they also promoted linkages between the landowners, scientists, managers and public and private institutions.

The Bahía Adair wetlands are under the jurisdiction of the Upper Gulf Biosphere Reserve's management program, so the Ramsar management plan must be consistent with the Reserve's plan. When the UGBR management plan is next updated (supposedly every 5 years), the Ramsar site plan can be incorporated into that instrument and therefore obtain legal status. Because of this overlap, management approaches for both instruments must incorporate both state and federal environmental authorities, in addition to local actors.

Bahía Adair is Mexico's first Ramsar site to have a management program and to include components of climate change. It is actively being used by local stakeholders even before it has been legally formalized within the context of the UGBR. In the Bahía San Jorge wetland the management planning process is following a similar approach as in Bahía Adair. Since this wetland is not located within an existing natural protected area it will require a different mechanism for legal formalization, a process which is still unclear. Nonetheless, when locals are actively engaged in the Ramsar site and developing a plan for its use, its implementation doesn't need to wait for formalization. Local environmental organizations can add value to the process by actively promoting its application by local actors and maintaining continuity in processes.

The strength of Ramsar designations as a legal tool for conservation of wetlands has not yet been tested. Without legal registry of some sort, its strength may be limited, making it important to use in conjunction with other instruments. In the case of Bahía San Jorge, where no other legal tools exist, the Ramsar designation gives CONANP a legal interest which facilitates their obtaining federal zone rights, as described below.

11.4.3 The Federal Maritime Zone

Coastal areas represent interfaces between marine and terrestrial ecosystems of biological and economic importance. Changes in land use in coastal areas for urban developments such as construction of ports, tourist and aquaculture infrastructure, and other uses, can disrupt the environmental services offered by dunes and wetlands, salt marshes, mangroves and other native vegetation. The construction of dams and deforestation of land for agriculture and livestock in the upper basin, create downstream problems by altering runoff sedimentation, provoking saltwater intrusion and contamination of aquifers. Coastal dynamics are also affected by the construction of piers, breakwaters and other marine infrastructure.

Tools are needed to reduce the pressure and impacts of these activities on wetlands, dunes, beaches and the coastal area in general. Although the legal framework of the coastal-marine area (the General Law of National Property or Ley General de Bienes Nacionales, LGBN) establishes that the coastal zone is federal property and of public interest (DOF 2004), economic interests complicate management of this zone, especially for conservation and environmental matters.

The LGBN empowers SEMARNAT to define the boundaries of the coastal zone and to exercise possession, control and management (including conservation). This law provides the basis for making collaborative arrangements with state and municipal governments, as well as with private entities so that they can use, manage, conserve and monitor the coastal zone. Embedded within the LGBN the legal concept of the Federal Maritime Zone offers a conservation tool for coastal ecosystems.

The Federal Maritime Zone (FMZ) is defined as the walkable strip of land extending inland 20 m from the high tide line along Mexico's coasts. The law provides for the use, development, exploitation and restoration of FMZ areas by individuals and institutions or other agencies through the following legal mechanisms:

- Concession Title: It is granted by SEMARNAT for exclusive use and exploitation of the FMZ, for a specific period of time, which can be extended.
- Permit: A document issued by SEMARNAT which covers the use of any body of seawater in the FMZ for a specific activity for a period of one year. It is not renewable, but reapplications can be made.
- Destination Agreement: document issued by SEMARNAT for the use and benefit of the FMZ by federal, state

or municipal governmental agencies. It has no expiration date and is not transferable. Those holding it cannot engage in any activity that is not specifically authorized. (DOF 2004)

Federal Zone Concession Titles are regularly requested for coastal zones because of their scenic value. This is especially true for tourism related activities and development of infrastructure which can have irreversible negative impacts on coastal ecosystems. Civil Society and Non-government Organizations or NGO's can also apply for concessions for conservation purposes. This tool is particularly useful for threatened wetlands where no other legal conservation instruments are in place, as it can limit adjacent land use. Concession Titles have a 15-year term and can be renewed. The use of FMZ Concession Titles as a conservation tool is very recent and as a consequence there are still many challenges to their successful implementation, but in the last 5 years efforts to increase their potential as conservation instruments have increased. Some of the limitations relate to the costs required to delimit areas, limited governmental capacity to process requests for large areas and the limited capacity of the beneficiary to comply with the obligations conferred by the Concession Title. It is essential to prioritize sites and to create effective linkages between Concession Titles and other management tools such as terrestrial and marine ecological zoning programs, NPA management plans or municipal development plans. To date, several conservation organizations have managed to secure concessions to protect and/or restore marshes, wetlands, beaches, mangrove sites, and areas for nesting birds and sea turtles.

Another way to protect the Federal Zone adjacent to wetlands is through the use of Destination Agreements. SEMARNAT may allocate the use, development or rights to benefit from federal zone property to another government entity or agency at the federal, state or municipal level. When the Agreement is given for conservation purposes, it can strengthen the conservation of adjacent ecosystems and help maintain the integrity of the environment. Destination Agreements can be co-managed by developing an agreement between the government entity that it has been issued to and another government entity or organization. The title holder can establish an agreement for use of a concession or permit by another party as long as the original purpose of the title is respected. Any modification of the title should be issued through a formal authorization. Although Destination Agreements for Federal Zone can only be assigned to government agencies, NGOs can play a key role in promoting agreements for conservation, integrating the necessary documentation and conducting field work to support the preservation of the Federal Zone, strengthening the conservation regime of coastal habitats within and outside Natural Protected Area (NPA). In addition, Destination Agreements are first in order of priority of all the Federal Maritime Zone instruments that grant titles to individuals, and the validity of the rights granted are not limited in time.

11.4.3.1 Destination Agreements in the Upper Gulf Reserve and Puerto Peñasco Municipality

The strategy selected to increase the conservation regime of Bahía Adair wetlands was to request Destination Agreements in the name of CONANP for the adjacent Federal Zone. This instrument was chosen because CONANP is the management body in charge of the Upper Gulf Biosphere Reserve where Bahía Adair is located. Also, because of the area's enormous size, it would be difficult for an NGO to obtain and properly manage Concession Titles.

Nine Destination Agreements totaling 153 linear kilometers (approximately 308.5 ha) adjacent to Bahía Adair were applied for jointly with CONANP and the DG-ZOFEMATAC (an office within SEMARNAT that sets the general scheme for granting rights of use of FMZ to individuals and corporations). In the Upper Gulf Reserve the FMZ overlaps with the boundaries of several private lands, so when the applications were processed, numerous conflicts over property rights ensued. It was decided not to include La Cholla and Cerro Prieto, the southern-most estuaries of Bahía Adair, in the application because they have already been heavily disturbed by the proximity of real estate projects and golf courses. The applications were submitted in July 2009 and were approved and published in the Official Journal of the Federation, DOF (Diario Oficial de la Federación) between November 2011 and July 2012, a 2 to 3 year process.

The Destination Agreement tool was also chosen to protect 68,257 m^2 in Estero Morúa in 2010, in this case on behalf of the city of Puerto Peñasco. The Federal Zone of interest consisted of a strip of land adjacent to the mouth of the estuary that represents important nesting habitat of the Least Tern (*Sternulla antillarum browni*). The Agreement was requested to protect the Least Tern, but it also serves as a preventive measure against potential development projects that could alter the hydrology of the estuary.

11.4.3.2 Concession Titles at Bahía San Jorge

The first step in the strategy to protect the Federal Zone at Bahía San Jorge was to obtain Concession Titles. This private-public (or mixed) tool was requested to conduct environmental education activities and at the same time to regulate the use of key areas in the wetland complex. Concession Titles were promoted for two areas that total 4.45 ha (2.22 linear kilometers of FMZ). Applying for a small area had the advantage that in the long term, the costs of monitoring and complying with the conservation and education requirements are lower than they would be for a larger area, and can be challenging for an NGO.

Concessions for some of the other sections of the Bahía San Jorge FMZ were already in the application process or had already been granted to several private and community groups with different objectives, including the development of oyster and shrimp farming activities. The procedure for a Concession Title application is the same as for a Destination Agreement. The process took 2 years with applications made in July 2009 and with notification of approval and the assignment of the Concession Titles granted in early 2011.

FMZ authorities began requesting that all of the NGOs that had been granted Concessions Titles in coastal zones pay the stipulated fees to retain them. The Federal Rights Law, (Ley Federal de Derechos, LFD) was contradictory on the issue of taxes: on the one hand the law states that FMZ concessions granted for conservation activities would be exempt from paying taxes and on the other hand, the conservation activities allowed, were limited to reforestation (DOF 1981). Since estuaries in the northern Gulf do not have mangrove forests that can be reforested, this seemed to limit the use of this tool for conservation. These discrepancies gave rise to a dispute between NGOs and the Federal Congress over expansion of the concept of conservation activities so as not to limit them to reforestation. Meanwhile NGOs who had been granted Concession Titles were advised by CEMDA and DAN (two legal environmental civil organizations) to not make such payments during the dispute. They also warned NGOs that even if the outcome of the dispute favored the exemption of taxes, they would still incur a debt that could not be cancelled, starting from the moment the Concession Titles were accepted. In December 2011 the dispute was resolved in favor of non-profit associations. The concept of conservation activities within the law, LFD, was redefined to include conservation activities in addition to reforestation, helping to exempt the tax payment for NGOs (DOF 2011).

The next steps in the strategy for conservation of Bahía San Jorge consisted of promoting its establishment as a Ramsar site (as described above) and then supporting CONANP to obtain Destination Agreements for the available Federal Zone. The area originally included in the application for a Concession Title was later incorporated into the application for a Destination Agreement for CONANP covering 254.7 ha (127 linear kilometers of FMZ) at Bahía San Jorge. Once this wetland became a Ramsar site, CONANP acquired direct legal responsibility for its management, which facilitated their application for the FMZ under a Destination Agreement. When the Agreement is obtained it will facilitate conservation and management of this bay.

11.4.4 Ecological Zoning Programs

Ecological Zoning Programs are a type of coastal management or planning tool that seeks to balance productive activities and the protection of natural resources. In the Upper Gulf of California there are two zoning programs in effect: the Gulf of California Marine Ecological Ordinance Program which was enacted in 2006, and the Regional Ordinance Program for the Sonoran Coast, which was published in 2009. These programs analyze the compatibility between different activities that have spatial overlap and that could be the source of potential environmental conflict, to determine the most suitable economic activity for a given area. These programs promote strategies and lines of action for use and conservation of resources. They propose zoning that is binding at all three levels of the government. If a project requires an alternative activity not permitted according to the zoning, the promoters must apply for a change in land use from the appropriate authorities, usually at the municipal level, where the zoning is much more specific, but only if the municipal government has a development plan. The zoning programs can be strengthened when coupled with existing legal documents. These government-led programs were created with extensive consultation with the various sectors of the communities involved.

Though these various tools have produced some positive results for conservation of marine and coastal resources, their implementation has not been easy. The designations and rules for their operation have limited traditional activities of local communities, restricting their use in time and space, and have occurred mostly without adequate development of alternatives to respond to the economic needs of local communities.

Some of these challenges are being addressed by promoting social participation processes that empower coastal communities in the northern Gulf of California with the knowledge and tools needed to create sustainable livelihoods in line with protection of the natural environment. Processes involve creating partnerships between coastal communities, fishers and authorities to engage them in the responsible management of resources, facilitating their active participation in the legal tools described above and providing them with alternative economic options for sustainable use of resources.

11.5 Social Participation Processes in the Upper Gulf of California

11.5.1 Background

When the UGBR was established in 1993, there was limited input from affected communities. At that time very little was known about small scale fishing activities; the sector was not very organized, and as a consequence they were not involved in any way. On the other hand, the proposal for establishing the Reserve was presented to industrial shrimp fishers at

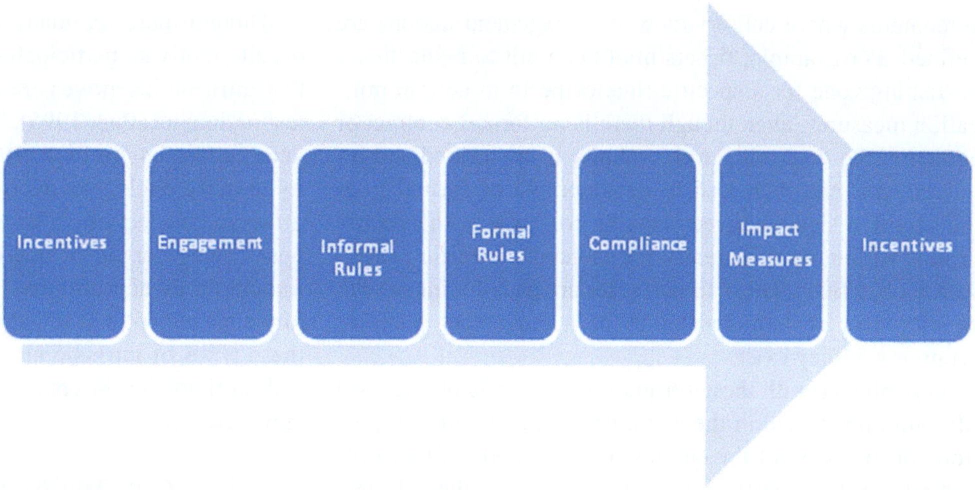

Fig. 11.3 Model social participation for resource management in the Northern Gulf of California

Puerto Peñasco and because of a crisis in shrimp production that began in 1990, this group agreed to the Reserve's establishment and signed the proposal, in hopes it would offer improved management for this important fishery. The industrial shrimp fishers also participated in a series of community meetings to develop the management plan for the Reserve.

In recognition of the need to engage small-scale fishers in management processes, a Community Action Program and research project was initiated in 1996 to engage this fishing sector. The result was publication of "Pescando Entre Mareas" (Cudney Bueno and Turk Boyer 1998), which was the first in depth description of coastal fisheries in the area. It involved fishers in sharing their experiences about fishing, describing what, where, and when they fish, as well as their concerns for management. Based on what was learned from this project, strategic initiatives were developed with this sector to help them develop sustainable fisheries.

As the pressure to protect the vaquita porpoise increased in the Upper Gulf Reserve and the fisheries regulations became stricter, conflict ensued (Turk-Boyer 2002). An emergent law was passed by SEMARNAT in 2002 that prohibited industrial fishers from entering the Reserve. After protests and highway closings the conflicts were resolved and industrial fishers were allowed to fish in the Reserve for a shorter duration (DOF 2002). The resolution also prohibited fishing in the area by non-local communities, and required industrial fishers to work under an Environmental Impact Study. Small-scale fishers were not required to present such a study until 2009.

In 2005, a forum known as Alto Golfo Sustentable (AGS) emerged that brought various stakeholders together to participate actively in management issues around the vaquita (NACAP Vaquita 2008). While this group was organizing to begin to address issues such as illegal fishing and shrimp management, the government declared a new area overlapping the Reserve, known as the Vaquita Refuge, where fishing would be prohibited. Though fishers were not involved in the declaration of the refuge area, they did get involved in establishing a management program and objectives for the area (DOF 2005c). The management program dictated the formation of a committee called "Órgano de Evaluación y Seguimiento" or OES that would be responsible for defining indicators and evaluating the effectiveness of management (Turk-Boyer and Barrera 2012). The committee was formalized in 2008 and eventually replaced AGS as a forum for discussion and organizing actions around vaquita and related fisheries issues. Today OES, a cross-scale forum, brings all the actors to the table and has elevated the dialogue and collective action to resolve conservation and fisheries issues.

11.5.2 Social Participation Model

A model of social participation and inter-agency collaboration was elaborated to generate solutions to the challenges of creating sustainable Upper Gulf fisheries and for coastal conservation (Fig. 11.3). In general, the model consists of a series of steps that begins and ends with incentives. At the beginning incentives such as in-kind assistance or financial support are offered to community groups as a way to engage them in the conservation or management of their own resources. Workshops and meetings are conducted to strengthen local knowledge and involve communities in decision-making and management, with the goal of giving them a sense of stewardship of their own resources. Often stakeholders are involved in projects such as monitoring of their fisheries or other natural resources, as a way of developing a common dialogue and understanding. Project development helps consolidate commitments from communities to actively participate in improving resource management.

If communities get engaged and broad participation and consensus is achieved, it facilitates the development and implementation of informal rules for the responsible use and conservation of resources. Informal rules are usually verbal agreements where conservation or management actions are defined. For example, fishers might commit to respecting a no-fishing zone for a specific time or perform certain mitigation measures, even though there is no formal contract or agreement to do so. If the community is committed and involved, this can then lead to development of formal rules that involve signing agreements or contracts with specific groups within the community for achieving specific conservation objectives. These formal rules are usually linked with development or implementation of management plans or programs, if they exist.

Compliance with these formal rules depends on the level of commitment within the community, thus monitoring actions of members of the community is important. Once the community is committed and complying with these formal rules, then, it is possible to establish indicators to measure and demonstrate the effectiveness or impact of management actions, such as recovery of commercial species. If management is successful as shown by these metrics it can be linked to new incentives, such as access to new international green markets, which can result in economic benefits.

11.5.2.1 Incentives

A variety of incentives is needed and these have been used to initiate and maintain community involvement in co-management processes in the Peñasco Corridor. They include knowledge, participative and interpretive incentives, and economic and legal incentives, as categorized by Jones et al. (2011). While NGOs can work with local community groups to encourage their active participation using the above incentives, economic factors and market forces are also a key driver for stakeholders whose livelihoods depend directly on the use of natural resources. Helping fishers and land owners access finances for initiating projects for better management can be helpful. Ultimately market rewards for responsible use of resources can be a strong and sustainable type of incentive, but care must be taken to prove compliance to make sure such rewards are merited. Promoting alternative economic activities that reduce pressure on resources is also important, and requires involvement of partners with capacity in business development. Without the appropriate legal incentives or disincentives it is impossible to formalize functional management processes and these must be supplied by the government. Examples include providing fishers with legal access to resources through permits, and enforcing laws that limit access, as well as enforcement of regulations and conditions of an Environmental Impact Study or Ordinance Plans. A balance of these different types of incentives is called for to not only stimulate and reward participation, but also reward compliance. Negative incentives are also important, and these usually come in the form of application of the law, or as sanctions imposed within a community group for not following informal or formal rules and commitments.

Though there are many extrinsic incentives that can be used to motivate participation and change, there is evidence that intrinsic incentives are more effective and have a more lasting impact (Pink 1995). The three greatest intrinsic motivators according to Pink (1995) are: (1) autonomy (the ability to make one's own decisions); (2) mastery (the ability to grow and perfect one's knowledge and skills); and (3) purpose (being part of or making a contribution to something important, meaningful and bigger than oneself). If community engagement and social participation strategies can offer these types of intrinsic incentives the chances of achieving full participation in creating a functional management system increases.

11.5.2.2 Community Engagement and Participation

One of the most critical and challenging steps in this process is obtaining the meaningful participation of members of the community in the development and implementation of solutions to their environmental problems. If a local group is successfully engaged, they are able to make decisions and assume responsibility for the agreements they make. Active participation of the community is a form of empowerment and gives participants stewardship over the use of their resources. These are powerful intrinsic motivators.

A key component to our work with communities and leaders is to speak with complete honesty and clarity about goals. This helps build trust and forms the foundation for future collaborations. Community groups have distinct socio-economic situations and as such they respond to different types of incentives. Finding the right incentives at the right moment can keep communities engaged and processes advancing.

The most successful approaches to community empowerment employed in the Corridor have used interactive participatory processes for engaging communities in decision-making processes for resource management. The interactive method considers participation as a right of the community, where groups contribute their opinions, are involved in analyzing and solving problems and in decision-making based on their knowledge of their resources.

The process involves visiting each community to identify the principal actors, although this does not guarantee their involvement. Because participation is a skill that must be learned and mastered by the community, we train community members on how to recognize opportunities and threats to their resources and thus make better decisions. We also teach them how to organize and how to express their interests and concerns so they may be incorporated into management plans, programs or projects that are being promoted for their communities or groups.

In order to exercise their rights and be able to negotiate agreements, members of the community need to understand existing laws, regulations and rules that are required to achieve long term environmental benefits. They also must learn how to adopt, appropriate, comply with and enforce the final decisions they make. Additional training in planning, negotiation, and resolution of conflicts can strengthen the cohesion among members and increase their ability to work in groups. Once they are trained, organized and committed to participate and work in sustainable environmental projects they can then share their knowledge and experience with other groups.

We use a variety of participatory techniques in this training process, such as meetings, workshops, discussion groups, focus groups, semi-structured interviews, and participatory mapping. We also engage local users in monitoring their resources, as this gives them direct socio-ecological feedback and helps initiate in depth discussions and exchanges on the state of the resources (Cudney-Bueno et al. 2009). We participate in the internal meetings organized by ejidos and co-operatives to help organize their community, identify issues of concern, and identify key players and potential leaders in the management of coastal and marine resources. This offers a venue to explain our vision for a given project and for sustainable use of resources.

Since the adoption of participatory processes depends largely on the level of commitment and ability of communities, these processes often require a great deal of time. The participants, in this case, members of a community or group, pass from being potential actors to becoming active players in resource management. They become proactive, responsible and capable of establishing alliances and agreements for co-management of resources, sharing benefits and responsibilities effectively and equitably with other actors, whether institutions or government agencies or other community groups. The trust and credibility these community participants develop, strengthens and legitimizes the community engagement process (Kelleher 1999; IMM 2008).

11.5.2.2.1 Education and Training

A strong commitment to environmental education for community groups and for the communities at large has fortified the participation processes described in this model, creating informed and capable citizens. High quality publications have been disseminated to the region's stakeholders including fisheries species summaries (i.e. PANGAS 2012), bilingual field guides, informative brochures on endangered species and management instruments, such as Ramsar sites. Workshops with well-designed curriculum to teach about the ecosystem and management processes have also been used extensively. Radio spots and social media are used to remind fishers and landowners of good practices and commitments to responsible behavior in a timely manner. Specific targeted training programs are also given to stakeholders for their participation in monitoring programs, for example. Training programs have been given in subtidal, bird, sea lion, waste, human use and fisheries monitoring.

Participation of the community at large in support of responsible resource use can consolidate a commitment to responsible behavior by those directly involved in resource use. Watching children learn about resources, communicate their concerns, and actively search for solutions can be highly motivating. It gives stake-holders a sense that they are part of something important for their entire community.

A 5 year program with fifth graders in the communities of the Corridor reached over 5,600 children, many who are relatives of the community groups engaged in these management processes. Children visited local wetlands to observe first-hand the natural environment and learned about sustainable fisheries. An annual environmental contest implemented in the Corridor has also involved hundreds of youth each year in researching and proposing solutions to environmental issues. Themes have covered such topics as conservation of wetlands, migratory birds in wetlands, climate change and its impact on wetlands, and fishing for the future. These contests involve community leaders as judges. Teachers and parents also get involved in helping kids, who often conduct community-wide campaigns. Contest winners enjoy field trips to the region's protected areas and follow-up on their projects. Summer camps and ecology clubs offer opportunities for youth to continue developing their environmental interests and skills. Such programs and interchanges promote a community-wide environmental ethic and contribute to integration of the entire community towards a common vision of sustainability.

The involvement of women has been an extremely helpful vehicle for community outreach. Women are sometimes more open than men to reflect on conservation issues and they also show a genuine interest to live in better environmental conditions rooted in their concern for their children and families. Working with mothers is also one of the best ways to reach children. The relationships established with the mothers and women in the communities of the region has helped us better understand the structure of the community and has given us better access for achieving conservation goals. Oyster farming in the region's estuaries is done predominantly by women. Programs to involve women at the Campodónico community of San Jorge Bay have led them to create an innovative business for developing handcrafts with recycled materials and contribute to economic diversification and stabilization for their community.

11.5.2.2.2 Capacity Building for Community Leadership

Through different encounters with community groups two types of natural community leaders can be identified: administrative leaders and moral leaders. Identifying natural leaders of the fishing cooperatives, ejidos and women's groups has been a cornerstone for the success that has been achieved in the Social Participation Processes developed in the Peñasco Corridor. When leaders open up to talk about their visions, concerns and interests related to their land or sea, they give us a better understanding of the socio-economic structure of the community, which opens up opportunities to address environmental issues and to achieve conservation goals. Working with the identified leaders helps us analyze the interests and abilities of each group and determine their compatibility with other ejido groups and fishing cooperatives in their area. To obtain positive long-term results, natural leaders need to be involved in capacity building and leadership training.

11.5.2.2.3 Project Development and Formalization of Groups

Developing projects together can be a key step for consolidating leadership and group participation, as it gives focus to collective efforts and helps a group advance towards larger management goals. We offer training to leaders by involving them directly in the implementation of projects. It has been important to (1) offer project management advice, (2) assist in identifying consultants and trainers, (3) provide technical training, (4) advise on equipment needs, (5) arrange for tools and support materials and (6) accompany the group throughout the process.

Training generally covers two topics: "strengthening social organization" and "administration and management of productive projects." With this focus, groups then start to develop, budget and find funds for their own projects. Finally we strengthen leadership and empower groups through the exchange of experiences with other groups in the region and other parts of the country. The idea is to encourage them to share achievements, lead by example and participate with others in a continual learning process. Strengthening ties with government institutions and civil society also helps build leadership capacity. Community leaders meet with officials to learn how to best express their views, negotiate and engage in actions for conservation.

Sharing experiences with other like-minded groups and developing collaborative projects or partnerships has the effect of broadening horizons and increases potential for collective action. This has been observed at the two Ramsar sites, where community members at Bahía Adair have formalized a network among themselves, have begun sharing experiences with Bahía San Jorge ejido members, and are participating in national exchanges with a wetland network from Oaxaca.

Project development can also help in the consolidation of groups. In the Peñasco Corridor the Bahia Adair Wetland Network that participated in the Ramsar management process has now formalized as a civil association and is managing monitoring and ecotourism projects. Two informal groups were created at Bahia San Jorge during similar processes, with organization around projects. Mujeres Trabajando (Women Working) is a women's collective that is creating an innovative handcraft business and the Los Lobos (The Sea Lions) is a group of fishers at Bahia San Jorge that have committed to monitoring human use and sea lion populations at San Jorge Island, with the objective of advancing a marine conservation agreement on the use of the island.

As small projects evolve into larger ones that involve collaborations between groups and/or communities, new structures and forums are needed to address collective concerns. This will be one of the greatest challenges for achieving an integrated coastal management scheme for the wetlands of the region and for the ecosystem-based fisheries management concept. The deep knowledge about community structure, leadership and relationships between groups and communities, as well as the connectivity between resource use and biological populations as described in Chapter 9 of this volume, will greatly facilitate creation of such forums.

11.5.2.3 Developing Informal and Formal Management Rules and Instruments

Once groups are consolidated that share a common vision, progress can be made towards formalizing project or management objectives. Solidifying agreements by documenting them in meeting minutes and having members sign such agreements is powerful mechanism for solidifying their commitment.

When groups have the opportunity to participate directly in formal management processes, it serves to help develop their capacity, and empower them in their own learning processes. This can happen through participation in advisory boards, such as the Upper Gulf Reserve, or the Vaquita Refuge, or Subcommittees for managing fisheries resources. Through such forums local fishers or community members gain direct access to government authorities and interaction with NGOs and they gain a broader perspective on the issues at hand. Often only formally organized groups are allowed to participate in representation of their sector. There are few such cross-scale forums in existence in the region, but where they do occur, management issues are resolved and progress is made (Cudney Bueno et al. 2009), such as in the Vaquita Refuge (Turk-Boyer and Barrera 2012).

11.5.2.4 Compliance and Measuring Impacts

The lack of compliance with the basic legal tools that are in operation such as permits is prevalent. Several processes must work together to achieve compliance. First and foremost the rules must be clear and known by the actors involved. It is especially powerful if local stakeholders are involved in setting the rules, as this gives them a vested interest from the beginning. Secondly, authorities must enforce the rules. Proving compliance becomes an important factor for determining whether management efforts are having the desired impact or result and can provide powerful feedback to all the actors involved in the management process. A set of indicators have been proposed for monitoring in the Peñasco Corridor, see Chapter 9, Sect. 9.12, Table 9.6, and baseline data has been obtained for most of these.

11.5.2.5 Economic and Market-based Incentives

Market-based incentives can help drive compliance with management regulations and reduction of impacts on resources and move this participation model forward. In support of community efforts in the conservation and management of wetlands in the northern Sonora we have promoted development of various alternative economic activities including alternative fisheries, ecotourism, and a project for production of handcrafts using recycled materials. These projects often involve (1) training, (2) construction and/or acquisition of infrastructure and equipment, (3) development of business plans and (4) development and promotion of brands. Such alternatives are important incentives for local communities and ultimately will enhance the capacity of local communities to use resources in sustainable ways.

The NaturArte Ecotourism Corridor was created with a focus on wetland conservation. It offers estuary landowners a way to participate in the local tourism economy by offering services that have a low impact on wetland ecological function. It was conceived as a network of community groups involved in ecotourism activities, tied together through a brand that would offer participating groups support for the development of their projects.

With a downturn in the tourism activity in the region from 2009 to 2012, NaturArte has been on hold, but the economic forecast is changing and the interest of local communities in tourism-related activities is partly a reflection of that. The original concept for NaturArte was as a network of businesses with shared goals and shared governance structure. As a network NaturArte can economize promotion and marketing of NaturArte products and, strengthen the voice of local community groups. NaturArte can help facilitate a business cluster around tourism for local communities.

Community development of sustainable tourism becomes possible when communities, governments and NGOs work in partnership to benefit such projects. This new formula seeks to replicate successful models of community-based nature tourism and to avoid the mistakes of those same models.

The first community projects that were integrated into NaturArte were three oyster farms in Estero Morúa. We worked with each of these groups to develop a project and service, helped access funds for improvement of infrastructure and gave training in topics such as identification of birds, business administration and English as a second language. Visitation to the region by tourists has been very low, so these projects are not currently very active.

In the meantime, another group of ecotourism enterprises has been growing, as part of the alternative economic projects promoted with the Bahía Adair wetland conservation and management program. The community group REHBA has formed their own corridor, the Bahía Adair Ecotourism Corridor, CECBA, with the goal of promoting their wetland as a tourist destination, and offering a suite of services.

Other training offered to the community groups included team building, management and administration of projects. Funding support was acquired for development of infrastructure using green technology, and for purchase of equipment such as kayaks and life jackets. The final support was to link these community groups to a national business incubation process, which integrates development of a business plan, business skills, market research, corporate image design and access to funders.

11.5.3 Social Participation in the Peñasco Biological-Fisheries Corridor

Since 2007 there has been a strategic focus on promoting participation of landowners associated with wetlands, as well as with fishers, women and youth in the sustainable management of wetlands and fisheries resources along the north-eastern coast of Sonora (Table 11.1), with varying levels of advancement in these processes. The processes link to some of the tools and legal instruments described previously and are strengthened with comprehensive community outreach and education strategies. In some cases the impact on society, the economy and the regional environment is already noticeable. As multiple groups in a community become involved, we approach a threshold that can change the environmental ethic of an entire community, resulting in concrete management results. We summarize advances in these processes with eight community groups of the Peñasco Corridor that depend on healthy wetland systems. Next steps will be to bring community groups together for a shared vision for integrated coastal zone management.

11.6 Discussion

Successful conservation of the wetland habitats of the north-eastern coast of Sonora, in the Upper Gulf of California, requires an integrated and multi-pronged approach and is a long term process. Different approaches have been used to integrate distinct community groups into a process for achieving sustainable fisheries. The proposed model for Social Participation for resource management is adaptable and responsive to the needs of children, fishers, landowners and wives of fishers. We mold our methods to offer the

Table 11.1 Community groups involved in participatory processes within the Puerto Peñasco Corridor in northern Sonora, Mexico and their level of advancement in the process

Area of work	Community group	Management instruments and actions
Bahía Adair Wetland complex: Area: 42,430 ha	Ejidos landowners Bahía Adair 4 ejido members 6 youth	Ramsar site designation
		Ramsar management plan
		Wetlands surveillance and bird monitoring
		Monitoring of human use of wetlands
		Monitoring of solid wastes and cleaning
		PROFEPA surveillance certification and training
		Creation of REHBA wetland network group
		Ecotourism project: CECBA
Bahía San Jorge wetland complex Area: 12,198 ha	Campodónico fishing coop 10 fishers 1 housewife 1 teacher	Ramsar site designation
		Management plan (MP) for blue crab
		Fisheries catch monitoring
		Monitoring human use, sea lions, Isla San Jorge
		Wetland surveillance and bird monitoring
		PROFEPA surveillance certification
		Establishment of Grupo Lobos
	Campodónico Mujeres Trabajando 15 fishers' wives	Handcrafts made with recycled waste
		Establishment of Mujeres Trabajando group
Rocky and muddy seafloor in the Puerto Peñasco Biological-fisheries Corridor	Puerto Peñasco 5 diver coops. Aprox. 45 dive fishers	Species management plans: geoduck clam, murex snail, rock scallop
		Environmental impact studies: rock scallop
		Monitoring blue crab and octopus
		Biological and fishery monitoring organization into management groups
		Subcomittee for blue crab
		Integrator for geoduck clam
Puerto Peñasco Upper Gulf biosphere reserve	Fishing coops 13 permit holders 272 fishers	Environmental Impact study UGBR
		Community decision-making committee formed
		13 mitigation measures identified
		Good practices for 9 fisheries
		Logbook and onboard monitoring program
		Education program
	Punta Jagüey and Sto. Tomás Aprox 7 fishers	Fisheries catch monitoring
		Monitoring blue crab fishery for MP
	Desemboque: 2 fishing coop. Aprox, 90 fishers	Fisheries catch monitoring
		Monitoring blue crab fishery for MP
		Monitoring winged oyster fishery-voluntary closed season
	Puerto Lobos 4 fishers	Fisheries catch monitoring
		Monitoring octopus fishery for MP
Puerto Peñasco Biological-fisheries Corridor	Peñasco, Campodónico, Desemboque and Puerto Lobos 716 students and teachers in 2011–2012	Environmental contest for youth: "Fishing for the Future"

right level of capacity building, assistance in organization, establishing commitments and dissemination to assure effective community engagement and empowerment. The model guides us as we seek to understand the needs and interests of different community groups and offer unique solutions to the problems they face. With fishers, for example, our first approach for engaging them often uses biological monitoring as an incentive, but ultimately we try to transform this into market based incentives that support their participation in management of sustainable fisheries.

The proposed model is founded in community-level social participation which is understood as the organization of communities, through a process of continuous reflection and consensus, to enable them to make decisions to achieve better living conditions (Mulet-Robello and Castanedo-Rojas 2002).

Environmental education is a basic tool for social participation. Education can change thoughts, attitudes and relationships with the environment in a community (Medina and Jiménez 2001). Education should support community initiatives and lead to the participation and organization of communities in search of solutions for environmental and social problems. Environmental conservation is a mutual responsibility of both community and government agencies. Community Social participation provides a new vision of sustainable development as a practice of groups seeking to improve their quality of life through coexistence, respect and conservation of natural resources (Medina and Jiménez 2001).

The concept and necessity of community social participation in decision-making has been recognized in various forums both at international levels, within the context of the United Nations (UNEP 1992) and Agenda XXI, and in Mexico's National Development Plans (DOF 1995, 2013). In Mexico social participation is a fundamental component of the constitution, supported by the Law for Transparency and Access to Public Information and the Ecology Law, though its application is not achieving the desired results, primarily due to the inability of local communities to respond. Civil society organizations can play an important role in promoting community participation, through partnerships, political influence and by guiding actions that lead to community participation in the design of environmental policies.

The incentives and the formal and informal rules developed through the Social Participation Process are linked to a high level of compliance from the community groups that are most engaged in these processes. The more knowledgeable and better trained the community is, the higher is the rate of participation in decision-making on environmental policies, which in turn results in a better quality of life.

Though important advances have been made in these processes, bumps in the road have occurred and when analyzed offer important lessons. In the past local stakeholders, especially in the Upper Gulf Reserve, have been subjected to top down decisions with limited opportunity to participate. This has bred mistrust and contributed to the breakdown of local relationships and processes. Partnerships with stakeholders are very delicate, especially when they are in their formative stages and transparency is not as evident. Forums that bring all actors to the table can help clarify participants' positions. It is much better for stakeholders to hear directly from government in matters regarding regulations set by government. The distance of northern Gulf communities from Mexico's central government challenges direct communication across these different scales, but where it occurs progress in reducing conflicts and securing engagement of stakeholders is evident, such as in the OES forum for the Vaquita Refuge. Local actors have learned the value of expressing their concerns directly to their government. This was evidenced when fishers organized a meeting with national senators in their community before they took a vote on the new shrimp law in the spring of 2013. Local NGO actors can and should make the effort to bring these diverse voices together.

When trust is questioned, it is essential to actively work to renew confidence. This can be done through face to face presence, integrity of words and actions, consistency, and a willingness to help advance the agenda of partners as well as your own organizations. Building community trust can help fortify the overall process when for some reason trust is shaken with one particular group. Earned trust is very powerful. When trust is earned, stakeholders begin to seek out relationships and partnerships to advance on management issues together. They then become active and responsible stewards and leaders in the processes that affect their lives and use of resources. Such paradigm shifts are occurring in the Upper Gulf, where local stakeholders are contracting with conservation partners to collaborate for advancing management.

The isolation and lack of connectivity of coastal populations of the northern Gulf leaves them vulnerable in many ways. Top down decisions, lack of connection to markets, limited education opportunities, and lack of basic services all contribute to instability and insecurity. Connectivity and economic diversification are key factors that can enhance the adaptive capacity of communities. National climate change agenda stresses the need to develop adaptive capacity of communities and the natural resources they depend upon. The vision proposed here contributes to both. By promoting management processes that empower local communities in the protection of their resources and connecting them with their government to do so, local communities become less vulnerable. Through maintenance of connectivity between resources, habitats and ecological services communities can achieve greater security for the environment they depend upon. By creating economic opportunities that help incubate and integrate local businesses and give them access to both local and international markets, we reduce vulnerability of coastal communities and increase their adaptability in the face of constant change in both political and climate arenas.

11.7 Conclusions

The conservation of the wetlands in the Peñasco Corridor, essential habitat for the region's fisheries, has been achieved using a variety of legal tools including the review of environmental impact statements, the designation of Ramsar sites, the development of management plans, and protection of adjacent federal maritime zone providing one or more layers of protection to each wetland. These have been coupled with training of community leaders and groups, the implementation of innovative projects and education and community awareness activities to achieve concrete conservation goals.

To identify the most appropriate tools to implement in a given site a comprehensive site analysis is needed which includes not only the definition of environmental targets, but also identification of social and economic development opportunities that help ensure the integration of landowners and stakeholders into the conservation and management process. Successful long-term conservation depends on this type of analysis that identifies threats and actively develops strategies to prevent possible ecological damage.

The process must be integrated with existing environmental laws, plans, programs and strategies that define lines of action to protect and conserve the environment. Environmental agencies are often underfunded and understaffed and lack adequate financial resources and the skilled human capital to advance programs.

This work highlights the role played by non-governmental organizations who help fill in gaps not covered by laws and where the capacity of environmental authorities to protect resources is lacking. The implementation of community participation processes such as those described here for the Peñasco Corridor requires a deep commitment with adequate human, physical and financial resources, often exceeding the capacity of a single organization.

The proposed social participation process reflects a participation model that has been successful for wetland conservation because we were able to get meaningful involvement and commitment from community groups. Without a sensitive and committed community it is difficult to achieve long-term conservation. These processes are only a guide for the strategies to be implemented and should be tailored to suit many variables, the most important being community involvement. This process highlights the importance of involving not only fishers in wetland conservation, but landowners, who can be key to protecting the integrity of these habitats essential for fisheries.

To further fisheries management in the coastal waters associated with the region's wetlands, it will be important to bring together individual fishing communities with shared interests in the use of their fisheries resources. We have identified the connectivity between communities in terms of fisheries, fishing zones, and related this to connectivity between marine species (Chapter 9, Sects. 9.10 and 9.11 this issue). By creating community groups that can work together on management of their resources, we expect to advance the broader context of integrated management that is called for in ecosystem-based fisheries management.

References

Aburto-Oropeza O, Ezcurra E, Danemann G et al (2008) Mangroves in the Gulf of California increase fishery yields. Proc Natl Acad Sci U S A 105:10456–10459. doi:10.1073/pnas.0804601105

Addison-Sorey A (1989) History of the Puerto Penasco area. Part 3: prehistory. CEDO News 2(1):16–17

Alcock J (1992) Golondrinas en peligro en el estero Morua. CEDO News 5(1):8–10

Arriaga L, Espinoza JM, Aguilar C et al (2000) Regiones terrestres prioritarias de México. Comisión Nacional para el Conocimiento y uso de la Biodiversidad, Mexico

Astrálaga M (2006) La Convención Ramsar y los ecosistemas de Manglar. 6. http://www.ramsar.rg/speech/speech_campeche_manglares2006.pdf. Accessed 15 July 2013

Avila-Forcada S, Martínez-Cruz AL, Muñoz-Piña C (2012) Conservation of vaquita marina in the Northern Gulf of California. Mar Policy 36:613–622

Barbier EB, Hacker SD, Kennedy C et al (2011) The value of estuarine and coastal ecosystem services. Ecol Monogr 81(2):169–193

Barlow GW (1961) Gobies of the genus *Gillichthys*, with comments on the sensory canals as a taxonomic tool. Copeia 1961(4):423–437

Brusca RC, Cudney-Bueno R, Moreno-Báez M (2006) Gulf of California esteros and estuarios analysis, state of knowledge and conservation priority recommendations. Final Report to the David and Lucile Packard Foundation by the Arizona-Sonora Desert Museum

Calderón-Aguilera LE, Marinone SG, Aragon-Noriega EA (2003) Influence of oceanographic processes on the early life stages of the blue shrimp (*Litopenaeus stylirostris*) in the Upper Gulf of California. J Mar Syst 39:117–128

Castillo-López A, Rosemartin A, Morzaria-Luna HN (2007) Estudios previo justificativo para el establecimiento de la zona de conservación ecológica para el charrán mínimo *Sterna antillarum*, en estero Morúa. Centro Intercultural de Estudios de Desiertos y Océanos AC, Puerto Peñasco

Cinti A, Shaw W, Cudney-Bueno R et al (2010) The unintended consequences of formal fisheries policies: social disparities and resource overuse in a major fishing community in the Gulf of California, Mexico. Mar Policy 34:328–339

CITES (2011) Checklist of CITES species—2011. CITES, Geneva

CONANP (2007) Programa de conservación y manejo de la Reserva de la Biosfera Alto Golfo de California y Delta del Río Colorado. Comisión Nacional de Áreas Naturales Protegidas, Mexico

CONANP (2012) Comité Nacional de Humedales. Comisión Nacional de Áreas Naturales Protegidas. http://ramsar.conanp.gob.mx. Accessed 3 Feb 2013

CONANP (2013) Los humedales en la legislación ambiental internacional y nacional. Comisión Nacional de Áreas Naturales Protegidas. http://ramsar.conanp.gob.mx/legislacion.php. Accessed 25 Aug 2013

Crowder LB, Hazen EL, Avissar N et al (2008) The impacts of fisheries on marine ecosystems and the transition to ecosystem-based management. Annu Rev Ecol Evol Syst 39:259–278

Cudney-Bueno R, Turk Boyer MJ (1998) Pescando entre mareas del Alto Golfo de California: Una guía sobre pesca artesanal, su gente y sus propuestas de manejo. Centro Intercultural de Estudios de Desiertos y Océanos AC, Puerto Peñasco

Cudney-Bueno R, Bourillón L, Sáenz-Arroyo A et al (2009) Governance and effects of marine reserves in the Gulf of California, Mexico. Ocean Coast Manag 52:207–218

De la Cruz-González FJ, Aguilar-Ramírez D, Hernández-Flores A (2011) Indicadores económico-financieros para la captura de camarón y otras especies utilizando chinchorro de línea y red de arrastre prototipo RS-INP-MEX. Instituto Nacional de Pesca, México

DOF (1955a) Acuerdo que establece vedas y reglamenta la pesca de las especies conocidas con los nombres de totoaba y cabalucho y establece normas para la pesca del tiburón en el Golfo de California. Diario Oficial de la Federación, Mexico
DOF (1955b) Acuerdo que establece veda y fija zona de refugio para la protección de todas las especies de pesca en el Golfo de California. Secretaría de Marina, Poder Ejecutivo, Mexico
DOF (2002) Norma Oficial Mexicana de Emergencia NOM-EM-139-ECOL-2002, que establece las medidas de protección de los ecosistemas marinos y costeros y de las especies sujetas a protección especial en aguas de la Reserva de la Biosfera del Alto Golfo de California y Delta del Río Colorado. Secretaría de Medio Ambiente y Recursos Naturales. Diario Oficial de la Federación, México
DOF (1981) Ley Federal de Derechos. Secretaría de Hacienda y Crédito Público. Diario Oficial de la Federación, Mexico
DOF (1988) Acuerdo por el que se decreta la Ley General del Equilibrio Ecológico y la Protección al Ambiente. Secretaría de Medio Ambiente y Recursos Naturales y Pesca. Diario Oficial de la Federación, Mexico
DOF (1993) Decreto por el que se establece se declara área natural protegida con el carácter de Reserva de la Biosfera, la región conocida como Alto Golfo de California y Delta del Río Colorado, ubicada en aguas del Golfo de California y los municipios de Mexicali, B.C., de Puerto Peñasco y San Luis Río Colora-do, Son. Diario Oficial de la Federación, Mexico
DOF (1995) Decreto por el que se aprueba el Plan Nacional de Desarrollo 1995–2000. Secretaría de Gobernación. Diario Oficial de la Federación, Mexico
DOF (2000) Reglamento de la Ley General del Equilibrio Ecológico y la Protección al Ambiente en materia de Áreas Naturales Protegidas. Secretaría de Me-dio Ambiente y Recursos Naturales y Pesca. Diario Oficial de la Federación, Mexico
DOF (2004) Ley General de Bienes Nacionales. Diario Oficial de la Federación, Mexico
DOF (2005a) Acuerdo mediante el cual se establece el área de refugio para la protección de la vaquita (*Phocoena sinus*). Diario Oficial de la Federación, Mexico
DOF (2005b) Programa de Protección de la Vaquita dentro del Área de Refugio ubicada en la porción occidental del Alto Golfo de California. Diario Oficial de la Federación, Mexico
DOF (2005c) Programa de Protección de la Vaquita dentro del Área de Refugio ubicada en la porción occidental del Alto Golfo de California. Diario Oficial de la Federación, Mexico
DOF (2007) Acuerdo por el que se decreta la Ley General de Pesca y Acuacultura Sustentable. Secretaría de Agricultura, Ganadería, Desarrollo Rural, Pesca y Alimentación. Diario Oficial de la Federación, Mexico
DOF (2010) Norma Oficial Mexicana NOM-059-SEMARNAT-2010 para la Protección ambiental—Especies nativas de México de flora y fauna silvestres—categorías de riesgo y especificaciones para su inclusión, exclusión o cambio—Lista de especies en riesgo. Secretaría de Medio Ambiente y Recursos Natura-les. Diario Oficial de la Federación, Mexico
DOF (2011) Decreto por el que se reforman, adicionan y derogan diversas disposiciones de la Ley Federal de Derechos. Secretaría de Hacienda y Crédito Público. Diario Oficial de la Federación, Mexico
DOF (2012) Acuerdo por el cual se establece una red de zonas de refugio en aguas marinas de jurisdicción federal frente a la costa oriental del Estado de Baja California Sur, en el corredor marino San Cosme a Punta Coyote. Secretaría de Gobernación. Diario Oficial de la Federación, Mexico
DOF (2013) Decreto por el que se aprueba el Plan Nacional de Desarrollo 2013–2018. Secretaría de Gobernación. Diario Oficial de la Federación, Mexico
Downton-Hoffmann CA, Turk-Boyer PJ, Thaxton B et al (2013) Corredor biológico pesquero Puerto Peñasco-Puerto Lobos, Sonora. Centro Intercultural de Estudios de Desiertos y Océanos AC, Puerto Peñasco
Drinkwater KF, Frank KT (1994) Effects of river regulation and diversion on marine fish and invertebrates. Aquat Conserv Freshw Mar Ecosyst 4:135–151
Erisman BE, Aburto-Oropeza O, Gonzalez-Abraham C et al (2012) Spatio-temporal dynamics of a fish spawning aggregation and its fishery in the Gulf of California. Sci Rep. doi:10.1038/srep00284
Felger RS (2000) Flora of the Gran Desierto and Rio Colorado of Northwestern Mexico. The University of Arizona Press, Tucson
Fletcher S, Kawabe M, Rewhorn S (2011) Wetland conservation and sustainable coastal governance in Japan and England. Mar Pollut Bull 62:956–962. doi:10.1016/j.marpolbul.2011.02.048
FONATUR (2003) Proyecto de Escalera Náutica del Mar de Cortés. http://ccds.semarnat.gob.mx/regiones/r-no/2002-2004/sesiones_ordinarias/1a_sesion_2003/documentos_presentados_pdf/ e-nautica-mardecortes-1a-ext-no.pdf. Accessed 16 Aug 2013
García-Frapolli E, Ramos-Fernández G, Galicia E et al (2009) The complex reality of biodiversity conservation through natural protected area policy: three cases from the Yucatan Peninsula, Mexico. Land Use Policy 26:715–722. doi:10.1016/j.landusepol.2008.09.008
Glenn EP, Nagler PL, Brusca RC et al (2006) Coastal wetlands of the northern Gulf of California: inventory and conservation status. Aquat Conserv Freshw Mar Ecosyst 16:5–28
Guido S (2006) Red de monitoreo de los desarrollos turísticos costeros del Noroeste. Alcosta, Ensenada
Houde ED, Rutherford ES (1993) Recent trends in estuarine fisheries: predictions of fish production and yield. Estuaries 16(2):161–176
IMM (2008) Sustainable Livelihoods Enhancement and Diversification (SLED): a Manual for practitioners. IUCN, Gland; CORDIO, Kalmar; ICRAN, Cambridge
Iris-Maldonado CA (2011) Estructura de la comunidad de peces de dos esteros en el norte del Golfo de California. Dissertation, CICESE
IUCN (2004) International Union for Conservation of Nature, IUCN 2004 Red list. http://www.redlist.org. Accessed 14 Aug 2010
Jaramillo-Legorreta AM, Rojas-Bracho L, Gerrodette T (1999) A new abundance estimate for vaquitas: first step for recovery. Mar Mamm Sci 15:957–973
Jones PJS, Qiu W, De Santo EM (2011) Governing marine protected areas: getting the balance right. United Nations Environment Programme, Nairobi
Kelleher G (1999) Guidelines for marine protected areas. IUCN, Gland
Marinone SG, Ulloa MJ, Parés-Sierra A et al (2008) Connectivity in the Northern Gulf of California from particle tracking in a three-dimensional numerical model. J Mar Syst 71:149–158. doi:10.1016/j.marsys.2007.06.005
Martínez I, Turk-Boyer P, Sanchez CA et al (2011) Corredor biológico-pesquero Peñasco. Punto Borrascoso a Puerto Lobos, Sonora, Norte del Golfo de California: Una propuesta de ordenamiento pesquero de la pesca ribereña. Centro Intercultural de Estudios de Desiertos AC, Puerto Peñasco
Medina JA, Jiménez I (2001) Guía para la gestión integral de los residuos sólidos municipales. SEMARNAT, Mexico
Mellink E, Palacios E (1993) Note on breeding coastal waterbirds in north-western Sonora. West Birds 24:29–37
Mulet-Robello C, Castanedo-Rojas I (2002) La participación comunitaria y el medio ambiente. Rev Cubana Enfermer 18(2):125–128
NACAP Vaquita (2008) Plan de acción de América del Norte para la conservación de la vaquita. Comisión para la Conservación Ambiental. http://www.cec.org. Accessed 13 March 2010
Okin GS, Mahowald N, Chadwick OA et al (2005) Impact of desert dust on the biogeochemistry of phosphorus in terrestrial ecosystems. Glob Biogeochem Cycl. doi:10.1029/2003GB002145

Palacios E, Mellink E (1996) Status of the least tern in the Gulf of California. J Field Ornithol 67(1):48–58

PANGAS (2012) Jaiba *Callinectes bellicosus*. Ficha Informativa de la Pesca Ribereña del Norte del Golfo de California. Centro Intercultural de Estudios de Desiertos y Océanos AC, Puerto Peñasco

Pikitch EK, Santora C, Babcock EA et al (2004) Ecology: ecosystem-based fishery management. Science 305:346–347

Pink DH (1995) Drive: the surprising truth about what motivates us. Rivergead Books, New York

Ramsar (1996) Ramsar convention on wetlands. http://ramsar.org/. Accessed 21 April 2009

Ramsar (2008) Strategic framework and guidelines for the future development of the list of wetlands of international importance of the convention on wetlands. Ramsar Convention Bureau, Iran

Ramsar (2009) Information sheet for Ramsar wetlands Humedales de Bahía Adair, Sonora, México. http://ramsar.org/. Accessed 21 April 2009

Ramsar (2010) Information sheet for Ramsar Wetlands Humedales de Bahía San Jorge, Sonora, México. http://ramsar.org/. Accessed 14 May 2010

Rodríguez-Quiroz G, Bracamontes A (2008) Pertinencia de las ANP como política de conservación y mejoramiento de la calidad de vida. Análisis de la percepción en la reserva de la Biosfera del Alto Golfo de California y Delta del Río Colorado en Estudios Sociales, Mexico

Ruiz-López DM (2009) Diagnóstico Socioeconómico y evaluación de una estrategia de compensación al sector pesquero del Golfo de Santa Clara, Sonora. Dissertation, Instituto Politécnico Nacional

Rupnow J (2008) Recreational fisheries and the ecology and management of targeted resources in the Northern Gulf of California, Mexico. Dissertation, Arizona State University

Sáenz-Arroyo A, Roberts CM, Torre J et al (2005) Using list naturalists' observations and grey literature to reassess marine species at the Gulf grouper in the Gulf of California, Mexico. Fish Fish 6:121–133

SECTUR (2007) La inversión privada identificada en el sector turismo. Secretaría de Turismo, Mexico

SEMARNAP (1995) Programa de manejo de la reserva de la biosfera Alto Golfo de California y Delta del Río Colorado. Secretaría de medio Ambiente, Recursos Naturales y Pesca, Mexico

SEMARNAT (2008) Programa de acción para la conservación de la especie vaquita (*Phocoena sinus*). Secretaría de Medio Ambiente y Recursos Naturales, Mexico

Spackeen J (2009) Analysis of food web dynamics in the Northern Gulf of California using stable isotopes. Dissertation, University of Miami

Travieso-Bello A (2009) La legislación aplicable a los sitios Ramsar en México. Letras jurídicas 20:1–14

Turk-Boyer PJ (1989) Estero Morua: the formation of an estuary. CEDO News 2(1):6–7

Turk-Boyer PJ (2002) Shrimp trawling industry protests new fishing regulations for the upper Gulf of California—Colorado river delta biosphere reserve at Puerto Peñasco, Sonora. Centro Intercultural de Estudios de Desiertos y Océanos AC, Puerto Peñasco

Turk-Boyer PJ, Barrera JC (2012) Indicadores y mecanismos para la evaluación de la efectividad del programa de protección de la vaquita dentro del área de refugio ubicada en la porción occidental del Alto Golfo de California. Centro Intercultural de Estudios de Desiertos y Océanos AC, Puerto Peñasco

UNEP (1992) Conferencia de las naciones unidas sobre el medio ambiente y el desarrollo Río 1992. http://www.unep.org. Accessed 15 July 2013

Uribe P, Moguel S, Torre J et al (2010) Implementación de reservas marinas en México. Carambola Films, CEMDA, COBI, México

Villacaña-Yepéz GL (2012) Manifestación de impacto ambiental modalidad regional para el proyecto "Home Port Turístico de Puerto Peñasco", promovido por la Comisión de Fomento al Turismo del Estado de Sonora, Gobierno del Estado de Sonora. QV Gestión Ambiental, San Carlos

Zuria-Jordan I (1996) Disponibilidad de alimento y reproducción de *Sterna antillarum* (Aves: Laridae) en dos colonias del Noroeste de México. Dissertation, CICESE

Zuria I, Mellink E (2005) Fish abundance and the 1995 nesting season of the Least Tern at Bahia de San Jorge, northern Gulf of California, Mexico. Water birds 28:172–180

Erratum

"Marine Areas of Responsible Fishing": A Path Toward Small-Scale Fisheries Co-Management in Costa Rica? Perspectives from Golfo Dulce

Luc Fargier, Hans J. Hartmann and Helena Molina-Ureña

Erratum to:
Chapter 10 in: F. Amezcua, B. Bellgraph (eds.), *Fisheries Management of Mexican and Central American Estuaries*, Estuaries of the World,
DOI 10.1007/978-94-017-8917-2_10

The Publisher regrets to inform that in the current chapter, the image and references related to appendix 1 is missing. The correct version of the appendix is as follows:

Appendix 1 Co-management potential evaluation matrix from a literature synthesis of case studies analyzing fisheries participative processes. Numbers in the references column correspond to the following publications: **1**, Chuenpagdee and Jentoft (2007) ; **2**, Carlsson and Berkes (2005) ; **3**, Govan (2008); **4**, Nielsen et al. (2004); **5**, McConney and Baldeo (2007); **6**, Geoghegan and Renard (2002); **7**, Geoghegan et al. (1999); **8**, Renard (2001); **9**, Cumberbatch (2001); **10**, Mahon and Mascia (2003); **11**, Ravndal (2002): **12**, Renard (1991); **13**, Govan (2003); **14**, Brown and Pomeroy (1999); **15**, CARICOM-CFRAMP (1995); **16**, Almerigi et al. (1999); **17**, White et al. (1994); **18**, McConney (1999); **19**, Begossi and Brown (2003); **20**, (Renard, 1991); **21**, Pomeroy et al (2003); **22**, Pomeroy et al. (2001); **23**, Pomeroy and Carlos (1997); **24**, Jentoft et al. (1998); **25**, Noble (2000); **26**, Fonseca-Borrás (2009); **27**, Luna (1999); **28**, Nuñez Saravia (2000); **29**, (2003); **30**, Ostrom (1990); **31**, Nuñez Saravia (2005); **32**, Pinkerton (2005); **33**, Pomeroy et al. (2011); **34**, Gutiérrez et al. (2011); **35**, Kikuchi and Hayami (1980); **36**, Hayami and Kikuchi (1981); **37**, Pomeroy and Berkes (1997); **38**, Weitzner (2000). Ind.: Individual.

The online version of the original chapter can be found under
DOI 10.1007/978-94-017-8917-2_10

L. Fargier (✉) · H. J. Hartmann
UMR 7266 LIENSs (Research Center for Coasts, Environment and Societies), Université de La Rochelle (ULR), 2 rue Olympe de Gouges, 17000 La Rochelle, France
e-mail: luc.fargier@univ-lr.fr

H. J. Hartmann
e-mail: hhartman@univ-lr.fr

H. Molina-Ureña
CIMAR (Research Center in Marine Sciences and Limnology), Universidad de Costa Rica (UCR), San Pedro de Montes de Oca, 2060-11501 San José, Costa Rica
e-mail: hmolina@rsmas.miami.edu

F. Amezcua, B. Bellgraph (eds.), *Fisheries Management of Mexican and Central American Estuaries*, Estuaries of the World, DOI 10.1007/978-94-017-8917-2_12, © Springer Science+Business Media Dordrecht 2014

		Factor		References
Ex-ante	Cause	Resource decrease, conflicts		35, 36, 37
Ex-ante	Initiative	Local		1, 3, 26
Ex-ante	Initiative	Mixed (local + external agent)		1
Ex-ante	Initiative	Government		1
Implementation	Supra-communautary	Backing	Legal	6, 11, 12, 13, 18, 19, 20, 21, 22, 27, 28, 29, 33, 37
Implementation	Supra-communautary	Backing	Financial	11, 13, 15, 18, 19, 21, 22, 27, 28, 29, 31, 33
Implementation	Supra-communautary	Backing	Technical	11, 13, 15, 18, 22, 27, 28, 29, 31, 33
Implementation	Supra-communautary	Rights to organize		30, 37
Implementation	Supra-communautary	External agents		1, 3, 5, 22, 33
Implementation	Communautary	Organized groups		9, 11, 15, 17, 22, 33
Implementation	Communautary	Leadership		21, 22, 32, 33, 34, 38
Implementation	Communautary	Clearly defined boundaries	User group	8, 21, 22, 24, 25, 30, 33
Implementation	Communautary	Clearly defined boundaries	Management area	3, 5, 21, 22, 24, 25, 30, 33, 34
Implementation	Communautary	Clearly defined boundaries	Resource	2, 21, 22, 24, 30, 32, 33, 34
Implementation	Communautary	Clearly defined roles		2, 3, 5, 6, 7, 8, 9, 10, 13, 14, 17, 21, 26, 28, 31, 33
Implementation	Communautary	Participation		6, 7, 8, 9, 12, 14, 17, 20, 21, 22, 23, 26, 33
Implementation	Communautary	Local knowledge		3, 4, 5, 14, 21
Implementation	Communautary	Collective-choice arrangements		30, 32
Implementation	Communautary	Conflict-resolution mechanisms		21, 22, 26, 30, 33
Implementation	Communautary	Local will		27, 28, 29, 31, 32, 33
Implementation	Communautary	Graduated sanctions		27, 30, 33
Implementation	Communautary	Monitoring		30, 33
Implementation	Communautary	Nested enterprises		30, 33
Implementation	Ind.	Benefits > Costs		22,30, 33
Values	Supra-communautary	Equity		8, 19, 20, 21, 26
Values	Supra-communautary	Transparence		5, 6
Values	Supra-communautary / Communautary	Trust		5, 21, 26, 27
Values	Supra-communautary / Communautary	Respect		5, 18, 21, 26
Values	Supra-communautary / Communautary	Cooperation		28, 29, 31
Values	Communautary	Ownership		3, 21, 26
Values	Communautary	Social cohesion		34
Values	Communautary	Empowerment		3, 4, 26
Values	Communautary	Legitimity		8, 24, 26
Values	Ind.	Topophily		26, 32
Values	Ind.	Right to participate		26

Almerigi, S, Mahon, Alleyne, Y, Atherley, K, Cumberbatch, J, Mahon, S (1999) Barbados Coastal Conservation Programme (Phase 1), Demonstration Projects. Participatory coastal zone management in Barbados. . (ed C. C. Z. M. Unit), pp. 30

Begossi, A, Brown, D (2003) Fisheries co-management experiences in Latin America and the Caribbean. *The Fisheries Co-management Experience. Accomplishments, Challenges and Prospects* (eds D C Wilson, J R Nielsen, P Degnbol), pp. 348. Kluwer, Dordrecht

Brown, D N, Pomeroy, R S (1999) Co-management of Caribbean Community (CARICOM) fisheries. *Marine Policy*, 23, 549–570

CARICOM-CFRAMP (1995) Report of the strategic planning mission on community involvementand education sub-project. (ed C F M Unit) Belize

Carlsson L, Berkes F (2005) Co-management: concepts and methodological implications. Journal of Environmental Management 75 (1):65–76

CBM (2003) Estado Actual del Comanejo de Áreas Protegidas en Mesoamerica. Proyecto para la consolidación del Corredor Biológico Mesoamericano. CCAD/PNUD/GEF, Managua, Nicaragua

Chuenpagdee R, Jentoft S (2007) Step zero for fisheries co-management: What precedes implementation. Marine Policy 31 (6):657–668

Cumberbatch, J (2001) Case study of the Folkstone Marine Park and Reserve, Barbados. (ed CANARI). CANARI Technical Report, St Lucia

Fonseca-Borrás M (2009) Comanejo de areas silvestres protegidas en Costa Rica: Analisis comparativo de las experencias del parque nacional Cahuita y el parque nacional Marino-Ballena. Universidad de Costa Rica, San José, Costa Rica

Geoghegan, T, Renard, Y (2002) Beyond community involvement: lessons from the insular Caribbean. *Parks*, 12, 16–25

Geoghegan, T, Renard, Y, Brown, N, Krishnarayan, V (1999) Evaluation of Caribbean experiences in participatory planning and management of marine and coastal resources. *CANARI Technical Report* (ed CANARI). CANARI, St. Lucia

Govan, H (2003) Co-management of natural resources in Central America: The road to "equitable distribution of the benefits of biodiversity" or another empty discourse of the technical elite? *Ecosystems, Protected Areas and Peoples Initiative: Lessons learned in community-based management and co-manegement of natural resources and protected areas: focus on coastal and marine resources and Indigenous People's communities in Central American and the Caribbean.* (ed CMWG-IUCN), pp 36 IUCN, San José, Costa Rica

Govan H (2008) Overview: Reclaiming "Protected Areas" as a livelihood tool for Pacific Island people. In: Cohen P, Valemei AD, Govan H (eds) Annotated bibliography on socioeconomic and ecological impacts of Marine Protected Areas in Pacific Island countries. WorldFish Bibliography, vol No. 1870. The WorldFish Center, Penang, pp 1–5

Gutiérrez NL, Hilborn R, Defeo O (2011) Leadership, social capital and incentives promote successful fisheries. Nature 470 (7334):386–389

Hayami, Y, Kikuchi, M (1981) *Asian Village Economy at the Crossroads: An Economic Approach to Institutional Change*, University of Tokyo Press, Tokyo

Jentoft S, Mc Cay BJ, Wilson DC (1998) Social theory and fisheries co-management. Marine Policy 22 (4-5):423–436

Kikuchi, M, Hayami, Y (1980) Inducements to Institutional Innovation in an Agrarian Economy. *Economic Development and Cultural Change*, 29, 21–36

Luna R (1999) El comanejo de áreas protegidas en Centroamérica. In: PROARCA/CAPAS/AID (ed) Memoria del Taller Centroamericano sobre Comanejo de Areas Protegidas, realizado en Panamá, entre el 22 y el 24 de noviembre de 1999. Panamá

Mahon, R, Mascia, M (2003) The Barbados Marine Reserve: a late bloomer? *Gulf and Caribbean Research*, 14, 171–180.

McConney P, Baldeo R (2007) Lessons in co-management from beach seine and lobster fisheries in Grenada. Fisheries Research Evaluation and Management of Coastal Fisheries in Latin America and the Caribbean, 1st CoastFish Conference 87 (1):77–85

McConney, P A (1999) Organising fisherfolk in Barbados without completing a clean round. *Proceedings of the Gulf and Caribbean Fisheries Institute* 52, 290–299

Nielsen, J R, Degnbol, P, Viswanathan, K K, Ahmed, M, Hara, M, Abdullah, N M R (2004) Fisheries co-management--an institutional innovation? Lessons from South East Asia and Southern Africa. *Marine Policy*, 28, 151–160

Noble BF (2000) Institutional criteria for co-management. Marine Policy 24 (1):69–77

Nuñez Saravia, O M (2000) El comanejo y la participación de la sociedad civil en las áreas protegidas de Centroamérica. (ed F D N-T N C-. PROARCA/CAPAS/AID), pp. 46 Guatemala

Nuñez Saravia OM (2000) El comanejo y la participación de la sociedad civil en las áreas protegidas de Centroamérica. Guatemala

Ostrom E (1990) Governing the Commons, The evolution of Institutions for collective action. The Political economy of institutions and decisions, 22d ed. Cambridge University Press, New York

Pinkerton E (2005) Les partenariats de gestion. In: Cochrane KL (ed) Guide du gestionnaire des pêcheries: Les mesures d'aménagement et leur application, vol 424. Document technique sur les pêches. FAO, Rome, p 235

Pomeroy RS, Berkes F (1997) Two to tango: The role of government in fisheries co-management. Marine Policy 21 (5):465–480

Pomeroy RS, Carlos MB (1997) Community-based coastal resource management in the Philippines: A review and evaluation of programs and projects, 1984-1994. Marine Policy 21 (5):445–464

Pomeroy RS, Cinner JE, Raakjær J (2011) Conditions for Successful Co-management : Lessons Learned in Asia, Africa, the Pacific, and the wider Caribbean. In: Pomeroy R, Andrew NL (eds) Small-scale Fisheries Management Frameworks : Frameworks and approaches for the Developing World. CABI Publishing

Pomeroy RS, Katon BM, Harkes I (2001) Conditions affecting the success of fisheries co-management: lessons from Asia. Marine Policy 25 (3):197–208

Pomeroy RS, McConney PA, Mahon R (2003b) Comparative analysis of coastal resource co-management in the carribean. CCA guidelines project.

Ravndal, V (2002) Community co-managed park system for Belize. *Final Project Evaluation* UNDP/GEF, Belize

Renard, Y (1991) Institutional challenges for community-based management in the Caribbean. *Nature and Resources*, 27, 4–9

Renard, Y (2001) Case of the Soufriere Marine Management Area (SMMA). *Technical Report No. CANARI, St Lucia* CANARI, St Lucia

Weitzner V (2000) Probing the Powers at Play in Co-Management from teh Bottom Up: The case of Cahuita National Park, Costa Rica. In: Vlllth Biennial Conference of the International Association for the Study of Common Property "Constituting the Commons: Craning Sustainable Commons in the New Millennium", Indiana University in Bloomington, Indiana, May 31-June 4, 2000.

White, A, Hale, L Z, Renard, Y, Cortesi, L (1994) *Collaborative and community-based management of coral reefs*, West Hartford, Connecticut, USA

Epilogue: Fisheries Management of Estuarine Systems of Mexico and Central America, what we know and what we need

The aim of this book is to contribute scientific information at an interdisciplinary level that will help to attain adequate fisheries management of the estuarine systems of Mexico and Central America. As pointed out in the introduction and chapters of this book, fishing activities within ecosystems in this region of the world are generally undertaken by small-scale fishers within an inherently complex framework. Complexity arises from the various types of fishing gears employed; the number and characteristics of a myriad of exploited species; the socioeconomic, educational and cultural context of the fishers exploiting these systems; as well as the environmental and anthropogenic factors. However, proper management and regulation of these small-scale fisheries is inhibited by a lack of physical and biological information in the area being fished, inadequate legislation for the protection and conservation of these ecosystems, and a paucity of strong governmental institutions with an ecosystem perspective to ensure effective governance.

It is generally recognized that successful conservation and management of fisheries requires adequate data about the exploited species and the ecosystem as a whole, as well as information about the socioeconomic context of the fishers and others that directly affect the exploited resources, as was indicated in the introduction. Unfortunately, this biological and socioeconomic knowledge is scarce or nonexistent for the majority of the estuarine systems in Mexico and Central America, and for those ecosystems where some information is available; the results are generally treated as independent or unrelated to other processes occurring in these systems. In this sense, this book collected relevant available information useful for adequate management of these ecosystems including the ecology of the exploited species and fisheries-related issues, results and information on physicochemical aspects and their relation to the flora and fauna inhabiting the estuarine systems, and socioeconomic aspects.

The first part of the book dealt with naturally occurring and anthropogenically caused physicochemical issues of estuaries throughout the region, and how these characteristics are relevant for the conservation and management of these ecosystems. In Chap. 1 and 3, De La Lanza et al. and Calvario et al., respectively, highlight the importance and negative effects of untreated waste effluents discharged into estuarine and lagoon systems that impact estuarine habitat, fisheries, and water quality by increasing organic matter and nutrients such as ammonium, total nitrogen and phosphorus. In both studies, it was concluded that the studied systems (one in the Gulf of Mexico and the other on the Pacific Coast) were heterotrophic, and that although the level of pollutants discharged into these systems was high, the water quality was of sufficient quality to maintain productive fisheries. In both cases, the hydrological period and short water residence time helped to sufficiently eliminate the discharged pollutants. However, both studies also highlighted the importance of controlling urban discharges because as human population abundance increases, so will the amount of anthropogenic discharges. In turn, additional discharge could affect water quality and cause eutrophication, which could provoke decreased biodiversity and lower fisheries yields.

In Chap. 2, Ruelas et al. discussed how the pollution of estuaries could increase the presence of Mercury (Hg) in edible parts of fish and invertebrates, which could have an effect on human health through bioaccumulation. After analyzing the available information on Hg levels in fish (elasmobranchs and teleosts) and some of the most important edible invertebrates (shrimps, clams, mussels and oysters) landed in estuarine systems from Mexico, they concluded that only scalloped hammerhead sharks (*Sphyrna lewini*) present a clear risk to humans due to the level of Hg and methyl Hg in edible muscle tissue. However, the authors also indicated that the available information regarding Hg concentration of many exploited fish species, and the rates of fish consumption in Mexico, is scarce. Further, it is important to incorporate this information into fisheries management programs to limit or preclude consumption of effected species by size, sex, or during specific seasons when the risk of health effects is greatest.

F. Amezcua, B. Bellgraph (eds.), *Fisheries Management of Mexican and Central American Estuaries*,
Estuaries of the World, DOI 10.1007/978-94-017-8917-2, © Springer Science+Business Media Dordrecht 2014

In Chap. 4, Jara-Marini et al. studied the food web of an estuarine system in the Gulf of California that received a considerable amount of sewage from a nearby urban area (this system was also studied in Chap. 2). They documented a food web with 5 trophic levels in this estuary, which is consistent with previous studies in coastal environments, and indicates that although this system has been affected by anthropogenic activities, the food web still functions adequately. However, the results also showed enrichment of $\delta^{15}N$ in the food web, which suggests nutrient enrichment by anthropogenic discharges into the ecosystem. The consequence of nutrient enrichment is that it can cause structural changes to coastal lagoon ecosystems, such as a reduction in species diversity and changes to the top-down and bottom-up regulating forces in food webs. Another consequence of nutrient enrichment is the stimulation of bacterial activity that can result in benthic oxygen depletion, which may cause long-term changes in the structure of benthic assemblages. Further changes could, in turn, affect fish abundance, which is strongly dependent on benthic organisms as this study demonstrated. An artisanal fishery that primarily targets finfish is also present in this estuary and thus, removal of fishes from the system without proper management guidelines could affect the long-term nutrient budget of the estuary by disrupting the linkage between coastal and marine processes.

The four chapters describing physicochemical characteristics of Mexico estuaries represent a baseline for evaluating the current state of estuarine water quality in Mexico and Central America and indicate a need for proper future management. It is apparent that these estuaries are not yet at their breaking point and that conditions could be much worse; however, the consequences of anthropogenic pollution are becoming more evident. Estuaries are reaching their nutrient carrying capacities and nearing eutrophic status, and bioaccumulation of Mercury in human-consumable tissues of the top piscine predators has been documented. Overall, these studies highlight an urgent need for proper waste water management before the effects to fisheries and other exploited species are irreversible.

The second section of this book dealt with general- and fisheries-ecology topics from estuarine systems in Mexico. In Chap. 5, Flores et al. discussed the effects of hydrological regimes on estuarine systems. Using a case study from the most extensive mangrove region in the American Pacific, "Marismas Nacionales" (National Floodplains), these authors described how the opening of a water-connection channel from one coastal lagoon to the open sea triggered a major ecological disaster. A channel that was originally planned to be 40-m wide by 2-m deep grew uncontrollably until it reached a width of 700 m and a depth of 20 m due to strong ebbing currents that provoked bank erosion. The consequence of this was the mortality of more than 15,000 ha of mangroves and the affectation of 33 % of the total mangroves in the system. Counterintuitively, local artisanal fisheries improved as landings of finfish and shrimp increased dramatically once this channel was formed. It has been typically believed that mangroves are a key habitat for sustainable fisheries, but in this chapter the authors indicate that environmental conditions in some areas of mangrove forests can be extremely difficult for the survival of aquatic resources due to oxygen depletion. They concluded that water inputs to mangrove forests are required for adequate fish habitat, and to sustain a productive fishery. In this sense, only certain types of mangrove forests are beneficial to fisheries, whilst other types of mangrove forests play other different ecological functions than supporting fisheries. This chapter is controversial in that it challenges the widely held belief that extensive areas of mangrove forests are needed for high fishery yields; however, the authors provide evidence that this paradigm may not always be true, particularly for Marismas Nacionales.

In Chap. 6, Vera and Salas reviewed the link between fish productivity and the variation in physical and chemical processes, which directly affect the abundance and distribution of zooplankton in areas of freshwater influence. Their goal was to describe how changes in environmental factors affected zooplankton, and they documented functional relationships that could be used to predict changes in zooplankton biomass and abundance, which in turn has an effect on the biomass of exploited species. In fact, the authors concluded that there is a direct link between zooplankton biomass and abundance of fish, and that the distribution pattern of fishes is similar to that of the zooplankton. Therefore, it is important to understand these processes in order to establish management tools that take into account the phenomena affecting the early life history stages of fish.

In Chap. 7, Ramirez et al. dealt explicitly with fisheries management issues of small-scale fisheries and proposed a management plan for the artisanal fisheries in the State of Sinaloa, Mexico, which has some of the highest fisheries landings among all of this country. In this paper, the authors agreed that the information needed for proper management of such a complex activity is lacking and they highlighted the urgent need for a management plan. They proposed a management strategy based on the use of zones as management units so that fishing processes could be understood by region, and to identify possible spatiotemporal changes of the marine communities between zones. Thanks to this zonation, the targeted species were identified and categorized per region, and it was discussed how using this method could allow the identification of catch trends for adaptive management.

Turk-Boyer et al., in Chap. 8, showed that an ecosystem-based approach to fisheries management is possible when the necessary information is available. They exemplified this using a case study in wetlands from the upper Gulf of California. Using data from trophic studies, oceanographic-biological models on larval dispersion, genetic data from commercial species, and information on the patterns of human use along the coast including fisheries, tourism and coastal development, the authors defined essential habitats for target species and identified their trophic interactions. It was then determined that these areas can be protected to ensure sustainability of the fisheries resources.

The last section of the book dealt with socioeconomic topics, which are an essential part of the management of small-scale fisheries, but are usually overlooked by stakeholders, the government, and decision-makers. Chapter 9, by Fargier et al., analysed participatory management processes of small-scale fisheries in Costa Rica, a country in which the fisheries data and the biological information of the exploited species is scarce, and with management plans that are not adequate for the existing conditions. Using the "Marine Area of Responsible Fishing" model, they described how the inclusion of small-scale fishers organizations into the development of management plans can help to achieve long-term success in fisheries conservation through a set of measures outlined by the authors.

In the 10th and final chapter of the book, Turk- Boyer et al. outlined a management strategy for a coastal area in the upper Gulf of California that has been developed significantly due to tourism. This strategy included the designation of Natural Protected Areas, Federal Zone Concessions and Ramsar Sites by the federal government, as well as the active participation of fishers, local communities and civil society organizations in order to guarantee long-term protection of essential coastal habitats. Government and civic organizations have conjointly developed a series of programs of different themes (educational, tourism development, handcrafting) that include the fishers and their families, and are aimed to help them solve environmental and financial problems. Thanks to these programs, fishers are participating in management initiatives for individual species, and a vision for ecosystem-based management is growing. A wetland conservation ethic is also emerging and could be as important in the long term as other tools that have been utilized.

Overall, this book presents a much-needed diagnosis of the anthropogenic factors affecting estuarine systems, and the potential medium- and long-term impacts that a lack of management could have on exploited resources. These impacts vary from threats to human health by the consumption of organisms with bioaccumulated toxins, to the loss of fishery yields as a consequence of coastal habitat destruction. Furthermore, information on the current condition of coastal areas of Mexico and Central America is presented, such as the effects of abiotic factors on zooplankton and the need to understand these processes due to their direct influence on the distribution and abundance of exploited fish species. Information was additionally presented about the relationship between mangrove-forest types and fisheries yield, and several case scenarios were presented to exemplify practical management tools. Conclusively, this cumulative knowledge of estuaries will assist decision makers in managing and restoring coastal ecosystems. Although more information is needed on the biology and ecology of Mexico and Central American estuaries, this book offers a glimpse of hope, and a first step, in achieving proper management of estuaries and estuarine fisheries.

Felipe Amezcua and Brian Bellgraph

Index

F. Amezcua, B. Bellgraph (eds.), *Fisheries Management of Mexican and Central American Estuaries*, Estuaries of the World, DOI 10.1007/978-94-017-8917-2, © Springer Science+Business Media Dordrecht 2014

MIX
Papier aus verantwortungsvollen Quellen
Paper from responsible sources
FSC® C105338

If you have any concerns about our products,
you can contact us on
ProductSafety@springernature.com

In case Publisher is established outside the EU,
the EU authorized representative is:
Springer Nature Customer Service Center GmbH
Europaplatz 3, 69115 Heidelberg, Germany

Printed by Libri Plureos GmbH
in Hamburg, Germany